New Challenges for Data Design

David Bihanic
Editor

New Challenges for Data Design

 Springer

Editor
David Bihanic
University of Valenciennes
 and Hainaut-Cambresis
Valenciennes
France

ISBN 978-1-4471-6595-8   ISBN 978-1-4471-6596-5 (eBook)
DOI 10.1007/978-1-4471-6596-5

Library of Congress Control Number: 2014949332

British Library Cataloguing in Publication Data
A catalogue record for this book is available from the British Library

Springer London Heidelberg New York Dordrecht

© Springer-Verlag London 2015
This work is subject to copyright. All rights are reserved by the Publisher, whether the whole or part of the material is concerned, specifically the rights of translation, reprinting, reuse of illustrations, recitation, broadcasting, reproduction on microfilms or in any other physical way, and transmission or information storage and retrieval, electronic adaptation, computer software, or by similar or dissimilar methodology now known or hereafter developed. Exempted from this legal reservation are brief excerpts in connection with reviews or scholarly analysis or material supplied specifically for the purpose of being entered and executed on a computer system, for exclusive use by the purchaser of the work. Duplication of this publication or parts thereof is permitted only under the provisions of the Copyright Law of the Publisher's location, in its current version, and permission for use must always be obtained from Springer. Permissions for use may be obtained through RightsLink at the Copyright Clearance Center. Violations are liable to prosecution under the respective Copyright Law.
The use of general descriptive names, registered names, trademarks, service marks, etc. in this publication does not imply, even in the absence of a specific statement, that such names are exempt from the relevant protective laws and regulations and therefore free for general use.
While the advice and information in this book are believed to be true and accurate at the date of publication, neither the authors nor the editors nor the publisher can accept any legal responsibility for any errors or omissions that may be made. The publisher makes no warranty, express or implied, with respect to the material contained herein.

Printed on acid-free paper

Springer is part of Springer Science+Business Media (www.springer.com)

Preface

Given the "information overload"[1] erupting from an exponentially growing amount of data being produced, exchanged, and generated electronically,[2] Data design is now, more than ever, paramount in not only addressing the material nature and behavior of such piles of unstructured raw data, but also in ascertaining a spectrum of new user scenarios there within. A cross between the arts and design science, its focus is on deploying both practical and semantic types and solutions that not only represent data flows (data production, transactions, and generation) and data clusters (semantic data-mapping), but also visualize volume changes in line with a number of parameters and methods.

Doing away with traditional design models oriented around information-fed presentations, data design opts for another vision, striving to devise spaces for both display and viewing purposes that couple context with environment. In turn, greater visibility of the relational dynamic inherent in the myriad facets of the data itself[3] results, ultimately giving shape, substance, and stance to the myriad changes, "trajectories,"[4] and shifts in data.[5]

What then does it take to make this creative endeavor[6] possible? What design-bred steps or approaches are needed to display and view these genuinely complex, relationship-based data ecosystems, while at the same time digging up new user relationships[7] lodged there within?

Within the wake of these questions lies the research in question. While drawing attention to alternative ways of processing colossal volumes of data in motion, the research aims at shedding light through the ever-evolving eye of a designer on data

---

[1] This data massification, or "Big Data," has reached critical status in a multitude of areas.
[2] Current volume worldwide is nearing three (3)-zettabytes.
[3] The opportunities for data abstraction are on the rise.
[4] With regard to lines and curves drawn in space by various groups of data in motion.
[5] The realization that the nature of data is both immaterial and intrinsic.
[6] Based on the assumption that data is physical matter.
[7] This refers to the triggering of new applications and uses.

display and viewing methods. It only then makes sense that the spotlight be directed at designers whose expertise and expert eye on the matter unravel as broadly as possible the complexity inherent in the relational dynamics of data components[8] through unprecedented formal and graphic arrangements. With an approach whose axis rotates simultaneously around the senses as it does around sensations, designers illustrate that a visual and graphic take can and does factor into a better grasp on organizational rationale involving data sets and clusters, and, in turn, facilitate greater visual reasoning.

The work divides the designers' contributions into eight parts, which comprise articles and written interviews. The first seven parts of the book outline the scope of data design. Each designer-author has been assigned the task of writing a chapter based on a theme, topic, or issue of his choice.[9] What follows is a line-up of "viewpoints"[10] that bring the main fields and areas of interest[11] under the discipline's umbrella to the surface and offer an in-depth look into practices boasting both foresight and imagination. The eighth and final part features a series of interviews with data designers and artists whose way of working, as well as what results from it, embody originality and marked singularity. Walking a fine line between testimony and trust, declaration and demand, and perception and projection, these designer-artists offer their readers, on the one hand, access into the intricacies of their personal journey and commitment, and on the other, the most contemporary principles in data design.

Each designer-contributor also strives to provide new answers to the question, "What challenges await Data Design?" To avoid falling into too narrow a mindset, each works hard to explain the breadth carried today by design and its widespread application across a mix of business sectors. With end users in mind, designer-contributors lift the curtain on the myriad purposes for which the field was originally intended, forging the bond even more between data design and the aims and intentions of those who contribute to it. As a result, a number of enlightening concepts and bright ideas unfold within the confines of this book to help dispel the thick fog around this new and still relatively unknown discipline. A plethora of equally eye-opening and edifying new terms, words, and key expressions also unfurl. Informing, influencing, and inspiring are just a few of the buzz words belonging to an initiative that is, first and foremost, a creative one, not to mention the possibility to discern the ever-changing and naturally complex nature of today's datasphere.

David Bihanic

---

[8] In other words, what shapes, articulates, and establishes the guidelines specific to relational data.

[9] In relation to his/her own practical, conceptual or theoretical interests.

[10] Fueled by intuition, these viewpoints are backed by a solid foundation in critical thinking and judgment.

[11] These stem from the choice of themes, topics, or issues addressed.

Acknowledgments

With the support of the CALHISTE laboratory (Culture, Arts, Literature, History of Societies and Foreign Territories—EA 4343).

We would like to personally thank Krista Schmidtke and Ramona Bourhis for the translation and editorial support.

Contents

Part I   Depicting Data in Graphical Form

1   Show, Don't Tell.................................... 3
Ben Willers

2   Giving Shape to Data 23
David Bihanic

Part II   Visually Representing and Explaining Data

3   The New Aesthetic of Data Narrative................. 57
Giorgia Lupi

4   A Process Dedicated to Cognition and Memory 89
Wesley Grubbs

5   Graphics Lies, Misleading Visuals 103
Alberto Cairo

Part III   Mapping and Visualizing Data

6   atNight: Nocturnal Landscapes and Invisible Networks 119
Mar Santamaria-Varas and Pablo Martínez-Díez

**7   Visualizing Ambiguity in an Era of Data Abundance
     and Very Large Software Systems**...................... 139
Ali Almossawi

Part IV Interacting with Data

8 Living Networks 159
   Santiago Ortiz

9 Epiphanies Through Interactions with Data 175
   Dino Citraro and Kim Rees

Part V Exploring and Manipulating Data

10 Sketching with Data 189
   Fabien Girardin

11 Information Visualizations and Interfaces in the Humanities 207
   Giorgio Uboldi and Giorgio Caviglia

Part VI Translating and Handling Large Datasets

12 Big Data, Big Stories 221
   Richard Vijgen

13 *Dispositif* Mapping 235
   Christopher Warnow

Part VII Experiencing Data Through Multiple Modalities

14 Sustainability: Visualized 253
   Arlene Birt

15 Encoding Memories 283
   Sha Hwang and Rachel Binx

16 Changing Minds to Changing the World 293
   Scott Murray

Part VIII Interviews

17 Beauty in Data 315
   Jonathan Harris

18	**Tracing My Life**.. Nicholas Felton	341
19	**Multidisciplinary Design in the Age of Data**................. Stephan Thiel, Steffen Fiedler and Jonas Loh	353
20	**Designing for Small and Large Datasets**.................... Jan Willem Tulp	377
21	**Process and Progress: A Practitioner's Perspective on the How, What and Why of Data Visualization**...................... Moritz Stefaner	391
22	**The Art & Craft of Portraying Data**...................... Stefanie Posavec	405
23	**From Experience to Understanding**....................... Benjamin Wiederkehr	423
Author Index		443

Contributors

Ali Almossawi Mozilla Corporation, San Francisco, CA, USA

David Bihanic CALHISTE Laboratory, University of Valenciennes and Hainaut-Cambresis, Valenciennes, France

Rachel Binx Walnut, CA, USA

Arlene Birt Background Stories, Ames, IA, USA

Alberto Cairo University of Miami, Coral Gables, FL, USA

Giorgio Caviglia Stanford University, Stanford, CA, USA

Dino Citraro Periscopic, Portland, OR, USA

Nicholas Felton New York, NY, USA

Steffen Fiedler Studio NAND, Berlin, Germany

Fabien Girardin Near Future Laboratory, Sierre, Switzerland

Wesley Grubbs Pitch Interactive, Berkeley, CA, USA

Jonathan Harris New York, NY, USA

Sha Hwang Walnut, CA, USA

Jonas Loh Studio NAND, Berlin, Germany

Giorgia Lupi Accurat, New York, NY, USA

Pablo Martínez-Díez Tech (UPC), Barcelona, Spain; BAU Design College of Barcelona, Universitat de Vic (UVIC), Barcelona, Spain

Scott Murray University of San Francisco (USF), San Francisco, CA, USA

Santiago Ortiz Moebio, Buenos Aires, Argentina

Stefanie Posavec London, UK

Kim Rees Periscopic, Portland, OR, USA

Mar Santamaria-Varas School of Architecture of Barcelona (ETSAB), Universitat Politècnica de Catalunya, Barcelona, Spain

Moritz Stefaner Lilienthal, Germany

Stephan Thiel Studio NAND, Berlin, Germany

Jan Willem Tulp TULP Interactive, The Hague, The Netherlands

Giorgio Uboldi Politecnico di Milano, Milan, Italy

Richard Vijgen Richard Vijgen Studio, GT, Arnhem, The Netherlands

Christopher Warnow Wuerzburg, Bavaria, Germany

Benjamin Wiederkehr Interactive Things, Zürich, Switzerland

Ben Willers Lincoln, UK

Part I
Depicting Data in Graphical Form

Chapter 1
Show, Don't Tell

Ben Willers

Abstract Too often, I see visual representations of data used as a means to an end, reinforcing or adding creditability to an argument, or worse still an attempt to decorate or distract. Those experienced with handling large quantities of raw data will know of the pleasures that can be had from uncovering hidden truths buried within. These may be unusual and unexpected, sometimes even controversial. They provoke thought, spark conversations, and encourage further exploration. By carefully selecting data and display methods, we are able to communicate these discoveries without explicit exposition, allowing the reader to explore the rich data landscape that lies before us. In this article, I will discuss the challenges and opportunities when showing data in my work. Building on recent projects, I shall try to demonstrate how visualizations may not only reveal the meaning of data in a graphical manner, but also produce meaning through new styles of visual language.

1.1 Between Worlds

I usually find myself caught between two extremes. During my studies, I became aware of many hostile disputes between established practitioners working at the pragmatic and artistic ends of the data visualization spectrum, each displaying a stubborn intolerance for opinions other than their own. Convinced there was value in either approach; I began to flirt with styles influenced by both arguments. The intention was to experiment and develop visual depictions of data that were clear and informative, yet also enigmatic and captivating, opening communication opportunities with a diverse audience. The balancing act continues in my work to this day, although this has developed somewhat from those feuds relating to form and function.

B. Willers (✉)
12 High Street Reepham, Lincoln, LN3 4DP, UK
e-mail: ben.willers@yahoo.co.uk

© Springer-Verlag London 2015
D. Bihanic (ed.), *New Challenges for Data Design*,
DOI 10.1007/978-1-4471-6596-5_1

Much has been written about storytelling possibilities within data visualization, and there are opposing opinions here also. A rather popular choice is the narrative-driven approach, presenting the audience with a linear path to follow which is dictated entirely by the author. If we were to compare visualizations to video games (something else I share a deep enthusiasm for), these would be similar to on-rail shooters, a genre once popular in amusement arcades and designed to be experienced in short sessions, delivering an immediate feeling of satisfaction. These games deliberately impose limitations on the player, only permitting progression along a predefined route to deliver a series of highly scripted set pieces.

Alternatively, there are exploration-led visualizations where the audience is encouraged to look more deeply and uncover stories at their leisure. The gaming equivalent of these are known as open world or 'sandbox'-type experiences. These often forgo traditional video game goals, instead allowing the user to explore and create their own styles of play within the game world. While perhaps somewhat overwhelming from the outset, these titles often prove extremely addictive and a player can invest a significant amount of time exploring the possibilities on offer.

Out of these two aforementioned visualization styles, it is the exploration-based approach which I find more alluring. These pieces usually demand a greater level of reader persistence, although I tend to find them more rewarding and engaging in process. Being able to extract meaningful information for oneself is inevitably going to be more interesting, more relevant, and more memorable than had this instead been prominently signposted. Examining the bigger picture may reveal compelling insights that could otherwise go unnoticed, leading us toward a more comprehensive level of understanding. They may offer alternative explanations or inspire further questions, stimulating us to investigate still deeper. I see this as a classic case of getting more out the more you put in.

I am fully aware that not everyone happens to share this view. Exploration and discovery yield great rewards, but only to those who invest significant time and effort, something I fear the 'Google Generation' has little tolerance for. Let us consider the recent rise in popularity of the infographic. Designed to appeal to those whose time or patience is limited, they often present an excessively simplified view that can be consumed with minimal cognitive processing. A similar phenomenon also exists in the gaming world. Many titles have recently adopted an obtrusive handholding approach, prompting the player with hints or instructions when confronted with even the simplest of challenges, almost eliminating the need for these games to be 'played' at all. While perhaps appealing to more casual users, I usually feel this leaves us with a significantly diluted gameplay experience. It is not all doom and gloom; however, some more successful examples have managed to effectively combine tightly designed level structures with open world-style exploration. The path taken through these games is generally very similar for all players; however, they are encouraged to deviate from it and embark on optional side quests, rewarding the most thoughtful and observant players in the process.

Much of my recent attention has focused on combining these two user experiences in my visualization work, revealing patterns in data by gently guiding the reader toward stories of interest while also encouraging them to venture beyond the obvious to make discoveries for themselves. I recognize that in visualization there are no silver bullet solutions that can truly satisfy all parties, nor do I believe there should be. The depth and variety of work out there is something that should be celebrated, and my preferred approach is just one that make up this rich tapestry. My methodology is outlined across the pages which follow, along with some recent examples to show the efforts in my quest.

1.2 My Life in Data

I accidentally stumbled into the world of data visualization while studying on the MA Design program at the University of Lincoln. Without a clear vision in the beginning, I proposed to continue where my graphic design studies concluded, investigating consumer attitudes to brands. After conducting a short survey, I began using charts and diagrams to present the results. I also started documenting my daily progress and other activities using a visual logbook, featuring a small information graphic to represent each day. It soon became apparent to myself and others around me that these aspects were more interesting than the project itself. Already a third of the way through the course, I rapidly resubmitted my proposal to make the visual display of data the focus instead.

Further research and exploration in this area left me feeling torn. I sympathized with the views of visual perception experts, but also with those working on more sublime images. Tufte's theory of data–ink ratio was especially appealing to me, yet so was McCandless's playful approach. I began experimenting with styles influenced by both sides, hoping to appeal to the widest possible audience, although fearing my efforts would ultimately fail to please all concerned.

During the major project phase, I created a series of experimental visualizations using data I collected through daily self-monitoring. These included charts showing my eating, walking and sleeping patterns, as well as my body weight, bank account activity, and television viewing. A final piece (Fig. 1.1) summarized my life to date, displaying more subjective information by mapping my general mood against various events during my 27 years of existence. The MA experience provided me with an opportunity to handle data for the first time and experiment with various visual encoding methods, laying the foundations for much of the work I have produced since.

I was initially hesitant to include any titles or legends in my work, remaining quite content with the abstract visual forms that were emerging from the data. As the weeks progressed, I continued to monitor and record my activities, and with each day that passed, I eagerly compared the newly gathered information with that which preceded it. Soon, I began to observe some startling patterns. Although only evident to myself at this stage because of my familiarity with the data, others around me were intrigued once I began to explain my discoveries. Rather than telling others about what I had observed, I was keen to invite them to see whether they could identify the same patterns I had become aware of. Eager to avoid any undue

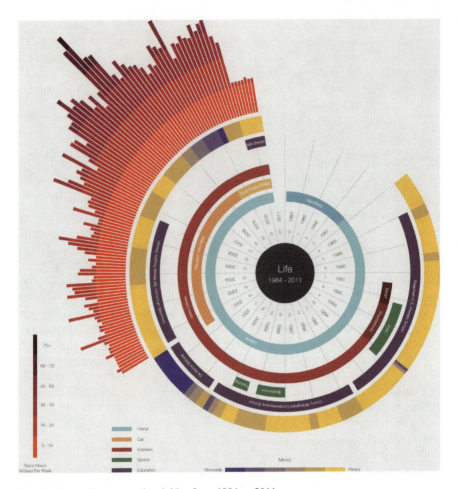

Fig. 1.1 Personal events and activities from 1984 to 2011

exposition, subtle labeling was gradually introduced so that others could explore these relationships without interference from myself.

Considerable efforts were made to eliminate visual noise by removing visual elements that did not enhance our ability to interpret the data. I later came to appreciate this approach does not necessarily deprive a visualization of its esthetic charm. In the same way that many find Bauhaus designs alluring because of their radically simplified forms, visualizations may also benefit from a clean and honest visual appearance.

Unfortunately, the most effective and widely understood visualization formats like bar charts are now so ubiquitous that they rarely possess an ability to mesmerize or enchant us. While this should be of little concern to those designers working on more pragmatic pieces for an already captivated client, these display methods are unlikely to entice many casual observers. A more experimental approach can be advantageous in this respect, luring unsuspecting eyes toward the data, ensnaring the reader as they attempt to decipher these enigmatic patterns. Conversely, excessively obtuse visualization techniques may potentially leave the reader confused and frustrated. It is important to consider for what purpose the data will ultimately be used and to design accordingly. A more artistic approach would unlikely be appreciated if the intention was for rapid decision making.

My intention throughout my studies was test the middle ground, seeing how far I could flex the conventions before they began to snap. Using this experimental approach, I hoped to discover display approaches that would be exciting to read, yet still manage to make sense on an intuitive level. Each piece was also designed to be visually distinct and reflect the theme of the data through the inclusion of visual metaphors. Opposed to superfluous decoration, I decided these should be reflected directly through the encoding methods used. For example, a visualization of my credit card activity in 2010 (Fig. 1.2) was inspired by barcode patterns, while a piece showing time spent asleep over a 4-month period (Fig. 1.3) was influenced by diagrams showing phases of the moon. These references are of course open to interpretation and are not required to be detected at all, yet they will hopefully serve as a pleasant surprise, rewarding the most attentive viewers.

An encoding technique notably absent throughout my MA work was the use of area to represent quantitative information. Seemingly beloved by infographic and data journalists, I could not bring myself to employ a display method as fundamentally flawed as this. It has been demonstrated through numerous scientific studies that the brain is a poor judge of area, typically underestimating the size of shapes larger than the standard. The use of circles is particularly troublesome in this respect, as Willard Cope Brinton discussed a century ago in his pioneering book,

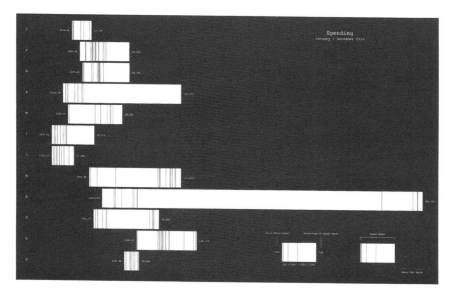

Fig. 1.2 Credit card purchases in 2010

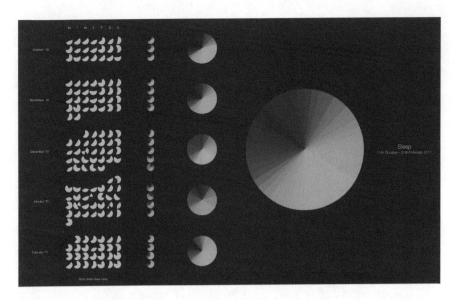

Fig. 1.3 Time spent sleeping from October 2010 to January 2011

Graphic Methods For Presenting Facts (Brinton 2007). Although detested by a few, the encoding of data using the area method remains a popular visualization choice among many designers, probably due to the ease with which they can be created and the flexibility they bring to layouts.

Pie charts were deemed another cause for concern, especially when inviting a reader to compare segments across multiple examples. Evaluating angles and arc lengths introduces additional workloads on the brain, impairing our ability to perceive patterns with ease. My dissertation focused on the challenges, limitations, and merits of these controversial approaches, helping me to make more informed decisions in my work which followed.

1.3 Money Troubles

A graphic published by *The New York Times* (Marsh 2011) in October 2011 on the prevailing global financial crisis caught my attention. Rather than attempting to demystify, it instead appeared to celebrate the complexity of the situation by presenting an incredibly convoluted flowchart with countries represented as circles scaled by area according to their gross domestic product. It was certainly eye-catching, perhaps even hypnotizing. A cascade of bubbles was shown scattered across the page connected by a sequence of arcs, evoking images in my mind of electrons caught in a series of bizarre orbits. The graphic was accompanied by extensive editorial copy, suggesting to me that the journalists were unconvinced that the graphic could communicate effectively alone. I spent some time attempting to

1 Show, Don't Tell

unscramble it, but came away not much wiser. Frustratingly, the part which I felt was most interesting (gross government debt as a percentage of GDP) was not even attempted to be represented visually; instead, this was simply included as a percentage figure inside each bubble and struck me as being somewhat of an afterthought.

When tasked to produce a similar visualization on the financial crisis in Europe (Fig. 1.4), I was keen to avoid much of the confusion caused by the aforementioned piece. Most visual perception experts will proclaim that line length and two-dimensional positioning are the encoding methods the brain can most easily comprehend. Unfortunately, as I discussed previously, many people simply do not consider bar charts to be very alluring. My response was to consider alternative layouts that would appear fresh and exciting, while also clearly demonstrating the net result from two opposing factors. GDP per capita and government debt per capita from 2000 to 2010 are represented by a series of bars sprouting in opposing directions to reflect the positive or negative effects each can have on a county's economy. As the two exert away from each other, a further series of bars between the two reveals government debt as a percentage of GDP, effectively allowing us to see which is leading the tug-of-war battle. Using a baseline of 100 % (where the two are in equilibrium), these bars either grow upward or downward toward the force which is greater. To further accentuate the notion of growth or contraction, positive and negative aspects are colored green or red accordingly.

Countries with the highest debt-to-GDP ratio in the most recent year (2010) are displayed on the left of the page (Fig. 1.5), while those with the lowest proportion of debt are shown toward the right. When presented like this, it is easy to see why some countries like Greece, Italy, and Ireland were focused on so heavily in the news around this time. The two least worrisome countries (at least according to this method of analysis) appear to be Luxembourg and Estonia, the first because of an extraordinarily high GDP per capita, the second because of virtually no reported government debt.

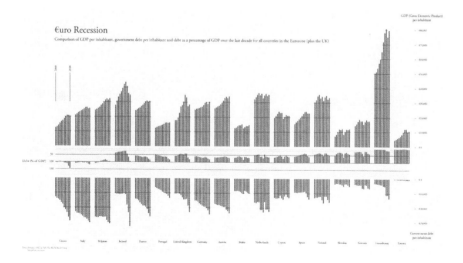

Fig. 1.4 Effects of the early 2000s' recession in Europe

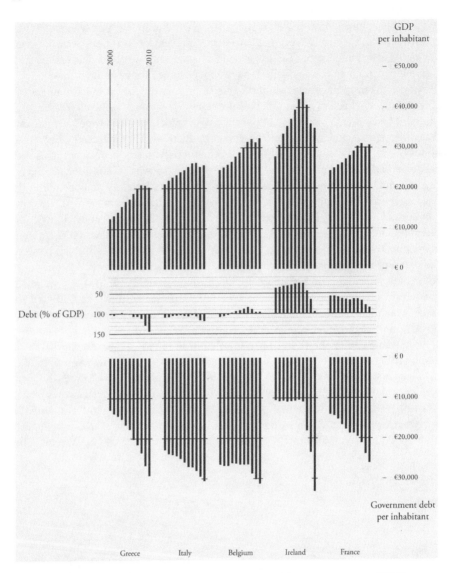

Fig. 1.5 GDP per capita, government debt per capita, and debt as a percentage of GDP

Showing data from the previous 11 years allows the reader to not only see how each country currently ranks among their European neighbors, but also how successfully each has weathered the financial storm. Some like Germany have remained fairly constant throughout the last decade; meanwhile, others like the United Kingdom have experienced a roller coaster-like journey, with GDP levels climbing dramatically in the early 2000s, then plunging severally in the years which followed. Using per capita figures is essential if we are to fairly compare countries of different populations. The reader is provided with an opportunity to freely peruse the data in a manner of their pleasing, allowing everyone to walk away having gained a different, yet equally valid, level of insight.

Efforts were made to remove unnecessary clutter from the design. Even the gridlines which would normally accompany a piece such as this were reduced to small dashes at points intersecting with the bars, hopefully making the graphic appear less congested and more inviting to casual onlookers. While experimenting with layouts during development, I noticed the piece begin to adopt a city-like form, one which I thought was especially appropriate for a visualization based on economic performance, and something many observers have commented on since. The graphic was generally well received, winning the competition set by the Information is Beautiful group for which it was created. This was the first opportunity I had to visualize data beyond the confines of the MA, and judging from the response generated, my first real evidence there was validity in my approach.

1.4 A Story to Behold

A criticism I sometimes hear concerning my work is the lack of story or clarity of message shown. I once asked a client to clarify this a little more and what came next astonished me. 'More text' was the reply. Clearly, we had different opinions about what a story should entail, my theory being that explanatory copy hardly takes advantage of the unique qualities visualizations bring.

Journalists, novelists, and historians will typically embrace storytelling in very different ways within their respective fields. This ambiguous term can even be interpreted in various ways within the data visualization community. Infographics typically borrow heavily from more traditional forms of written media, directing the reader in a linear fashion using text to drive the narrative while dispensing nuggets of data in a logical sequence to support a core message. Exploration-led visualizations may not present such an obviously structured narrative to follow, but I believe these are no less capable of revealing captivating hidden insights.

There are some that may argue that these pieces are divest of 'story' and that any observed patterns are purely coincidental. I, however, appreciate that these are often the result of careful data selection and meticulous planning. Some of my favorite visualization examples are those which manage to convey a sensation of thought-provoking exploration without the audience realizing they are being lured down well-trodden paths. Stories in these pieces do not occur organically through smart presentation alone, and it would be a mistake to assume that these visuals simply present all the information which is available and hope the reader is able to extract meaning from it. The designer adopts the role of an author by carefully selecting the data, not only determining how a story is to be shown, but also what stories are featured to begin with. Display methods are then selected that may gently nudge the user toward areas of particular interest by considering the path the eye will naturally wish to follow. The impression acquired will be dependent on the user's own choices and judgment, and also how successfully they are able to capture the hints left by the designer. Those adventurous readers will hopefully choose to continue down a series of alternative routes, uncovering further stories along the way, prompting further questions and hopefully providing an incentive to adventure deeper still.

Pacing is an important factor we must consider in any form of storytelling. By enticing the reader to peel away layers from the top, we can influence the reading flow, allowing them to survey the lay of the land before adventuring deeper into parts unknown. The first impression received could be from a macro perspective showing something readily known or expected. Although perhaps not immensely enlightening, slowly introducing the reader to the data in this way allows them to build familiarity with the subject, while also providing them with the confidence to question and analyze other areas more intensely later on. As they continue to drill down, they could be presented with a series of microviews, or perhaps an alternative perspective on the data already shown. If all goes well, the users will feel themselves growing in confidence and ability alongside the increasing complexity of the data, providing an escalating feeling of progression and satisfaction throughout their journey of discovery.

Visualizations which go beyond the ruthless objective of communicating a specific narrative can be construed in a variety of ways by different readers, and each interpretation may be as equally as valid as the next. There is a risk, however, that the impression obtained will be at odds with the sentiment of the author, a concern if the intention of the visualization is to persuade. These alternative impressions also risk confusing those readers who seek clear and definitive answers. Clearly, there are situations where this approach would not be desirable; however, there is sometimes merit in exploring the wider picture and recognizing that the world is not always as black and white as others sometimes try to make us believe.

1.5 Playing with Data

Whenever embarking a new project, I begin by becoming intimately familiar with the data available. This will inevitably lead me to developing a series of hunches and questions which I will proceed to investigate carefully. Additional datasets are sometimes introduced so that I can test further theories, or the information may be distilled to focus on specific aspects. Spreadsheet programs like Microsoft Excel are capable of generating some fairly generic looking charts; however, I find these tools indispensable, allowing me to rapidly digest the information while experimenting with various data configurations. This process continues until a series of compelling patterns begin to emerge that were not immediately apparent through studying the numbers alone.

Sometimes, the most intriguing stories are those which I never set out to discover. While recently researching average body mass index values for adults around the world (Fig. 1.6), I instinctively wanted to test a theory that people living in countries with a high Human Development Index would have a greater BMI than those in less developed areas. This indeed turned out to be true in many cases; however, the trend was not as strong as I had expected. Further experimentation led me to something I had not anticipated and far more curious. It appears that in more

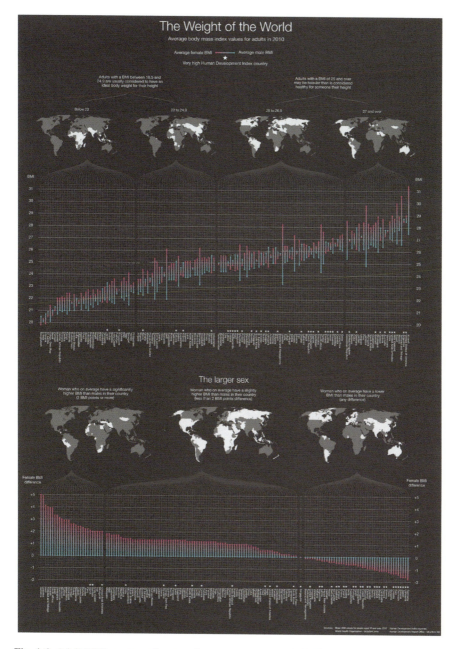

Fig. 1.6 Adult BMI average of men and women per country in 2010

developed countries (particularly in Europe) the men usually have a higher BMI value women, while the opposite is true within many poorer regions, sometimes by quite a considerable margin.

The questions I instinctively ask myself when first laying eyes on the data are probably going to be similar to those asked by the reader who surveys my work. By reflecting on my own initial impressions, I hope to get a peek inside the heads of the audience, anticipating how they will approach the data and what theories they would likely wish to test. This is why I generally prefer visualizing subjects I have no prior knowledge of, or starting from an experience level I anticipate the reader to initially be at. The process of acquitting myself with the data allows me to view it more objectively, perhaps noticing patterns I may have overlooked had I approached it with preconceived ideas. This becomes increasingly challenging as my familiarity with the data grows throughout the project, so I make a habit of recording my early reactions in a sketchbook, consulting these notes throughout the development to ensure I am able to satisfy my initial queries, and in turn hopefully those of the audience.

While an infographic will typically distill the information right down so that only the core message remains, visualizations by their very nature tend to be more data rich. One of the greatest challenges I face in any project is deciding how much data to feature. Show too much and I risk complicating the structure and overwhelming the reader. Show too little and the audience will be deprived of opportunities to see stories within a wider context. I always endeavor to present information as impartially as possible, and certainly never resort to trickery or deceit. Being completely honest implies that nothing is left behind; however, I usually find that being selective to some degree is inevitable. Using the BMI piece again as an example, featuring every country would have almost doubled the width of the page, resulting in excessive horizontal scrolling on all but the highest resolution displays. In this case, I decided to remove all countries with a population of less than one million, leaving a still respectable 153 examples to compare. This has the potential to dramatically alter the shape of a story, and decisions must be rationalized carefully. Remaining completely neutral is often more challenging than it may outwardly appear.

Once familiar enough with the data, I begin to sketch ways that I feel would organize it most clearly. The process begins with paper and pen, advancing to crudely drawn screen prototypes using samples of real data. There will inevitably be successes and failures along the way, as is the nature of the design process. I cwholeheartedly endorse this organic approach. Forcing data to fit predetermined display models rarely produces satisfactory results in my experience. I found the early stages of my MA incredibly frustrating as the data I had rarely cooperated with my artistic vision. I later realized I was placing too much emphasis on the visual style from an early stage. Consideration to esthetic qualities should be secondary to those of content, for a visualization is admired for the insights we extract from within, not those patterns which are merely appealing to the eye.

1.6 Shifts in Power

The United Kingdom, if you believe the tabloids, is currently on the brink of an energy crisis. Rising fuel costs and dwindling resources are forcing us to look elsewhere to satisfy our thirst for power. Much of the media's attention has recently focused on the uptake of renewable energy, but how significant are these changes in the grand scheme of things?

Energy mix data are traditionally visualized using the stacked bar method, displaying total energy consumed, plus the type and amount of each fuel utilized to generate this. The stacked bar is, however, an inherently flawed display method if the intention is to encourage comparisons within. The only values which we are able to reliably compare are the total, and those bars sat directly upon the baseline. All others are displayed haphazardly, their elevation determined by the height of any bars below. To make judgments, the eye must laboriously meander from one to the other, a task which makes direct comparisons challenging to say the least.

My solution was to split the total and part-to-whole relationship comparison tasks into two (Fig. 1.7). Using energy consumption data from 1970 to 2010 in the UK, I displayed total energy using tall gray bars, grouping these into four distinct decades. A series of narrower color-coded bars were layered on top of these to show a breakdown by fuel type per year. Crucially, though there is no stacking involved, all elements are aligned to the same baseline so that trends are hopefully easier to follow.

I am extremely fond of visualization techniques which allow readers to explore the same data from a variety of angles. This is an area where interactive visualizations really excel, allowing the user to switch and compare views at will. Unfortunately, this is a luxury that static designers like myself do not get to work with. An approach I have used on many occasions is to repeatedly display the same information over and over, but from a series of alternative perspectives. In this case, the page is divided into three sections. The first is intended as a gentle introduction and offers the most straightforward view, focusing purely on the total energy consumed per year. The second section grows in complexity, again showing energy usage but breaking this down by sector, showing the total amount used by industry, transport, domestic, and service areas. Additional layers are also introduced here to reveal a further breakdown by fuel type, allowing us to see within each sector how much was generated by solid fuels, gas, renewables, etc. The final section repeats the second but flips it on its head, displaying energy consumption by fuel type first, followed by a breakdown by sector.

Explicit signposting has intentionally been omitted, allowing the reader to freely explore and interpret the information without any exposition from myself. Although strictly only displaying energy consumption in the UK, the data help paint a much wider picture. Industry appears to have slumped dramatically, while other areas

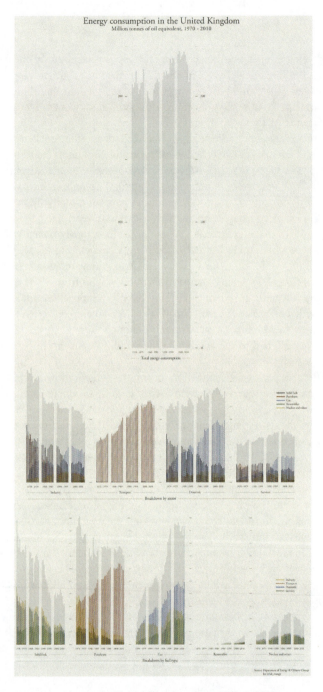

Fig. 1.7 Energy consumption in the United Kingdom from 1970 to 2010

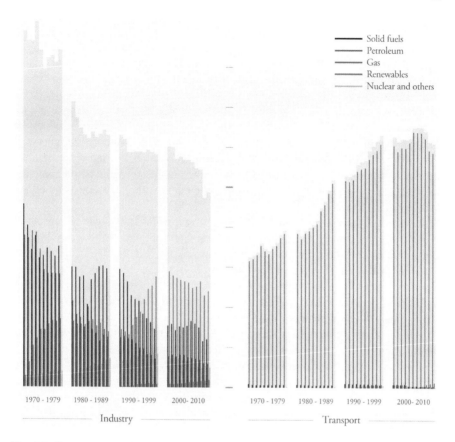

Fig. 1.8 Energy consumption trends in industry and transport sectors

continue to soar. All sectors are powered from a wide mixture of fuels, except for transport which remains predominately petroleum (Fig. 1.8) (it will be interesting to see how we adapt once this resource runs dry). Renewables are slowly on the rise, but still appear as a tiny blip on the radar when compared to gas and others.

An attentive reader will acquire even more as they delve deeper still, perhaps discovering multiple layers of interwoven stories. I find the year 1984 particularly intriguing as energy produced from solid fuels dipped dramatically, while the amount generated from petroleum temporarily rose to fill the void. The graphic offers no explanation for this, and does not even highlight it in a pronounced way. Most viewers will likely overlook this detail, or fail to make a connection between this anomaly and the miners' strike from this time. Those who are knowledgeable and thoughtful enough to connect the links and observe something beyond the obvious will hopefully gain a small sense of satisfaction from their discovery.

1.7 A Familiar Tune

The Eurovision Song Contest has for some time been one of my guilty pleasures. Not because of the quality of the music of course, I am endlessly fascinated by the blatant bloc voting between certain members. Whether these alliances are politically motivated or a result of shared musical tastes remains a subject of debate, but the statistics clearly show a series of like-minded clusters dotted across the continent. The most prominent example is Cyprus and Greece, each awarding the other with that elusive maximum of 'douze points' year after year. Following these trends is becoming quite a challenge, especially since the number of participating countries swelled to forty-three in 2011, each awarding their 'favorite' acts between one and twelve points.

Rather than simply highlighting a few key examples, I was keen to present an expansive view, hoping that signals would emerge from within the noise. Initially concerned there would be few notable patterns beyond the famous examples; I devised a system of grids and colored boxes to test the data for myself. I was relieved to see so many rhythms begin to emerge from the data as the visualization took shape, and this method of display was deemed so successful that I applied it directly into the final design (Fig. 1.9).

Fearing that this incredibly large and complex-looking graphic would appear intimidating on first inspection, I decided to add an additional section serving as a gentle introduction. The reader is initially greeted with a more familiar-looking series of bar charts displaying total points won by each country in the last decade. Although interesting to a certain degree, particularly if you wish to look up how successful your own country has been in recent years, the intention was to provide a simple mental warm-up exercise, preparing the brain for the more interesting task to come. It also has the benefit of establishing the 10-year period that is used across the visualization.

I deliberated quite extensively on how many years worth of data to feature, considering options for five or eight, but eventually settling on the previous ten to provide an even wider view. The scale of the graphic was already beyond anything I had attempted before, leaving me concerned that those viewing on low-resolution displays would not be able to appreciate the full picture. Unwilling to compromise on the scope, I made great efforts to eliminate redundant information, squeezing everything together into a tightly condensed package (Fig. 1.10).

The reader is able to scan the entire image and quickly identify those areas of particular interest. Voting countries are displayed along the vertical axis of the page, while those receiving points are shown along the horizontal axis. A series of blocks where these intersect reveals the number of points awarded in competitions from 2002 to 2011. These are arranged in ten columns from left to right, one for each year accordingly. Points awarded are shown in orange, and points not awarded are shown in white. Years displayed in gray show that either one or both countries were absent from the competition on that occasion. Scanning the image for large clusters of similar colors makes it abundantly clear who has historically scored others

1 Show, Don't Tell

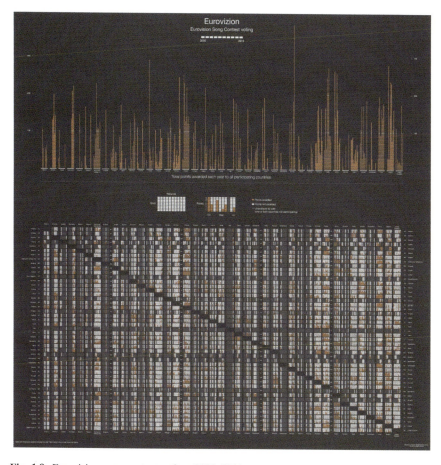

Fig. 1.9 Eurovision song contest voting, 2002–2011

highly, and interestingly which countries have generally been unfavorable toward each other, allowing us to make informed predictions about competitions in the future. I took great pleasure pouring over this piece during the televised coverage of Eurovision 2012, successfully foreseeing many of the scores before they had even been declared.

1.8 The Missing Link

My approach to visualizing data stems from my own curiosity and a desire to explore. While some established narrative techniques often provide only a blinkered view, multilayered and nonlinear graphic representations of data allow us to make more informed judgments and comprehend on a much deeper level.

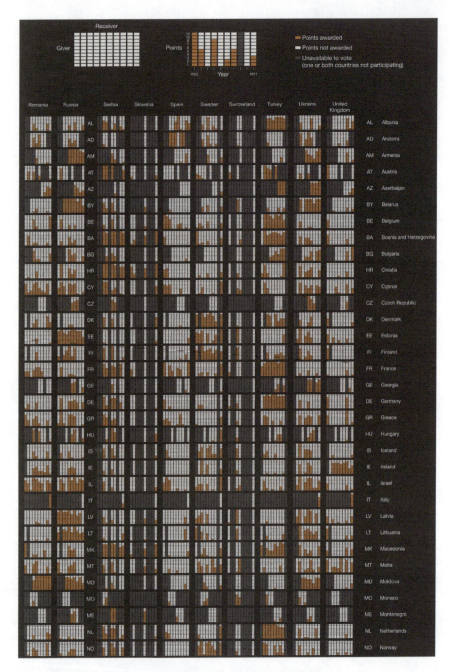

Fig. 1.10 Points awarded and received by each participating country

One of the greatest challenges I face in my work is finding a balance that does not undermine the richness of the data, yet is still accessible and engaging to the reader. Many of us have been conditioned to read in a particular way, along a linear path in a one-dimensional fashion. Unfamiliar territory is inevitably going to cause many readers, and perhaps some designers with a degree of anxiety, which may explain why I see so many displays of data that simply mimic the traditional reading style. I always assume my reader is highly capable of judgment and reasoning, possessing a desire to learn but lacking knowledge within a particular area. My heart always sinks whenever a client asks me to simplify the scope of a design and turn up the visual 'cool factor' when a multidimensional approach would be my preference. It is my hope that as audiences become more accustomed to these alternative methods of reading, designers will grow in confidence and deliver thought-provoking visualizations on a more regular basis.

To my eyes, the most exciting graphic representations of data are those which offer a fresh perspective, opening windows and broadening our outlook. These may be used to explain, or allow us to explore. Those who endeavor to explain do so by answering five fundamental questions: what, who, where, when, and why. My approach, perhaps controversially, often involves providing answers to only the first four in this list. I believe stories in data should be built on a solid foundation of facts, preserving impartially which allow the audience to arrive at their own conclusions based on the indisputable evidence provided. Introducing opinions into the mixture risks clouding our judgment, and so the question of 'why,' I feel, is sometimes best left for the reader to resolve. Inviting audiences to participate in this manner allows for a more engaging and credible stories to be shared, while offering the reader an unparalleled level of understanding in the process.

Writers will sometimes use open endings to mentally involve their readers, a technique that can prove effective even after they have stepped away from the text. Rather than passively accepting a story at face value, the audience internally questions the deeper meaning and the implications this may have on the future. Exploration-led visualizations may sometimes leave matters equally unresolved, prompting further questions and sparking conversations. Rather than attempting to leave the viewer completely satisfied, these pieces encourage us to venture onward to broaden our horizons. The quest for knowledge is one which I hope is never truly complete.

References

Brinton WC (2007) Graphic methods for presenting facts, New edn. Kessinger Publishing, Whitefish

Marsh B (2011) It's all connected: a spectator's guide to the Euro crisis. The New York Times, URL, January 11, 2014, http://nyti.ms/1lZ748j

Author Biography

Ben Willers is a freelance Graphic Designer who enjoys exploring and creating data rich visualizations. After graduating from the University of Lincoln in 2012 with a MA in Design, Ben's work featured on the cover of IdN Extra 07, a special edition of the international design journal dedicated to infographics and the visual display of data. He also won a competition set by the Information is Beautiful group and had work exhibited at ExpoViz 2012, a public exhibition held at the Campus des Cordeliers in Paris.

Ben currently works from London on a wide variety of projects for a diverse set of clients, including the Qatar based news group Al Jazeera, the US magazine Popular Mechanics, and marketing and research companies in Europe and the US. He primarily works on static visualizations for screen and print, however he has also collaborated with organizations and individuals on interactive projects. The Startup Universe, an exploration led piece Ben worked on with Visually and Accurat was shortlisted in the 2013 Information is Beautiful Awards interactive category.

For more information and contact:

http://www.benwillers.com
http://uk.linkedin.com/pub/ben-willers/27/986/274
https://twitter.com/b_willers

Chapter 2
Giving Shape to Data

David Bihanic

Abstract A key challenge for designers is to be able to present large amounts of data in very clear and simple ways. By this, it involves bypassing, if not eliminating entirely, sophisticated forms and convoluted graphics, as well as countless textual metadata (legends or notes). High-dimensional data space, requiring the use of advanced user tools to navigate inside, is also to avoid. What it boils down to is nothing other than data assimilated into straightforward and captivating forms. The aim is to design primitive or minimal shapes stripped down to their very essence for the purposes of delivering both a sensitive and meaningful experience of datasets—this kind of experience leads us to the fundamental principles of *Gestalt* perception. In this chapter, we shall attempt to prove that visual and graphic representation of "Big Data" needs to shift toward a new formalized approach using pure shapes to enhance the qualities of its aesthetic perception (and improve interaction with data representation and visualization)—note that the possibility to infer phenomena from data depends mainly on the aesthetic experience of the data itself. To do so, we will refer to some cutting-edge data design projects. It will also be an opportunity to introduce a new series of large-scale data representations currently underway, entitled *Data Shapes*, in which the simplicity of forms is addressed.

2.1 Introduction

Over 40 years ago, certain computer science laboratories specializing in human–system interaction set in motion significant research programs targeting new ways to display and describe information and knowledge. With this in mind, the majority of studies and experiments carried out unveiled a series of unique methods, techniques, and procedures designed to present metric- or statistic-oriented scientific data

---

D. Bihanic (✉)
CALHISTE Laboratory, University of Valenciennes and Hainaut-Cambresis, Le Mont Houy,
59313 Valenciennes Cedex 9, France
e-mail: david.bihanic@univ-valenciennes.fr

© Springer-Verlag London 2015
D. Bihanic (ed.), *New Challenges for Data Design*,
DOI 10.1007/978-1-4471-6596-5_2

(Kruskal 1977, 1972; Kruskal and Hart 1966; Freeny and Gabbe 1966; Freeny et al. 1969; Caroll and Chang 1970).

As time went on, research naturally evolved to a more holistic approach on how new systems built viewed information, not to mention the number of avenues[1] there within. From data-viewing to data-processing, these initial, theoretical findings were combined with more practical ones pertaining to how cognitive systems (human, artificial) work and interact. Approximately 20 years later, this crossroads of thought drew in a multitude of experts from around the globe aiming to create visual and graphic displays of information[2] for the fields of engineering and cognitive science (behavioral psychology). Among those fueling this initiative were Card et al. (1999), Robertson et al. (1989), Stasko (1993, 1996), Hollan et al. (1986), Hutchins et al. (1985), Hutchins (1995) and Furnas and Bederson (1995), as well as various laboratories affiliated with the Xerox PARC (Palo Alto Research Center), the University of California, San Diego (UCSD),[3] the University of Maryland, Georgia Institute of Technology (Georgia Tech),[4] Virginia Polytechnic Institute and State University (Virginia Tech),[5] and the laboratories of IBM, AT&T and Bell.

Aware of the infinite possibilities at its disposal, the field of Design decided to take matters into its own hands in the 1990s and seize the opportunity wherein to give rise to a creative entity that would embody its expertise and know-how better known as data design. With a focus on technically complex issues, the newly created field sought to provide more formal,[6] Enactment of rationale principles. "interface-based"[7] input from the angle of new facets specific to the physical image,[8] alongside consideration attributed to processes of perception and cognitive representation (see Denis 1989). As a result, the interface transitioned to one wherein both expression and representation of the informational complexity reached closure—for should the quality and accuracy of information-display systems depend on the interface, it would only be correct to say that formalizing the process and model paved the way for the meaning that ensued and the significance thereto related. Be they graphic shapes and symbols, colors, movements and sequences or *interactors*[9] of data manipulation, all of the above act also as influential parameters in an ongoing transcription pattern of informational reality or a partial viewing of it.

---

[1] In reference to various contexts (situations, tasks).

[2] E.g., treemap, touchgraph, and complex network graphs.

[3] Home of the Distributed Cognition and Human–Computer Interaction Laboratory overseen by James D. Hollan and Edwin Hutchins.

[4] Home of the Graphics, Visualization, and Usability [GVU] Center.

[5] Home of the Laboratory for Information Visualization and Evaluation [InfoVis].

[6] Creative process and artistic expression, as well as the enactment of shapes generated (designs, indications, illustrations, symbols, etc.) in a system-to-structure pattern.

[7] In reference to the intersection of informational (semiotic), visual and graphic (aesthetic) and functional (technical) patterns.

[8] In Ancient Greek, αἰσθητικός, aisthêtikós, meaning "who perceives through the senses, perceptible."

[9] Referring to components or interactive objects.

Due to an exponentially growing amount of data being produced, exchanged and generated electronically nearing the three (3)-zettabyte limit worldwide, new headway across technical, functional, ergonomic, and aesthetic fronts is now necessary to confront this epic phenomenon known as "Big Data." Traces of this "informational avalanche" are inadvertently swept under *virtually* endless piles of confidential data from government agencies of countries and nations, not to mention market-fed (stock exchange) financial data or "moneymaker" corporate data, and free/public data from the Internet (e-mails, social networks, blogs, forums, online groupware,[10] "Open Data," etc.), as well as from several other sources bursting with ever-growing quantities of data. The saying, "Desperate times call for desperate measures" never seemed more appropriate. A plague-like spreading of data on the loose is in urgent need of a new take on long-standing models and systems of information representation that can offer structure and substance. The stakes erupting from this revelation concern, first and foremost, those in Engineering Science, though it is clear that other fields could benefit from the revamp efforts, and namely that of design whose job it will be to invent and devise a new offering adapted to how data is presented and processed.

It is this latest contribution by the field of Design to data representation and visualization that we would like to examine more closely within this chapter. We will first take a look at how Data designers, while drawing attention to alternative ways of processing data, go about revisiting methods and techniques of its presentation. We will meticulously dissect designers' expertise and expert eye on the matter, and more importantly, how they unravel or rather *translate* as broadly as possible the complexity inherent in the relational dynamics of data components through unprecedented formal and graphic arrangements; in other words, what shapes, articulates, and establishes (on the micro/macro level) the guidelines specific to relational data. We will then focus on a number of highly theoretical, *Gestalt*-compliant approaches being unveiled by some Data designers. Pushing the limits of the notion to extremes, these designers are devising and applying a *new minimalist* approach to visual and graphic forms of massive dataset representation whose premise is on envisioning, understanding, and recalling the fundamentals. The emphasis now extends beyond the arbitrary shuffling about of meanings conveyed by datasets and clusters. It encompasses a sustainable movement that seeks to document the tucked away, secret, or hidden meaning there within. Having recourse to an arrangement of pure, uncomplicated, raw, and serene shapes, as if stripped of all graphically superfluous elements, the end result boasts an optimistic promise of new, data-derived aesthetic perceptions and insightful outcomes.

---

[10] Whose data are stored in public and/or private clouds.

2.2 Designing Data

Data design strives to place emphasis on the design of user-driven systems for data-processing purposes. For this to happen, the field calls upon various research done in cognitive science aimed at laying the foundation for "dynamic cognition" (see Dokic 2001; Abrioux et al. 1999) (in reference to a theory on cognitive dynamics; see Dokic 2010). This research assumes that a wide range of relationships exists not only between cognition and time (temporal perception), but also, and here lies the major point of interest, among cognition, perception and action (time-driven interactive sequence). This means that any change or any cognitive variation or evolution in an individual depends almost essentially on outside factors and phenomena. As Denis Brouillet and François Lang (Brouillet and Lang 2012) write, these changes, variations, and evolutions near "...the expression of a broader dynamic within which perception, cognition, sensation, emotion and action undergo circular and systemic causalities in a given environment and temporality."

In light of such assumptions examined more closely by George Lakoff and Mark Johnson (Lakoff and Mark 1980), Ronald Langacker in Cognitive Semantics (Langacker 1987, 1991), as well as Jean Piaget (Piaget and Chomsky 1979), Lev Vygotski (Vygotski 1986), David Rumelhart (Rumelhart and Norman 1975), Timothy T. Rogers, and James McClelland in Psychology (Rogers and McClelland 2004), Data designers have decided to put them into practice[11] by placing the user at the heart of interface-based data representation. The resulting developments require that a certain number of conditions be met, triggering an expanse in the decision-making or judgment process and offering a spectrum that spans from *relative judgment* or low decision-making ability to *absolute judgment* or high decision-making ability. The assumptions establish new visual and graphic possibilities able to view and process a growing number of items. These models of data display deploy a series of perception-driven tactics (see Latour 1985) that encourage discovery (visual data mining), judgment and analysis (visual synthesis).[12] How these tactics play out depends mainly on the user's perceptive–cognitive (underlying visual processing) and memory-based abilities. With this in mind and given that data representation models discern and decipher almost instantaneously basic characteristics (primitive perceptive processes) such as line orientation, length, thickness, size, curvature, cardinality, terminators, intersection, inclusion, color, flicker, direction of motion, stereoscopic depth, 3D cue, lighting direction, etc., it is easy to see how they reap all of the benefits specific to "preattentive" processing[13] as captured so well by Christopher Healey (Healey 1992).

---

[11] Referring here to *poiesis* (and not *praxis*), from Ancient Greek, ποίησις (poiêsis), meaning "doing, creating, bringing about, leading something into being"—Aristotelian distinction.

[12] In reference to isolating and differentiating structural, and at times, complex "motives," including trends, groups, "gaps", and isolated areas.

[13] Referring to a user's "low-level" visual capacity taken largely into account today by numerous designers and engineers.

From an interactive standpoint (and no longer here just for representation and visualization purposes; see Spence 2007), they call upon the user's faculty to isolate, distance, and thus differentiate certain elements, nonetheless, present within our visual field. Data display models rely on a handful of seasoned techniques such as "dynamic filtering" whose premise uses qualitative variables to first configure how items will appear, and in turn, stimulate and shape our differential analysis ability.

Consequently, and regardless of from where these representation models stem, be it in the research done by Jer Thorp, Santiago Ortiz, Benjamin Fry, Jeff Heer or Moritz Stefaner or that of Martin Wattenberg and Fernanda Viégas, all move ahead using the same dynamic visual programming or coding (in relation to graphic encoding) approach, which rewrites data in the form of graphic objects by pairing each data-derived variable with a graphic one, such as position, length, area, color, light, density, shape, texture, angle and curvature. How these objects evolve over time (referring, ultimately, to a dynamic modification of data-derived variables) can be compatible with "variometric"[14] data. On the interactive front, there seems to be a new series of related procedures designed for better cognitive understanding of data variability, scalability, extensibility, or even resiliency,[15] among which include a sound and comprehensive grasp on the *view* of the data, including regulation of the physical parameters in visual and spatial representation (see Bihanic 2003); the control of the *viewpoint*, be it "exocentric" or "egocentric"; the "multiple" viewpoints ("overview + details") that enable the user to possess both a macro- and microangle; and the "contextual" viewpoints ("focus + context") (Lamping et al. 1995) that categorize the details into the user's various areas of interest ("focus").

The myriad alternatives in terms of data representation, visualization, aggregation, extraction, and interaction pave the way for a better handle on the information fabric. Expanding outside the realm of data mining, monitoring, and supervision (through the graphic reconstruction of system-extracted information), these creative design approaches are banking on the human ability to perceive and process information. Repetition underlies the foundation and functioning of the data environment. It acts *in favor of* an ever-evolving and user-driven posture on data representation within a given system, as well as *toward* a mutually dependent trio consisting of situated perception, cognition, and action (see Dougherty 1985; Winograd and Flores 1986; Lave 1988; Cicourel et al. 1994; Laville 2000; Myin and O'Regan 2008) subjected to multiple unknowns and substantial changes within the environment (environmental dynamics).

---

[14] Periodic movement or lack thereof of data properties, increase in overall data volume, etc.

[15] In reference to the numerous changes in data shape and format.

2.2.1 Data Relationship Dynamics

As part of the study on dynamic cognition (Cognitive Variability; see Lautrey 2002), one of the major questions that has arisen from the creative proposals of Data designers pertains to action's supposed superiority within the cognitive process[16] or rather the supposed subordinate relationship of perception to action (beyond their universally known inseparability). In the absence of concrete answers, the experimental research in data design manages, nonetheless, to offer conclusive and convincing examples of a truly dominant action entity within all cognitive input. The superiority embodied by action is also apparent in "Situated Cognition" research (Vera and Simon 1993), which posits learning through a subject's experience or discovery rather than on rote, reception, or taught learning. This movement seeks to illustrate that the body (mostly through tactile or physical participation) is in no way withdrawn from the meaning of the action. On the one hand, its environment[17] weighs into the situations dictating the action (Cognitive Anthropology). From there surface a broadly empirical notion both *in* and *through* the action and acknowledgment of the value inherent in the experience (experiential learning) whose role is paramount in shaping meaning and the myriad facets there within.

2.2.1.1 Action and Perception

From the decisive role that experience plays in knowledge-building comes the idea that any cognitive event *hosts* itself in action. Should an idea such as this be approved, then action (mainly interface-based representation) measures and schemes would, in turn, not only occupy a crucial place, but also make up both the *means* and the *place* to understand meaning and its connotations. Henri Bergson wrote that if perception is to be understood as that which "[…] measures our eventual action on things, and conversely, the eventual action of things on us, then the greater the body's power to act and the more possibilities there are for perception" (Bergson 1934/2007). And as Leibniz (1996) wrote, well beyond our "small perceptions" is our apperceptive faculty, which entrusts us with a complete and comprehensive awareness of phenomena and things around us. This awareness opens up a number of avenues, allowing us to dig further below the surface of perception in search of reason, insight, and understanding.

Going forward, if we are able to acknowledge the suggestions offered by Data designers regarding advances that have already made significant headway, it is, first and foremost, the one in favor of lifting the veil (through application) on an action-oriented, apperceptive quality. We have just pointed out how much this suggestion

---

[16] Process of building skills and knowledge.

[17] From the space of the body caught in action toward exteroceptive sensitivity (not interoceptive or proprioceptive).

could prove vital when delving into the realm of significance. It would then make sense that the resulting advantage could trigger yet another that is tailored to taking perception to new heights (state of consciousness): from sensory-driven perception of things and phenomena to meaning-driven apperception of things and phenomena in action.

What is interesting to observe in certain data design projects is that from an apperceptive hypersensitivity such as this, the possibility to *visualize* information emerges: information that would, otherwise, not have been visible or apperceptive (see Tufte 1990, 2001) had it not been, up to now, entirely missing. The information in question pertains primarily to the *content* of data relationships (qualifying relationships), the *frequency* of its relationships with other data (quantifying exchanges), and several variables relating to data relationship dynamics that form the core of representation and visualization in large clusters of data.

Catching a glimpse of this dynamic means shifting from an immediate apperception of it to a physical intervention—in favor of an act or a sensitive gesture that handles and models data as a material and tangible entity: *feeling*, *perceiving*, and *apperceiving* (see Chouvel and Hascher 2012). The suddenly perceived data are viewed both as a material and product from the activity. It is the sediment of an environment boasting interface-driven models of representation and visualization and whose "cementing" (that of meaning) arises from its connections and combinations.

2.2.1.2 Enaction and Environment

Through some tweaking of the *Gestalt* principles of perceptual organization (proximity, similarity, good continuation, perception of movement, etc.), Data designers are able to devise unique proposals of visual grammar and organization, and cast a different look on Jacques Bertin's *Semiology of Graphics* (Bertin 2011). A variety of shapes ensues, which then trigger active data environments. Like true information ecosystems, these environments visually transform the associations, relationships, and transactions established between the data into phenomenological–aesthetic objects. As such, relational data dynamics are expressed and viewed through the way in which events unravel on the screen, including data effusion and scrolling, data concentration and grouping, data explosion, and dispersion. The more representation seeks to translate this dynamic, the more it appears as a "biocenosis" or ecosystem, uniting items, and biomorphic-like or cellular data.

With the focus being not only on metadata, but the data itself, these "semibiotic" representations demonstrate the irrefutable edge that space and all things related occupy in the polymodal representation of information. Backed by several studies oriented around more in-depth topological, dynamic, and morphodynamic models and patterns in cognitive science (cognitive grammar) (Petitot 2003; Petitot and Doursat 2011), these display models illustrate on an experimental basis that any intelligible conditions in the relationships seen here between the data are based on the coherence and adjustment of the system's spatial (naturally phenomenological) representation. Hence, without designating a material or formal aspect to this space and devoid of a primarily perceptive experience within it, no relational data dynamic is possible. Because a space is formed and structured around the interconnectedness of data (flow), cognitive understanding and data access are within reach. Therefore, the reason why these new data environments (data design) work so well is, without a doubt, related to their ability to shape new spaces of representation (space < > meaning).

The main question now nagging every designer is the following: What does it mean and through what means it is possible to create a visual and spatial representation of data capable of accounting for (given the experience) the information-specific properties and peculiarities via the experience? Let's not forget that merely and perceptively capturing the happenings does not enable one to grasp nor even visualize the dynamic addressed here. The user-subject needs to enter into an experiential relationship that is both sensory-based[18] and sensitive so that, one, this information makes its way to him, and two, its processing sparks new knowledge and skill development. Guided previously by the type of interface-based representation[19] chosen, it is the experience that presents itself to the user-subject, offering him (as the experimental subject) a plethora of opportunities. The experience is, in no way, forced upon him. Only the user-subject is at the helm, and only he can dictate the purpose, the direction, and eventually the results, pushing him to seek out even those less explicit. His motivation and drive in carrying out this exercise (perception-action) to the full extent of its potential will reward him with an apperceptive intuition.[20] Both heuristic (Depraz et al. 2003) and aesthetic (sensitive perception), the purpose fundamentally determines the quality of these interface-driven, spatial representations. Their very essence is multifold. They lay the groundwork for *seeing, exploring, penetrating,* and *roaming* these spaces (opening up the field to a number of interpretations[21] factoring into the experience).[22] The fundamental difference is here, which explains why we attribute a certain amount of originality to these active representation models. As "Big Data" continues to grow and spread, these representations are naturally inclined to resolve new data-processing challenges with the help of new and enlightening percepts (or affects, as necessary) and concepts that offer the possibility to grasp this new informational reality. One route in particular pursued by Data designers in an effort to shed some much needed light on the demand for new ideas comes from biomimicry (biomorphic origin), which transforms spatial data representations into agile and thriving environments comprised of graphic objects in perpetual

---

[18] For example, in the case of interactions with touch screen or multitouch technology devices (Lenay 2010), this experience triggers sensations specific to the haptic realm.

[19] In line with a selection of action and interaction modalities.

[20] In contrast with the relative agnosia to which the lazy and absent-minded perceiver will fall victim.

[21] To which inferences and deductions have recourse.

[22] Here targeting a hermeneutic dimension between Phenomenology and Cognitive Science.

movement. This same route adopts a new enactive notion of spatial perception (Froese and Spiers 2007), and at the same time, does away with "computationalistic" approaches that treat (and wrongly in our opinion) information processing as a logical and sequential calculation that allows little to no wiggle room when it comes to considering alternative meanings within the experience (interpretation). Alongside the ideas defended by the late and well-known neurologist, Varela (1989), Varela et al. (1993), Data designers count on giving *shape, substance, and stance* to the myriad *"trajectories"* and shifts in data. With all of this in mind, what then does it take for them to make this creative endeavor possible? What design-bred steps or approaches are most appropriate to display and view a genuinely complex, relationship-based "go-between" (interface) for massive datasets? Our aim in the next section will be to provide further insight and answers to these questions.

2.3 Giving Shape and Meaning to "Big Data"

2.3.1 The Appearance and Shape of Massive Datasets

Most of the testing done in data design combines *representation* and *visualization* within a comprehensive and formalized, "interface-driven" data initiative. As such, the interface lays out the framework for converting and transforming several piles of unstructured raw data into a system that allows for meaningful, compatible, and relevant[23] data flows and clusters to be displayed and viewed. It is then up to the Data designer to devise an original graphic script featuring a clearly defined and relatively exhaustive combination of terms with the ability to bring to life and execute a number of informational properties and applications. Among the qualities pertinent in transcribing any language,[24] one stands out with its very own unique purpose: that of formal expressiveness, whose definition here refers to conveying meaning of an entire group or a total volume of datasets. Given the large batches of extremely important information filtering through it, this degree of formal expressiveness takes on even greater dimensions, and namely with regards to the dynamic and frequency of data transactions and exchanges occurring, not to mention the total amount of information dispersed.

2.3.1.1 Perception and Sensation

Fully aware that an aesthetic grasp or take on data and the myriad ways in which it appears play a crucial role in how the meaning there within is extracted, some Data designers are now deploying new visual and graphic formalisms of massive dataset

---

[23] Formally ridding certain "critical mass" issues.
[24] For example, "distribution", referring to grapheme arrangement and association.

representation that round out or fill in, where necessary, those previously established, such as diagrams, networks, maps, symbols, or mashups. Rather similar to what Roger Brunet coined as "chorems" (Brunet 1986) or schematic representations of a geographic space or territory, these Data designers are shifting their creative talents into higher gear by not only positioning graphic primitives (point, line, and color), but also combining[25] visual primitives (X and Y plane or position, size, value, texture, color, orientation, and shape),[26] in addition to a number of other visual and graphic attributes such as *edge, contour, area* (empty-full relationship), *light, volume,* and *field* (all of which factor into giving rise to a graphic form of data representation and visualization).

All of these new formalisms (or adaptations of existing ones), a handful of which will undergo further speculation, conform to the same demand: that which removes any ornamentation, gimmicks or visual and graphic elements considered superfluous from the shapes depicting the data in an effort to retain but the crux, purity itself.[27] Simple and primitive, these shapes have been stripped and freed from all forms of sophisticated imitation and exaggerated virtuosity without surrendering the slightest bit of refinement or elegance. With nothing but geometry to define them, these shapes reduced to the barest form do not conceal, in any way, the qualities of the data they symbolize. Instead, they model them in accordance with their image and blend them into their aspect and appearance. Should a direction such as this have a legitimate place on an aesthetic level, thanks to the data's expressive added value, then it also has its place on a functional one. Enhanced performance (Data Interaction) relies on this perspective and arrangement (which we will examine later). In favor of absolute simplicity, the conditions of unmasking meaning from data (from simple or pure shapes to meaning exposure) forge stronger bonds. The same applies to an overall meaning (and not in what appears to be a reflection of meanings) for which a *path* may be found only through graphic representation and visualization and not in the work or handling of the data itself. In other words, an overall meaning could emerge from colossal volumes of massive datasets; however, it is likely to do so but through those shapes that have been assigned to them for a specific role or purpose. A more expeditious take on things would lead us to affirm here that the data do not, in any way, diffuse meaning; only the shapes, here of a symbolic nature, are meaningful. Isolated, the data do not contain any information that would allow it to choose a direction or orientation, define ahead of time, or hinder expansion of its meaning (excluding, of course, the semantic deviations that could arise among the various data elements). Architecture inspires the best analogy: These piles of data are like stones waiting to be put toward the erection of a building whose fate depends solely on the project and the work that ensues.

Shapes breed meaning. Here is an adage that Data designers could adopt, chant, and proclaim with the same enthusiasm as did those at the start of the twentieth

---

[25] According to distributions by *point, line* or *area*.

[26] In reference to the classification system defined by Bertin (1982, 2011).

[27] This demand also requires that any formal irregularities seen as accidental be corrected.

century in response to that uttered by famous architect Mies van der Rohe: "Less is more." If they have found it worthwhile to focus as much as they have on graphic formalization, it is precisely because there has been no other way, up to now, of accounting for and processing "Big Data" (whose magnitude remains a mystery to this day). Although computer systems are doing what they can to handle such an outpouring of data, both user and designer are still struggling to view the overall data ensemble (even a mental picture of it has proven challenging). Plus, as we have just mentioned, it will be up to the user and his interpretation of the data that will give rise to its meaning. Despite the performance levels and efficiency exhibited by systems, their ability to extract or describe the meaning from data has its limits. Therefore, it is necessary to reiterate the importance of not opposing what, here, appears as aesthetic data recovery versus truly perceptive data capture. The bimodality seen in the relationship pairing the graphic representation of data and the meaning attributed to it is inevitable, if not indispensable.

Hence, the primacy given to shapes (the plural usage being most appropriate here), each unconventional in its own way, is what gives these Data designers reason to bank on the creation of new formalisms. Paul Valéry[28] wrote, "Shape is everything" (Valéry 1957), with an understanding, here, referring to full and complete autonomy of shapes in the conception and expression of meaning. Keeping intuition intact (Intuitive Vision), some of the foremost qualities in target data will cross over into the apperceptive realm after discerning how shapes are aesthetically perceived and understood. Of the same opinion and enthusiasm as expressed in the *Gestalt* principles, these Data designers are committed to aesthetically illustrating massive datasets through a mix of pure sensation and perceptive capture. By taking advantage not only of the objective properties in data, but also the findings from data mining searches that reveal, more often than not, contrasting content from databases and ensembles, Data designers have at their disposal a rich breeding ground for experimentation and application. Their painstaking efforts have not been in vain, and from this preliminary analysis of the databases and datasets, other pertinent observations for the purposes of future creative syntheses will, no doubt, emerge.

If sensory-driven sensation or understanding marks the start of our awareness of these massive datasets, we are then left with figuring out how to incorporate either or both into our perceptive and cognitive faculties. In close, if the user is able to discern these shapes, then he is also able to discern their structure, both on a visual and perceptive level, in addition to their combinations and degrees of change,[29] while at the same time aesthetically judging the balance or imbalance of their proportions. This bimodal interpretation of shapes then fuels the origin of meaning whose interpretation is both perceptive (perceptive–cognitive stimuli and excitation) and aesthetic (impression, sensation). Banking fastidiously on the respect of various *sensory thresholds* (establishing the scope of the perceptive field based on a

---

[28] Quoting the poet, Frédéric Mistral.
[29] Specifically, in the case of generative or animated shapes.

given meaning) and limits of *differential thresholds* (detecting the smallest change in stimulation), the Data designer is also responsible for showcasing data through concrete graphic and aesthetic means.

2.3.1.2 The Role of Aesthetic Perception

In order to fully grasp the uncertain nature that perception (aesthetic data recovery and optical data capture) embodies, it is necessary to take a more in-depth look at the singularities of a purely aesthetic understanding of the shapes that stem from a sensitive awareness[30] of the world, its objects and their various states. Each aesthetically based data recovery makes room for a broader grip on things belonging to the worldly realm, while stressing our inability to examine them from all angles, narrow down the number of meanings there within, and penetrate right down to the essence itself (of an exclusive access within a phenomenological world). Contemplating a deeper, more aesthetic interpretation of a work, otherwise considered sterile by more short-sighted individuals, the observer, perplexed, will suddenly see the paths of perception unravel before his eyes (in line with Hegelian notions of "sense certainty" and perception). If perception is meant to instill in us a more thorough awareness of our surroundings, it is, no doubt, due to an intricate and detailed network of its bimodal functionality mentioned earlier, laying the very foundation upon which our perception takes root.

Determined to breathe new aesthetic life into how primarily massive dataset shapes are displayed (with high hopes that a new wave of graphic and more qualitative options surface for a variety of reasons), certain Data designers are committed to pushing through the idea and need to understand data on an aesthetic level, and furthermore, the phenomena that arise from it. One of those designers is Moritz Stefaner, who has recently authored *Wahlland* (Fig. 2.1) and *Stadtbilder*. The first project, *Wahlland*, offers the user a "hands-on" experience using a simplified map presenting the results from the 2013 German Bundestag elections. Extending well beyond its original function, the map becomes an original interface, breaking down and arranging the entire Germany territory (and its image, for that matter) by electoral districts according to voter behavior, attitude, and political party affiliation. The new data-mapping interface becomes the specially designed place and space for massive dataset exploration expressed through plain and simple geometric shapes serving as symbols, tools, and materials in a new version of data display for the purposes of visualizing and interpreting data. *Stadtbilder*, on the other hand, offers the opportunity to display what typically does not show up on a map; in other words, a city's shape and "living" infrastructure. Unpredictable and transient in nature, the shape is not indicative of geographically generated data retrieval, but rather a graphic transcription of a city in constant motion whose guise reflects the lifestyles of its residents, the "hotspots" they frequent and other outlets

---

[30] Prior to any new knowledge being acquired.

2 Giving Shape to Data 35

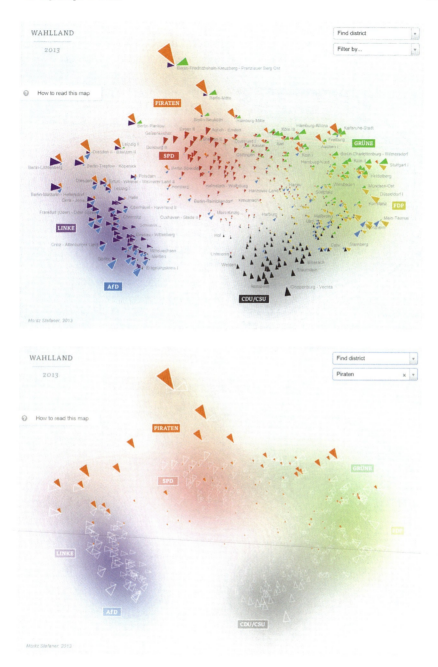

Fig. 2.1 *Wahlland*, Moritz Stefaner, 2013

contributing to the city's makeup. A multicolored, graphic mist hovers over the territory and keeps pace with the urban fabric and its enigmatic components. Spatial perception takes on a new dimension following a rather minimalistic graphic formalization of the multivariate data extracted from a portion of a city in motion.

In a rather different context, it is equally worthwhile to address electronic composer and visual artist Ryoji Ikeda's project, entitled *datamatics* (Fig. 2.2), whose experimentation with sound and image explores an infinite, or better yet, *transfinite* number of ways to render and visualize raw data via sonic and visual dimensions, layers, sequences, and patterns. Through a mix of graphic, often binary code-embedded textures in black and white (or empty and full), he exposes a wave of new environments whose success relies on an experiential dimension of aesthetic perception. Processing piles of data as substance, Ryoji Ikeda strives to simultaneously resolve how they materialize and are expressed.

There are some similarities between Ryoji Ikeda's work and that of designer Yugo Nakamura. The latter opts for a radically minimalist approach in his interface-based, graphic representations, which, in no way, exhibits any kind of graphic idleness, and which illustrates both the dynamic and interactive traits present in the interface's graphic shapes and objects. Through his unconventional and audacious, data-generated interfaces offering those present an epic visual, graphic, and audible experience, he gives life to an entirely new dimension of the experiential space and the relationships that arise within it.

Fig. 2.2 *data.path* audiovisual installation, Ryoji Ikeda, 2013—photo by Fernando Maquieira, courtesy of Espacio Fundación Telefónica

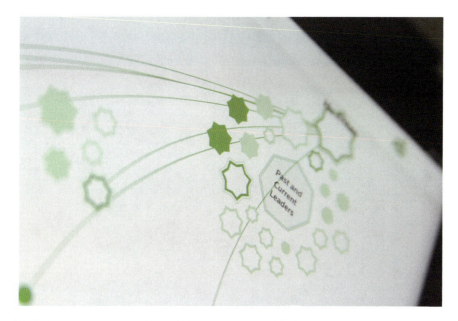

Fig. 2.3 *Connected China*, Fathom Information Design, 2013

Created by the Fathom Information Design studio, the *Connected China* project (Fig. 2.3) is a sound example of the operational (practical), cognitive, and aesthetic implications stemming from the use of symbolic and primitive shapes. Aiming to convey a clearer visual transcription of the overall meaning outside of the political and cultural factors shaping modern China (in reference to the false pretext of minimalist formalism), the interface-based representation also helps to devise the methods responsible for its understanding. In other words, how data appear greatly increases the user's ability to process it. Aesthetically perceived primary shapes and the indispensable presence of action (Data Interaction) spawn myriad possibilities in which to further dissect meaning.

The *Kindred Britain* (Fig. 2.4) project designed by Elijah Meeks and Scott Murray also falls within this school of thought. Here again, action plays a paramount role in the assimilation of the overall meaning. Calling upon a wide variety of both historic and social data from Great Britain that has been cross-checked and crossbred, the user is now in position to become fully immersed[31] within the interaction. Deploying a written form of basic shapes,[32] the interface-based representation underlies the action. The variety, breadth, and abundance of the clearly

---

[31] Theologically speaking, wholeheartedly and with practically all one's might.

[32] Any attempt or effort to radicalize the formal language of data representation and visualization involves bettering the conditions of user-driven processing (both perception- and action-related): toward its design. There is not enough time to get into the details of what this entails. It should be noted that we are currently working on other articles (to be published soon) that prove it. The projects already referenced offer sufficiently convincing examples.

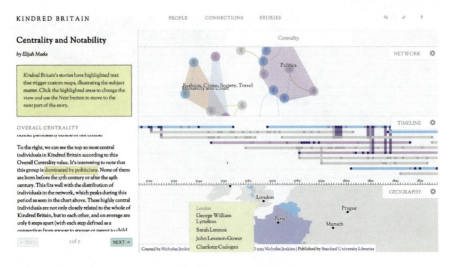

Fig. 2.4 *Kindred Britain*, Nicholas Jenkins, Elijah Meeks and Scott Murray, published by Stanford University Libraries, 2013

displayed data interactions will impel an overall meaning (consequence of data dynamics).

On a broader level, we would like to draw attention to the fact that Data designers' contribution to displaying and viewing large clusters of data is not to be taken lightly. The reason is that they propose a visual and interactive offering that boasts qualities of both an expressive and practical nature, ensuring that users have at their disposal what they need to not only perceive and understand the data-related connections and exchanges (interrelational data dynamics), but also take part in their processing. In an effort to strengthen the perception–action duo, Data designers have come to realize the influence and impact that action has on the kinetic perception and aesthetic grasp of information. In addition, they demonstrate that any graphic translation of large clusters of massive and in-motion datasets expressed as such leads to a rise in the ability to process them.[33] With this in mind and from a reception standpoint, the numerous effects (accentuation, intensity, etc.) and visual and graphic expression characteristics (a kind of formal, visual, and graphic "prosody") resort to impressions, emotions, and sensations that instill and "innervate" (figuratively speaking) the perceptive and cognitive process. With both percepts and affects, the user examines and reflects on the different directions that meaning and its interpretations can pursue as a result of the interrelated and visually expressed data shared. A sensitive and aesthetic experience such as this is viewed as a basis for crafting new perceptive and cognitive methods of information reception. It acts as a pivotal conduit for relaying certain elements of information to which we would have not had access otherwise.

---

[33] Favoring a wider variety and number of perceptive, interpretative, cognitive, emotional, intentional, and action-driven processing models.

Without completely relying on contemplative judgment for answers when in the presence of a work of art,[34] it would appear that the perceptive and cognitive mechanisms that fall under an aesthetic acumen of graphically processed information share a few things in common: A look or look back at Théodor Adorno's *Aesthetic Theory* (Adorno 1997) or Mikel Dufrenne's *The Phenomenology of Aesthetic Experience* (Dufrenne 1953) (both contemporaries) will be enough to remove all doubt. Other authors, such as Dewey (1934), Schaeffer (1966) and Shusterman (1992), have specifically focused their attention on the relationship between the "aesthetic experience" and "cognitive data capture (reception model)," and have arrived, individually, at the same observations and conclusions as we have. In sum, if certain properties, for example relational dynamics, in the entire data ensemble are within reach (both concretely and conceptually), it is precisely because there is a real push for interface-based representations (namely those devised by Data designers) to express these massive datasets, as well as their aggregation, appearance, and alterations (expressiveness). The expressions conveyed and transmitted through graphic objects and symbols[35] imply an aesthetic approach, which has now become a prerequisite in understanding the meaning within the data, be it, for our purposes, hidden or concealed.

2.3.2 Moving Toward Concrete Data Design

A number of Data designers today share a common vision and direction aimed at visually acknowledging the transformations, changes, and unknowns emanating from the current information hub. Tackling various piles of data, they strive not only to piece together an overview of the data, but also concoct an image of it that is genuine, powerful, and telling and, despite the multitude of interpretative facets generated, seeks, nonetheless, to transmit an overall meaning. Brought deliberately together by formalistic intentions, this twofold project embodies a new, broader "movement" that we have decided to refer to here as *Concrete Data Design*. In line (and on more than one account) with the declarations made by the late Theo van Doesburg regarding "Concrete Art" (van Doesburg et al. 1930), this emerging movement calls for a shape-driven government (of a sovereign and graphic plasticity) with infinite *re*-presentation powers or powers designed to convert reality (formalist realism). Banishing the lyrical, symbolic, and impressionist aspects oozing from a sentiment of excess subjectivity, Data designers behind this relatively

---

[34] A parallel development could help to specify how perception-targeted objects condition our approach and understanding of them. That said, not every object is a work of art (to put it bluntly). This does not mean that processing it (by the author) more formally is futile—on the contrary, for doing so would bestow upon this object truly aesthetic qualities that it is capable of ingesting thanks to the relatively similar processes of perception and cognition.

[35] Based on aspect—and behavior-based variables.

young movement advocate for the aesthetic clarity[36] of simple and adapted shapes in an attempt to gain objective ground (and, in turn, that of universal truth), and unanimously back the idea that an information hub such as this is not quite a lump of obscure data devoid of meaning.

2.3.2.1 Clarity is Aesthetics

New opportunities are likely to arise from the overwhelming trend in data recovery and capture today; however, their evolution is tremendously hindered should there not be any way to display or view them. As data prospers and accumulates while databases expand, these Data designers have set forth the objective to formulate even clearer, more *transparent* images[37] boasting unparalleled elegance and grandeur. Enlightening and eye-opening, these representations never lack relevance, and avoid falling into the arbitrary realm. Be they for the purposes of large sets of data[38] (informational hub) demanding a bit more distance or for lesser volumes (both are given the same attention), their goal is simplicity (see Maeda 2006). Of course, we perceive and interact better when things are orderly, balanced, and unified than when they are dismantled or disparate. What matters most is not how the shapes are arranged, but rather the quality of the shapes themselves. If composition (organization of shapes) triggers meaning, then pattern triggers not only the accuracy and precision in how it is expressed, but also the insight and judgment resulting from the sensitive perception and aesthetic experience of the data shown here. This neoformalist approach supported (and openly, at times) by certain Data designers involves reducing or simplifying shapes. It stresses their geometry and most fundamental forms (contours, angles, dimensions, curves, etc.). It even goes so far as to commit to greater simplicity and minimalism in shapes, as well as advocate an aesthetic ideal.

Therefore, complexity may find its way into all means and manners of data display, but never will the latter opt to sophisticate its shapes. It will, instead, look to radicalize them with increasingly more candor and meaning. A closer look at the notions of clarity and complexity reveals more similarities than differences and proves these seemingly "polar opposites" are anything but that. In some respects, these two notions complement one another, but they not merge. If clarity (bordering on simplicity) is to be understood as a conquest (of meaning via shapes), then complexity constitutes a victory or success, or better yet, an outcome paving the way to new paths of knowledge and insight. Like the ideas and concepts refuting all reductionist ideologies,[39] this revolutionary twist on complexity portrays it no

---

[36] Including accessibility and visualization.

[37] Referring here to both info graphic projects and visual and graphic interface environments of "Big Data" representation and visualization resulting from conceptual, formal, and creative investigation.

[38] In addition to the numerous flows and transactions present there within.

[39] These ideas and concepts are examined more closely by authors such as Morin and Le Moigne (1999; Morin 1999), Koestler (2011), Berthoz (2009), and Eco (1989).

longer as a burden or weight bearing forever down upon us, but rather as a chance, a resource, and an opportunity in itself. The time and energy put in by Data designers to reverse this long-standing and uncompromising view on complexity are finally paying off as the term now engenders an image and a meaning that are understandable, within reach, and sensitively perceivable. This newly defined complexity rhymes with curiosity (embedded with heuristic value) and a propensity to picture our ever-evolving and ever-revolving world from a multitude of angles and one in which our presence is non-negotiable. By depicting some of the data-protected phenomena (in response to new realities in the present world), Data designers are out to prove that the world does not only *not* fit into our concepts, but extends well outside of them. As a result, they are heading in pursuit of other means and opportunities wherein to understand this world, play a role within it, or take ownership of it.

2.3.2.2 Data Gestaltung

Differentiating simple, pure, and primitive shapes on an aesthetic level adds to the expressive power of representation. A power such as this tied to meaning and evolution makes reference to the *Gestalt* concept pointed out earlier, originating from the verb, "Gestalten," meaning "to shape or to attribute a meaningful configuration." In other words, if data representation and visualization boasts expressive properties now used to generate and liberate meaning, then it is due to the quality of the shapes themselves that cater to it. The shapes chosen and how they are arranged make for a well-rounded representation. An overall meaning with clarity and direction starts to break away from stark, chiseled, and polished shapes. Profound and panoptic, the meaning is not stuck amid the lonesome and isolated shapes, as if in aimless wander and devoid of affinity, but fully engaged in the exchange and relationship that they nurture and nourish. What takes shape makes sense to us. The same is true when designating a concrete entity or its shape, and a promising and productive whole or its representation, not to mention all of the other criteria that prompt them.

In keeping with this understanding, several contemporary designers, artists, and creative minds have opted to devise a skillful mix of primary shapes. By hypothetically giving *shape* and *content* to entities and wholes, respectively, in reality, these same individuals are able to come up with new, geometric-inspired models. Graphic languages, in turn, emerge, proving more convincing and accessible, and within which image (in its widest sense) does not glorify reality, but rather grants it another meaning. Image does not decorate, illustrate, or describe; it selects, recomposes, and graphically transposes certain objective data elements, which it then codifies and differentiates before presenting them in another light and from other angles. Here, image deploys a system of shapes and establishes a code made up of meaning(*s*). To remove, once and for all, any doubt regarding the abundance of such present in the language of primary shapes, a look at the work by design sisters Nicole and Petra Kapitza is all that is needed. Pooling patterns, fonts, and colors, they endeavor to pick apart nature-embedded geometry, and piece it back together with originality, tact, precision, and rigor. Armed with a multidisciplinary background in fine arts (see Albers 1963), book design, digital software, and illustration, both graphic designers prove that their creativity and combinations of simple shapes are teeming, more than ever, with promise and prospects. Furthermore, the potential range of meaning and scope is limitless.

Other designers, and not just any, have successfully managed to illustrate that all fundamental shape generation implies an even greater organization of the whole, be it rhythmically, logically, or systemically. The expressiveness in any representation depends upon this organization. Thus, as shapes become more subtle, the space they inhabit becomes more orderly and structured. The more the shapes mature, the more this space expands, and the more the meaning evolves, becomes clear, and intensifies. The approach guiding this shape-the-shape process is not focused on the selection or institution of the shapes alone (as explained), sitting proudly at the heart of a whole, but rather on the whole itself responsible for giving order to them. As a result, this orderly whole doubles as a catalyst, spurring the formulation of meaning, the development of interpretation, and the fulfillment of perceptive data capture and aesthetic data recovery. Let us take, for example, designer Alexander Chen's project, entitled *You Still Believe in Me*, which transcribes the musical expressiveness or musicality from the song originally interpreted by the famous quintet, The Beach Boys. Here, rhythmic pulsations and a vocals-ingrained color palette (chromesthesia) are paired to produce the transcription or "transliteration."[40] Underlying intentionally simple shapes stripped down the bone are a number of minuscule variables, parameters, and fluctuations. Disk-like shapes emit sensory-driven vibrations, and, by virtue of their rudimentary character and unity, keep intact the purity and beauty of the original musical arrangement. Let us look at another example. Developed by Matthew Epler, *The Color of Language* project examines more closely the expressive power of language. He says, "Language is more than words." It is, first and foremost, a system or organized framework made up of signs, glyphs, and shapes enabling an infinite number of combinations, and in turn, an infinite number of synopses and interpretations. Through this project, Matthew Epler invites us to experiment with this multiplicity, and in doing so, gauge the complexity in all language-based meaning construction.

The same conclusions may be applied to data. Formally translated, data also functions within an organized framework used to determine its visual structure. Let us recall that data alone are of little value and meaning, and subsequently, bear little interest just like the shapes entrusted with conveying a relatively small part of that meaning. From the organizing framework referring, here, to an all-encompassing representation or one embodying the shape-driven system of organization, ensues an overall meaning. If what is most important arises from the shapes in their purest form or from their pared anatomy or appearance, then it is their arrangement, assembly and connections that fuel the emergence of meaning. For example, designer Shahee Ilyas's project, entitled *Flags by Colours*, clearly demonstrates that

---

[40] Transcription of graphemes from one written alphabet to other graphemes in another system: a manner of transposing one system of signs to another without loss of meaning.

by incorporating a simple variable of an ensemble, and for our purposes here on a colorimetric scale,[41] and by ruling out any symbolic, iconic, or indexed trace, other methods of data display and interpretation may be established. *The Periodic Table* project by designer Alison Haigh also agrees with the above notion, which essentially defends the idea that the design of a meaningful display of data, calling for formal relationships to be perceived and interpreted, entails formally reducing data or giving it substance.

In his mind-altering project, entitled *Here is Today* (Fig. 2.5), Luke Twyman, like those mentioned earlier, puts his own spin on an interface-based representation tracing time, history, and geology. Delving into the history of mankind, the representation offers us the chance to visually grasp the degree of granularity separating the current moment from earlier periods, ages, and eras that have existed on earth. Starting from an extremely simple graphic to a simplified and user-friendly application (heightening both perception and action), a time-lapse rendering on a spectrum in motion unfolds. Bewildering and unsettling, the project is also educational as it attempts to chip away at the layers having trickled down and settled over time throughout the course of history.

There are similarities between Twyman's project and that of Carlo Zapponi and Vasundhara Parakh, entitled *Worldshap in* (Fig. 2.6). Here, a combination of basic shapes and models is used to visually compare countries throughout the world on the basis of criteria such as sustainable and human development. Each country is given a shape, whose form changes based on where it stands in relation to various statistical data indicators.[42] The sliding year bar and the option to add or remove countries and their respective results (resulting in an innovative and eye-catching superposition of shapes) determine the spatial coordinates, making it possible to display and view the data accordingly. It is, therefore, the entire interface that constitutes the system of data representation and visualization.

Formal and concrete data representation involves a broader notion of meaning. This is what Data designers are out to prove, including one, in particular, by the name of Genis Carreras. His aim is to graphically isolate various philosophical concepts. In doing so, representation takes on the notion of a strategy or concept image that shapes the meaning. Another example leads us to the initiative taken by creative advertising agency Bartle Bogle Hegarty (BBH-Labs) as part of the ad campaign, entitled *The Three Little Pigs*,[43] which, through a clever and compelling use of minimalist infographic designs disguised as basic shapes, underlines certain gaps in statistical data. Here again, the simplicity seen in the graphic picture helps to better visually compare the quantities represented. DavidMcCandless, MGMT studio designers, Caroline Oh and Young Sang Cho (Fig. 2.7) have devoted

---

[41] Calibrating and comparing the quantities of colors within each national emblem (gallery of the world's country flags).

[42] The statistical data presented are based on *The Human Development Index* (HDI) and have been extracted from a report published in 2011 by the United Nations, entitled *Sustainability and Equity: A Better Future for All*.

[43] Advertising campaign done for the British national daily newspaper, *The Guardian*.

Fig. 2.5 *Here is Today*, Luke Twyman, 2013

timeless energy to this very topic timeless energy. Understanding that all shapes of a whole are perceived not separately or individually, but distinctively, proportionately or comparatively, these creative specialists have their sights set on a concrete formalization of data. The key word here is *concrete*, and for the following three reasons: Firstly, the formalization process depends solely on data representation variables being assigned to a number of formal attributes, 15 of which figure among the most prevalent, including, *shape, form* (or design pattern, symbol), *line, edge, contour, color, value, area* (full-empty relationship), *texture, size, light,*

Fig. 2.6 *Worldshap in*, Carlo Zapponi and Vasundhara Parakh, 2013

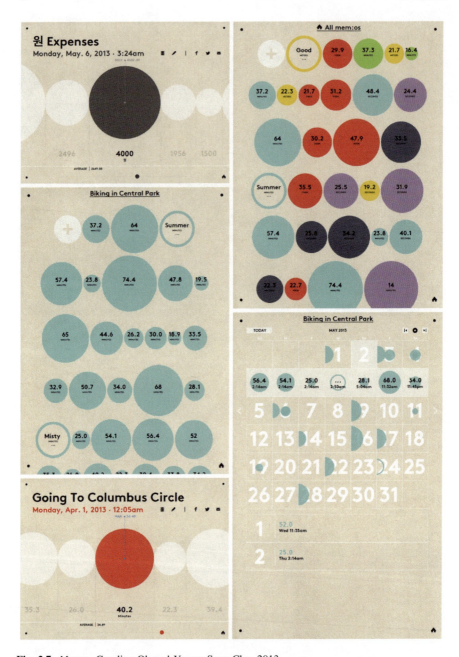

Fig. 2.7 *Mem:o*, Caroline Oh and Young Sang Cho, 2013

volume, plane, field (or space), and *format* (or orientation)—representation is then synonymous with graphic data code. Secondly, the term applies to the generation of pure and perfect shapes free of graphic imperfections or "anomalies" out to impede how the shapes are meant to be interpreted. These same shapes also exude honesty and transparency, exposing, in turn, but the lifeblood. Lastly, in the kingdom of pure shapes, creative expression is an infinite and exuberant realm, yet at the same time, one boasting measure and sophistication. As once proclaimed by concrete artists in their Second Manifesto, this creative expression breathes clarity and harmony into the artificial world in which we live. The deep-seated and indecipherable obscurity found within the information comprising databases, tables, and indices gives way to a clearer and more accurate conception of the images,[44] leaving no doubt, whatsoever, as to their meaning. Observable phenomena materialize from residual data. Although continually evasive and elusive, transforming this data via representation models would make it palpable and comprehensible. What was once beyond our faculties is now at our fingertips. As a result, we have all we need to not only assimilate our world, but also potentially alter it, and all this through the use of visual and graphic representations of data.

One of our latest projects, entitled *Data Shapes* (Figs. 2.8, 2.9), argues the idea of a new formal method that bestows upon massive dataset representation a concrete and explicit character. The project illustrates simple, yet fully formulated shapes of data extracted from the fields of economics, finance, and politics. Each of the shape's attributes is then directly associated with data and value means calculated in real-time and appearing as vector and spatial coordinates broken down by color. If the data and values are to be considered as indexed measurements of the working world, then the corresponding shapes are the proof. Here, the shapes do not represent the data. They do not communicate it nor divulge its numerous meanings. They are but its reflection, depicting a picture of the utmost accuracy, as well as one that could not be further from it. Rebelling against a more passive form of representation, the shifting nature and instability of the shapes avoid all forms of contact and comprehension. They try desperately to come together and form a whole, but to no avail. Our inability to ascertain the diverse meanings, not to mention the meaning of perception itself, result from these shapes and their volatility. They tap into our consciousness and push us to wonder about the things to which we have access and why they have been presented as such. This project is a sequence. Every single shape factors into its representation. Shapes are assigned means, or even a mean of means, calculated according to a number of criteria. Their fluctuations double as the pulse, and the range reading doubles as the heart.

The attempt to display and view data and the push to generate other aesthetically phenomenological data comprise the twofold purpose behind this project, and on a broader level, the creative approach whose definition we are committed to

---

[44] Visual and graphic representations of data, interface-based or not.

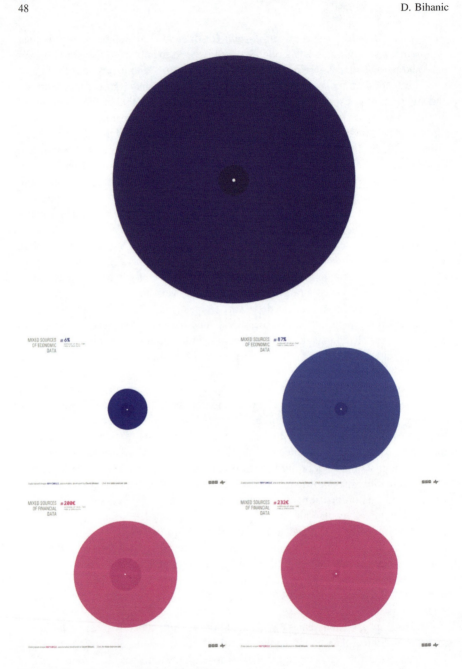

Fig. 2.8 *Data Shapes*, David Bihanic, 2013

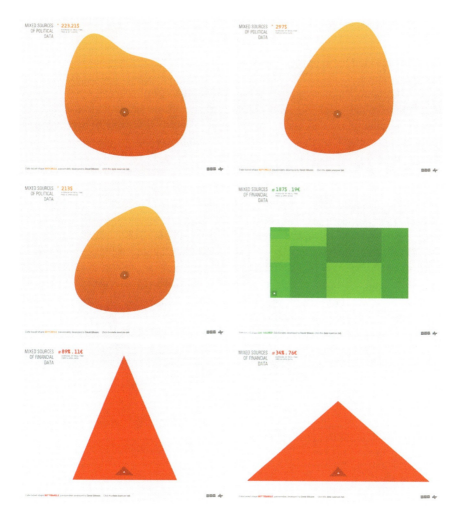

Fig. 2.9 *Data Shapes*, David Bihanic, 2013

elucidating.[45] Here, representation paints a more accurate picture of reality. By configuring shapes objectively, representation exposes itself to a genuine phenomenological data demonstration, ranging from the relationship with aesthetically oriented concrete data to the meaning of a newly formed aesthetic object and a new emergence of shapes.

---

[45] All things considered, the reasons driving this project are not completely foreign to those of *Listen Wikipedia* created by Stephen LaPorte and Mahmoud Hashemi or *Color Forecast* developed by Pimkie.

2.4 Conclusion

The approach of data design basically aims at creatively exploring and investigating new ways and methods of representing, visualizing, and processing computer data. To do so, Data designers are not refraining from forging other visions and paradigms that radically cut ties with tradition, rules, and designs, which, up to now, have been adamantly defended and accepted as true.[46] Their ultimate goal is to optimize all research, including those of a seemingly trivial nature, for the purposes of rebuilding the foundation of all knowledge. Through *exploration* and *exploitation* of massive datasets in motion, Data designers are committed to leaving no stone unturned, and exposing the *buried or hidden* (ideally *tried and true*) potential within disseminated data ensembles. By first forcing ourselves to clarify this handful of attributes and peculiarities[47] before delving further into the more formally aesthetic properties and expectations surrounding the display and view of "Big Data," only then will we be able to reach the conclusion that more emphasis needs to be placed on both the concrete and formal aspects of expressiveness in data representation and visualization, the benefits of which include more added value information-wise, embodying a greater sense of perspicacity and perceptive sensitivity of the user, in addition to a broader notion of meaning, and in turn, a more enriching experience.

References

Adorno WT (1997) Aesthetic theory. English edition: Adorno WT (1997) (trans: Hullot-Kentor R). University of Minnesota Press, Minneapolis
Abrioux Y, Varela FJ, Oulette P, Spolsky E, Schneider R, Chassay JF (1999) Dynamique et cognition: nouvelles approches. Théorie Littérature Epistémologie, Presses Universitaires Vincennes, no. 17
Albers J (1963) Interaction of color. Yale University Press, Connecticut
Bergson H (1934) Matière et mémoire. Felix Alcan, Paris, p. 61; e.g. (2007) Matter and Memory. Cosimo Classics, New York
Bertin J (2011) Semiology of graphics. ESRI Press, New York
Berthoz A (2009) Simplexité. Odile Jacob, Paris
Brunet R (1986) La carte-modèle et les chorèmes. Mappemonde 86(4):4–6
Bertin J (1982) Graphics and graphic information processing. Walter De Gruyter Inc., Boston
Bihanic D (2003) A complete system of tridimensional graphical representation of information: crystal Hy-Map. In: Proceedings of COSIGN'03
Brouillet D, Lang F (2012) Cognition environment. http://archive-fr.com/fr/l/lab-epsylon.fr/2012-10-09_401787_7/Cognition_environnement. Accessed 9 Mar 2014
Card SK, Mackinlay JD, Shneiderman B (1999) Readings in information visualization. Using Vision to Think. Morgan Kaufmann, San Francisco

---

[46] Wiping out, once and for all, all outdated and obsolete shapes.

[47] Inadvertently stressing how design's contribution to data representation and visualization is proving, now more than ever, indispensable.

Caroll DJ, Chang J-J (1970) Analysis of individual differences in multidimensional scaling via an n-way generalization of "Eckart-Young" decomposition. Psychometrika 35(3):283–319

Chouvel J-M, Hascher X (2012) Esthétique et cognition. Publications de la Sorbonne, Paris

Cicourel A, Conein B, Filippi G et al. (1994) Travail et cognition. Numéro spécial sur l'action située, Sociologie du travail 36(4)

Depraz N, Varela FJ, Vermersch P (2003) On becoming aware: a pragmatics of experiencing (Advances in consciousness research). John Benjamins Publishing, Philadelphia

Denis M (1989) Image et cognition. PUF, Paris, p 129

Dewey J (1934) Art as experience. Perigee Books, New York

Dufrenne M (1953) Phénoménologie de l'expérience esthétique. PUF, Paris

Dokic J (2010) L'architecture cognitive du sens esthétique. In: Borillo M (ed) Dans l'atelier de l'art. Expériences cognitives, Champ Vallon, Seyssel, pp 49–61

Dokic J (2001) L'esprit en mouvement. Essai sur la dynamique cognitive. CSLI, Collection Langage et Esprit, Stanford

Dougherty JWD (1985) Directions in cognitive anthropology. University of Illinois Press

Eco U (1989) The Open Work. (Trans: Cancogni A) Harvard University Press, Cambridge

Freeny EA, Gabbe JD, Michaels SA (1969) An experimental data structure for statistical computing. Statistical computation. Academic, London

Freeny EA, Gabbe JD (1966) Image of a Thunderstorm. AT&T Bell Laboratories (application)

Froese T, Spiers A (2007) Toward a phenomenological pragmatics of enactive perception. In: Proceedings of ENACTIVE'07

Furnas GW, Bederson BB (1995) Space-scale diagrams: understanding multiscale interfaces. In: Proceedings of CHI'95

Healey CG (1992) Visualization of multivariate data using preattentive processing. Master's thesis, Department of Computer Science, University of British Columbia, Vancouver

Hutchins E (1995) Cognition in the wild. MIT Press, Cambridge

Hutchins E, Hollan J, Norman DA (1985) Direct manipulation interfaces. Human-Computer Interaction

Hollan JD, Hutchins E, McCandless TP et al (1986) Graphical interfaces for simulation. La Jolla, Institute for Cognitive Science, University of California, Berkeley

Koestler A (2011) Le Cri d'Archimède: l'art de la découverte et la découverte de l'art. Les Belles Lettres, Paris

Kruskal JB (1972) Linear transformation of multivariate data to reveal clustering. In: Shepard NR, Kimball RA, Nerlove SB et al (eds) Multidimensional scaling: theory and applications in the behavioral sciences, 1 theory. Seminar Press, Oxford, pp 181–191

Kruskal JB (1977) Three-way arrays: rank and uniqueness of trilinear decompositions, with application to arithmetic complexity and statistics. In: Brualdi RA, Mehrmann V, Semrl P (eds) Linear Algebra and Its Applications, vol 18, pp 95–138

Kruskal JB, Hart RE (1966) A geometric interpretation of diagnostic data for a digital machine: based on a Morris. Ill Electron Central Office Bell Syst Tech J 45:1299–1338

Lakoff G, Mark J (1980) Metaphors we live by. University of Chicago Press

Lamping J, Rao R, Pirolli P (1995) A focus + context technique based on hyperbolic geometry for visualizing large hierarchies. In: Proceedings of CHI '95

Langacker R (1991) Foundations of cognitive grammar, descriptive applications, vol 2. Stanford University Press, Palo Alto

Langacker R (1987) Foundations of cognitive grammar, theoretical prerequisites, vol 1. Stanford University Press, Palo Alto

Latour B (1985) Les vues de l'esprit. Culture Technique, Éditions de l'École des Hautes Études en Sciences Sociales, 17, Paris

Lautrey J (2002) Le statut de la variabilité entre les individus en psychologie cognitive. Invariants et variabilités dans les sciences cognitives, Presses de la Maison des Sciences de l'Homme, Paris

Lave J (1988) Cognition in practice. Cambridge University Press, Cambridge

Laville F (2000) La cognition située : une nouvelle approche de la rationalité limitée. Revue économique, Presses de la Fondation Nationale des Sciences Politiques 51(6):1301–1331 (Paris)
Leibniz GW (1996) New essays on human understanding. Cambridge University Press, Cambridge
Lenay C (2010) It's so touching: emotional value in distal contact. Int J Des 4(2):15–25
Maeda J (2006) The law of simplicity. MIT Press, Cambridge
Morin E, Le Moigne J-L (1999) L'Intelligence de la complexité. L'Harmattan, Paris
Morin E (1999) Relier les connaissances. Seuil, Paris
Myin E, O'Regan JK (2008) Situated perception and sensation in vision and other modalities, a sensorimotor approach. Cambridge University Press, Cambridge
Petitot J (2003) Morphologie et esthétique. Maisonneuve et Larose, Paris
Petitot J, Doursat R (2011) Cognitive morphodynamics. dynamical morphological models of constituency in perception and syntax. Peter Lang, Bern
Piaget J, Chomsky N (1979) Théories du langage, Théories de l'apprentissage, Centre Royaumont pour une science de l'homme. Seuil, Paris
Robertson GG, Card SK, Mackinlay JD (1989) The cognitive coprocessor architecture for interactive user interfaces. In: Proceedings of SIGGRAPH'89
Rogers TT, McClelland JL (2004) Semantic cognition: a parallel distributed processing approach. MIT Press, Cambridge
Rumelhart DE, Norman DA (1975) Explorations in cognition. Freeman, San Francisco
Schaeffer J-M (1966) Les célibataires de l'art. Gallimard, Paris, p 135
Shusterman R (1992) Pragmatist aesthetics: living beauty, rethinking art. Blackwell, Oxford
Spence R (2007) Information visualization. Design for Interaction, 2nd edn. Pearson, Harlow
Stasko JT (1996) Future research directions in human-computer interaction. ACM Computing Surveys, 28(4es), Article 145
Stasko JT (1993) Animation in user interfaces: principles and techniques. In: Bass L, Dewan P (eds) Trends in software, special issue on user interface software, vol 1, Chap. 5, pp 81–101
Tufte ER (2001) The visual display of quantitative information. Graphics Press, Cheshire
Tufte ER (1990) Envisioning information. Graphics Press, Cheshire
Valéry P (1957) Variété. Victor Hugo créateur par la forme. Gallimard, Bibliothèque de la Pléiade, Œuvre I, Paris, pp 583–590
Varela FJ, Thompson ET, Rosch E (1993) The embodied mind: cognitive science and human experience. MIT Press, Cambridge
Varela FJ (1989) Autonomie et connaissance. Seuil, Paris
Vera AH, Simon HA (1993) Situated action: a symbolic interpretation. Cognitive Science 17:7–48
van Doesburg T, Carlsund OG, Hélion J et al (1930) The Basis of Concrete Art, Manifesto 1
Vygotski L (1986) Thought and language. MIT Press, Cambridge
Winograd T, Flores F (1986) Understanding computers and cognition. Ablex, New York

Author Biography

Bihanic David (Designer, Consultant & Researcher) While teaching as an Associate Professor at the University of Valenciennes and Hainaut-Cambresis in the north of France, David Bihanic (Saint-Nazaire, France, 1977) works as a Designer and Design Consultant; he is the founder and Managing Director of a creative agency called Fxdesignstudio (FXDS).

After obtaining a PhD from the Paris 1 Panthéon-Sorbonne University, he published numerous scientific articles relating to the new stakes of design; he then takes an active part in the evolution and transformations occurring in this field. Today, his research is mainly on data design, data visualization aesthetics, and creative informatics. His work spans the whole spectrum, from theory to implementation techniques and applications.

For more information and contact:

http://www.davidbihanic.com
http://fr.linkedin.com/in/dbihanic
https://twitter.com/dbihanic

Part II
Visually Representing and Explaining Data

Chapter 3
The New Aesthetic of Data Narrative

Giorgia Lupi

Abstract How can a data-driven visualization tell multiple interplaying stories and achieve a viable result in an abstract visual composition? How can we provide multi-levels of investigation of a certain phenomenon in a paper-based data visualization as if it were an interactive piece (article) where readers can lose themselves? This article relies on a series of exploratory data visualizations originally published for La Lettura, the Sunday cultural supplement of Corriere della Sera with my team at Accurat. Purposely, we here aim at delivering rich visual narratives able to maintain the complexity of the data but still making this complexity more accessible and understandable, publishing compound and complex stories told through data visualizations. The contribution will describe how we can imagine to open new perspectives in the newspaper-editorial field and how we can higher aim at educating readers' eyes to get familiar with new visual ways to convey the richness, the involvement and feelings (being engagement or concern) that we experience in our everyday lives rather than simplifications of the world. This contribution will outline the design process from the very first idea to the final results in different cases, presenting backstage materials such as sketches and intermediate versions and elucidating our personal aims, purposes, and expectations within this exploratory project. It concludes by tracing some final red threads to discover possible new approaches to the aesthetics of data visualization: focusing on how to get inspired from many different disciplines, how to build a personal method and set individual goals, and explaining how we hope our aesthetic and analytic choices can generate new possibilities for ongoing creativity and research in the field. The idea here is to open possible questions rather than providing finite and definitive answers: to ask ourselves how far can we go rather than delimiting the field.

G. Lupi (✉)
Accurat, 392 Broadway, New York 20013, USA
e-mail: giorgia.lupi@accurat.it

3.1 La Lettura, Visual Data

La Lettura[1] is the Sunday cultural supplement of Corriere della Sera, the highest circulation newspaper in Italy. The supplement is conceived as a long-read collection of articles about cultural and sociological phenomena, new media and communication-related topics. The aim of the issue is to provide readers with a product; they can read throughout all their week: with deep essays usually written by sociologists, professional writers, art and literature critics, historians, philosopher or modern thinkers.

As Accurat, we are regularly publishing in it, and we aim at revealing and advancing how data visualization can be used to provide new perspectives in the newspaper-editorial field: We publish compound and complex stories that each time are not told through an article, but through a data visualization (Fig. 3.1). We here try to think as journalists, rather than data analysts: Understanding in which contexts we should interpret the data we gather and analyze, and questioning ourselves about what it is interesting in these numbers and what possible correlations with other information we might experiment to unveil hidden stories. We aim at delivering rich visual narratives able to maintain the complexity of the data but still making this complexity more accessible and understandable through the visualization.

3.1.1 Story! Catchy, and Layered

For each story, we consider and pursue a topic we believe may be of particular interest to explore, ranging from current affairs to historical or cultural issues. Sometimes choices are driven by a fascination we have, sometimes by a compelling dataset we find and we would start from, other times we choose to present events and topics that are hot at the moment. We then analyze and compare different kinds of datasets trying to identify and reveal a central story, hopefully a not-so-expected one. We start from a question or an intuition we have and work from here, then try to put the information in context and find some further facts and materials to potentially correlate. Every time we aim at moving away from mere quantity in order to pursue a qualitative transformation of raw statistical material into something that will provide new knowledge: unexpected parallels, not common correlation or secondary tales, to enrich the main story with. In this respect, our work here cannot be considered data visualization in the pure sense; we are not just providing insight into numbers but into social issues or other qualitative aspects as well.

In addition, since we publish on a full-spread format within the cultural Sunday supplement of the highest circulation Italian newspaper, the leading narrative and the visual ways through which we display information have to be both catchy and

---

[1] La Lettura can be translated as "the very act of reading, of spending consistent time in the activity of reading."

Fig. 3.1 La Lettura visual data, Corriere della Sera. Accurat gallery of data visualizations

attractive: Once the first attention of the audience is "caught" by the aesthetic features of the image, the presentation of the information must be clear as might be expected.

The clarity does not need to come all at once; however, we also like the idea of providing several and consequent layers of exploration on the multiple dataset we analyze. We call it a "nonlinear storytelling" where people can get lost in singular elements, minor tales, and "last-mile" textual elements within the greater visualization.

3.1.2 Constraints, as a Resource

> Giving yourself a handicap to overcome will force you to think in a new and slightly different way, which is the prime goal of scratching.
>
> Twyla Tharp

If on the one hand, we are given considerable leeway design-wise, and in terms of topic choice, on the other hand the publisher establishes limitations on background color (and other color considerations), fonts (we are limited to two controlled font families), and, of course, format. Moreover, we work within a very short time span: We usually start and conclude the whole piece (e.g., proposing the topic, finding the data, analyzing them, and visualizing) in less than 5 or 6 days. Graphically speaking, we also give ourselves some extra rules using only one of the two fonts allowed and being very keen in avoiding any pictorial elements. We have found that all of these constraints are incredibly useful. First of all, they lay out the total of our compositions with a very deep consistency, creating what we can describe as a series if not a collection. And additionally, those limits and already fixed parameters help in staying focused on the very representation of contents, without losing time with the definition of a graphic layout.

3.1.3 Visual Models and Metaphors: Pushing Forward, Always

Since the goal here is neither to visualize data for decision-making processes nor are we representing information for scientific purposes, the opportunity to experiment with new visual metaphors is wide open and the exploratory nature of this work is clear. We can, every time, try to push forward how we can "compose" data visualizations that achieve (in our idea) aesthetic beauty and elegance through new visual metaphors, intentionally avoiding typical and already tested styles of representation (Fig. 3.2).

To us, elegance is not only beauty and prettiness, our intention is to make things not just understandable but also appealing, conveying the information in an effective way but also catching the attention (the eyes!) of the particular audience we are creating it for. Through these visual explorations, we constantly try to create harmonious compositions, where readers can look for, find and compare information.

By means of uncommon visual models, possibly different and experimental, we maintain and respect the complexity of the data we are analyzing. We do know that there is a science to data visualization or, at least, that recognized principles for representing information exist and are worth to be pursued in most of the cases.

Many times bar charts, scatter plots, regular time lines, and maps are the best way to convey data and messages indeed. That does not mean they are "an end," though. We here simply believe that keeping on exploring the realm of possibilities in the representation of information, could lead up to refining and perfecting the core of this "science," of this field, even passing through failures and mistakes. With this body of work, we truly try to test, to explore, and to understand how to push forward what is possible to imagine, and how we can even think of educating readers' eyes to some new visual metaphors and models.

3 The New Aesthetic of Data Narrative

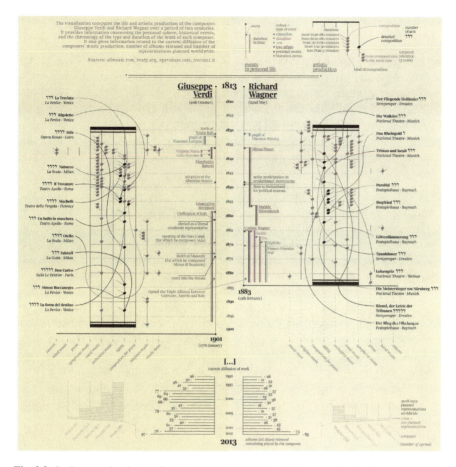

Fig. 3.2 La Lettura visual data, Corriere della Sera: *Verdi and Wagner visualized*, Accurat, 2013. Visualization of lives and artistic production of Giuseppe Verdi and Richard Wagner for their 2013 birth anniversary

Drawing the parallel with arts such as painting and music through centuries we know how much these disciplines have been able to constantly reinvent themselves even when a reinvention was not strictly necessary, opening new worlds and possibilities. The interesting question to us here is: How far we can go?

3.1.4 And, the Readers?

Since the newspaper has a very high circulation, it is exciting to observe if people like, understand, and spend time reading our pages and what they say, or if they just flip to the next page. We can do this just by looking around and looking for somebody with the newspaper, on Sundays. We found that our readers' reactions can be mainly grouped into two:

- People that would literally get lost and spend more than 5 or even 10 min looking at our pieces, talking loud about what they are discovering, commenting the topic, mentioning what they did not know before or did not expect at all, and being excited of what is surprising them while literally following their paths through the page with their fingers and heads.
- People that would look at that, try to get the big picture, quickly spanning their eyes through the elements, spending 1 or 2 min and then go to the next page (hopefully saving ours for later!)

A very well-done description of this complementary public comes from Shan Carter, the brilliant *New York Times* designer, which sometimes speaks about designing for both Bart Simpson and Lisa Simpson. The Bart Simpsons of the world might want a quick fix, a quick overview of the topic and then walk away, but a Lisa Simpson would surely desire to spend more time with the information, trying to understand and explore more in depth to gain further insights.

Here, we exactly aim at pleasing both a Bart Simpson and a Lisa Simpson, telling multiple and interplaying stories with several layers of information, the first of which should be immediately clear at a primary glance. Luckily, we never found anybody completely ignoring our pages. We never came across any readers acting with our visualizations as they would do with a piece of text or article they can spot they are not interested in just by seeing the title. We like to believe that the combination of recalling resonating or familiar images (in the way we frame the greater "architecture" of the visualization) plus our effort in composing unusual visual pieces through details and aesthetic choices plays its role in that.

3.2 A Method from Practice

Whether it's house or film or chair, it must have a structural concept.

Charles Eames

To achieve this multilayered storytelling with data even when visualizations are static and printed, everything depends on the concept of layering, establishing hierarchies, and making them clear: This is the case for both the data analysis (the stories we desire to tell), and the visual composition (i.e., the main architecture and the aesthetic value we desire to present), inviting readers to "get lost" within the narrative(s) and engage at deeper levels. Our design method, based upon layering multiple sub-narratives over a main construct, prescribes this specific phenomenon. The following is a dissection of our pieces, where the editorial process of selecting, analyzing, comparing, building hierarchies, etc., is in direct conjunction to the visual development of the layers (Lupi 2012) (Fig. 3.3).

3 The New Aesthetic of Data Narrative

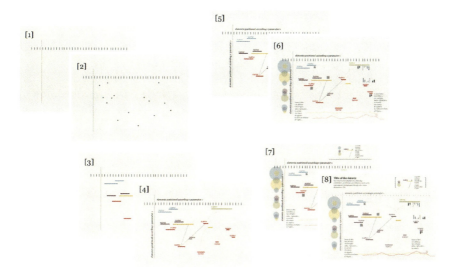

Fig. 3.3 Anatomy of the process

1. Composing the main architecture: This acts as the formalized base through which the main story will be mapped and displayed, upon this, one will see the most relevant patterns emerging from the story: The essential "map" that conceptually identifies where we are. This base is essentially a matrix or pattern that will serve as our organizer. It may be composed of cells, or distances, or other interrelated multiples.
2. Positioning singular elements (essentially dots) within the main framework. This process will test the effectiveness of the main architecture; the placement of elements reveals or confirms weaknesses and strengths, which may lead to modification of the main architecture.
3. Constructing shaped elements of dimensionality and form (essentially polygons) with quantitative and qualitative parameters and positioning these within the main architecture. As these elements have form, they must also be identified through colors according to opportunities to establish categorizations, thus advancing clarity and relationships that serve to enhance the story.
4. Elucidating internal relationships between elements. These links, directives, and qualifiers serve to give the story a comprehensive texture and correlate dependencies within the story.
5. Labeling and identifying through the addition of explanatory labels and short texts provides requisite last-mile clarity throughout the presentation.
6. Supplementing the greater story through the addition of minor or tangential tales elements. We consider this a very important step to contextualize the

phenomena in a wider world. These components link the story to external ideas, other times, or other places. Elements that are rendered here may come from very diverse sources—analysis that is undertaken once we have strongly established the core story. These elements, which may take the form of small images, textual components, graphic symbols, etc., are to be located where they best help to enrich the overall comprehension. They must not distract from the main story.
7. Providing small visual explanations such as a legend or key that assists readers and the general public who may not be familiar with norms of data visualization. These are composed to enlighten (reveal) the layered idea of the visualization, often constructed as miniatures of the layers themselves. The process usually involves simplification of the general architecture (e.g., the x and y axes, base time lines, or map components) as well as minimal explicit shapes, colors, and dimensions of singular elements. These explanations also provide units of measurement for distances and volumes, etc.
8. Fine-tuning and stylizing of elements' shapes, colors, and weights to make hierarchies pop out.

By visually highlighting the most relevant elements and lightening the other background layers of information, we should be able to allow information to be selectively and sequentially revealed, helping readers discover stories by themselves and recognizing the patterns or interrelationships from one element within the story to another.

> The details are not the details, they make the product, just as details make the architecture - the gauge of the wire, the selection of the wood, the finish of the castings - connections, the connections, the connections
>
> Charles Eames

The final fine-tuning of the piece is the necessary effort required to please readers' eyes: A well-balanced image where negative space and light elements play their role aesthetically. Is the process always so linear? Obviously, the answer is no. It is a constant iteration of explorations and a mixture of different approaches we can start to resolve the design problem with; with the constant goal in mind to allow people understand the stories, or, better said: see the stories.

3.3 Which Came First, the Question or the Dataset (or the Visualization Itself)?

I found it interesting to look back at our full set of data visualizations and start thinking how they came. I would macro-group them into two families that I can name as "visual-driven pieces" and "data-driven pieces."

The following is a description of the two different approaches with an in-depth elaboration on one of the most representative pieces for each category.

3.3.1 Visual-driven Approach

3.3.1.1 "Why Don't We Visualize This: Like This—?"

Sometimes, data visualizations come from a fascination or a visual intuition we might have in mind about possible ways a topic or a specific point of view on some information could be represented. We might even do not have the data yet, we know or suppose we can find them or we just discovered them and gave them a superficial look: The visual metaphor here is the lead element allowing us to tell the story(ies); an example follows.

Divergent Times: Atlases of World Histories

How distorted can our vision of past time be? This analysis (Figs. 3.4, 3.5 and 3.6) highlights how the perception of a historical period can depend on the way it is told. We have tried to find out the different importances given to half a century, centuries, and millennia on the basis of the latest editions of three main Italian historical atlases.

The main idea we had is relatively simple: How distorted can our vision of past time be? How does the perception of a historical period depend on the way it is told? Can we visualize it?

In Italy, we study history at school in a total linear chronological order. In 5 years of high school, for example, we would have the first year dedicated to Ancient times, the second year to the Medieval period, the third year focusing on the Renaissance, the fourth year covering nineteenth century and the beginning of the twentieth century, and a last year only expanding on the period from the first world war up to nowadays. We have tried to find out the different importance given to half a century, centuries and millennia on the basis of the latest editions of three main Italian historical atlases.

Each book has been analyzed to find out quantity, types, and geographical location of the contents, aiming at discovering possible differences in presenting events and periods. The result is an editorial reading of the three different editions through a visual rendering representing the main time, theme, and geographical distortions, thus enabling a comparison among texts. This time, everything was pretty clear to us, since the beginning.

Each atlas is illustrated with a real-time line merging with a distorted time line, altered on the basis of the space each book gives to the various periods of 50 years. At the bottom of each atlas, the visualization shows a chromatic comparison of different types of events (social, religious, political, economic, and warlike),

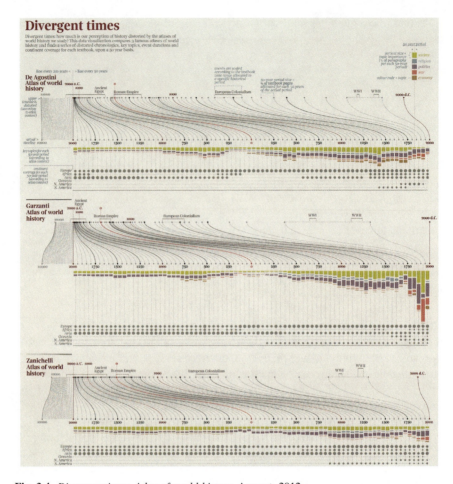

Fig. 3.4 *Divergent times: Atlas of world history*, Accurat, 2012

vertically positioned on the basis of their importance in the 50 years. The geographical areas mainly dealt with in each atlas are also highlighted.

We can spot a common interest for contemporary history, with an enormous space given to the twentieth century. It is curious to note that the last century takes a very different "space" in terms of contents: equal to 650 years for De Agostini, 1,000 years for Zanichelli, and even 1,600 for Garzanti.

A lot is told about society and politics, witnessing that history is not only made of wars (theme dealt with mainly in the last 150 years). Of the three, De Agostini proposes more economic contents, Garzanti highlights religious aspects, and Zanichelli shows the best balance of the various themes. Interesting differences also come up as to geographical localization.

3 The New Aesthetic of Data Narrative

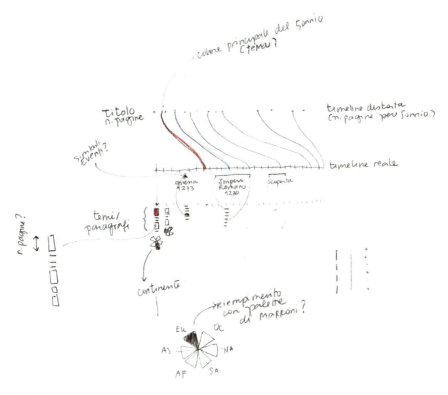

Fig. 3.5 *Divergent times: Atlas of world history*, first sketch

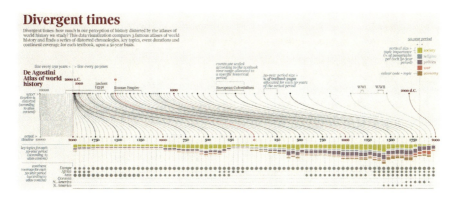

Fig. 3.6 *Divergent times: Atlas of world history*, zoom

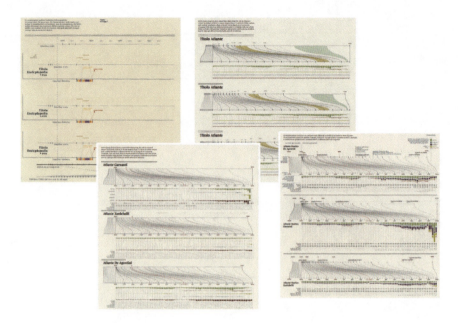

Fig. 3.7 *Divergent times: Atlas of world history*, intermediate stages

De Agostini is mainly European; Zanichelli is more focused on the East while Garzanti is the most global. All atlases, of course and as we expected, agree that the twentieth century is the key historical century.

Above (Fig. 3.7) are some intermediate steps, while we were trying to fit all the data into these three time lines, the main idea behind that is indeed very simple: visualizing a possible "mental distortion of time perception" given by how much we study different periods in books.

After this first intuition was satisfied, we spent time figuring out how to enrich this main story with other interesting comparisons. Other pieces that followed this criterion are highlighted in Fig. 3.8.

3.3.2 Data-driven Approach

Often, as you might expect, when this sudden intuition does not happen we would start from a topic worth to be explored, look for compelling datasets to analyze and cross and we would totally dig into the data material hunting for gripping narratives before puzzling out how to represent them.

This second group of visualizations, that I would call "driven by data" as an approach, can even further be divided into two.

In fact, sometimes datasets we look for, find, cross, and analyze seem to present a catchy story, or a compelling point of view to tell the story through at a very first glance. Some other times it happens that we choose a wide topic of general interest

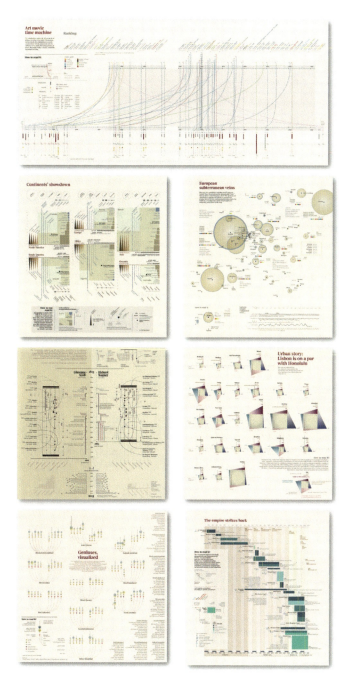

Fig. 3.8 *Empires strike again; Lisbon on pair with Honolulu; Subterranean underground veins of Europe; Continents' showdown; Time machine for art movies; Verdi and Wagner; Geniuses visualized*

and we start looking for data without knowing which specific point of view (narrative) we will end up representing. An example of the former follows.

3.3.2.1 Painters in the Making

Inspired by an article, we read in The New Yorker, in which Malcolm Gladwell tries to unravel rules underpinning age, creativity, and genius (The New Yorker, October 20, 2008, Annals of Culture, "Why do we equate genius with precocity?") comparing artists that scaled the pinnacle of their careers early and were young geniuses, like Picasso, with those like Cezanne who had to wait until their later years for their genius to reach full maturity (Fig. 3.9). The premise excited us. But everything was quite unclear to us at the beginning: We dived into data, we had intuitions about what could be potentially interesting to represent and how, but only when the visualization was complete we could really understand patterns.

First we had to set the criteria for selecting which artists to include in the visualization and also which of their paintings would be considered representative of the peak of their powers. After consulting the art experts at Corriere della Sera (main publisher for La Lettura), we chose the Italian edition of the Garzanti Art Encyclopedia as our reference. When we had our 90 artists and their representative paintings, we thought perhaps we could be more daring and choose a further painting from each artist but from a different source that would act as a counterpoint to the institutional perspective. We actually spent quite a fair bit of time on this, what would be an interesting parameter to explore?

Would it be an economic one, such as the one with the highest market value? Or perhaps, the value at which it was insured by each museum? We finally settled on something totally different. We decided to go for the "common people's choice," we would choose the first result that came up with each painters name on Google images! We were finally ready to fill some blank pages.

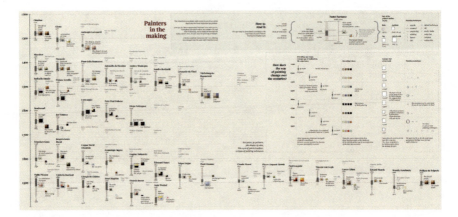

Fig. 3.9 *Painters in the making*, Accurat, 2012

3 The New Aesthetic of Data Narrative

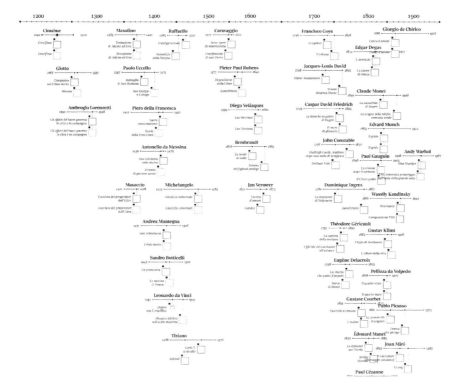

Fig. 3.10 First draft representing painters and paintings through time

Our first prototypes revolved around choosing how to represent each painter: We started with a line of length proportional to the length of his life and the two chosen pieces positioned along that life line at the marked time of the creation of the piece.

Before actually working with data we always try to mock up some sample data to see how all the elements fit together, and so we can begin experimenting with different graphic elements (Figs. 3.10, 3.11 and 3.12).

In our first sketches and drafts, we organized the visualizations around a horizontal time line. As we considered the initial drafts we realized, a vertical option would provide a more balanced overall composition and bring more clarity to all the elements. So, we rotated everything 90°.

At this point, we felt the urge to impose a more rigorous framework around how we represented each painter's life. We needed a more "immediate" visual strategy that would allow readers to more easily compare the different artists. We came up with the idea of dividing each artist's life into three defining periods: young (up to 35 yo), adult (36–60 yo) and mature (61 and over). We also considered how to deal with the fact that most of the artists lived in times when the average life span was a lot shorter: The three intervals for each artists' life was then normalized according to the average life expectancy for each century.

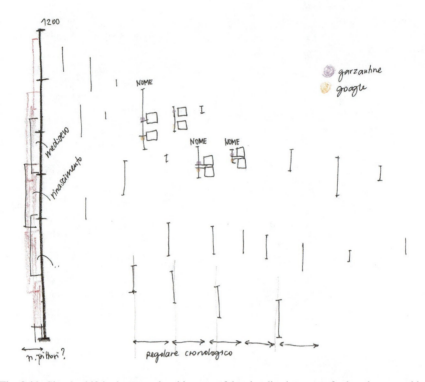

Fig. 3.11 Sketch, shift in the general architecture of the visualization, some further elements added

At this stage of the process, we were finally beginning to feel comfortable with the overall composition. It was now time to begin to add further layers of information to the visualization to bring greater depth and richness to it (Figs. 3.13 and 3.14).

Here, we decided to add the main colors of each painting to its representation, and a further layer of information was then added by making each square proportional in size to the painting it represented. The final touch was adding a small token, indicating the technique used by the painter.

While refining the content and visual strategies, we needed to make sure not to lose focus of the final version (Figs. 3.15 and 3.16). As the visualization took shape and more data were being added, we wanted to convey how the patterns were changing across time, to make sure we could safely communicate the unfolding big picture.

As it often happens, once the visualization had taken its final form, we realized our initial idea was really just a jump-off point, an insight that enabled us to begin to tackle the compelling potentialities as they came up.

In fact, the initial concept in which we were focused on the painter's age became just another element in what had become an enthralling visualization. And the more the visualization matured the more we started to conceive it as a sort of synoptic map, which would enable the viewer to explore the multilayered information

3 The New Aesthetic of Data Narrative

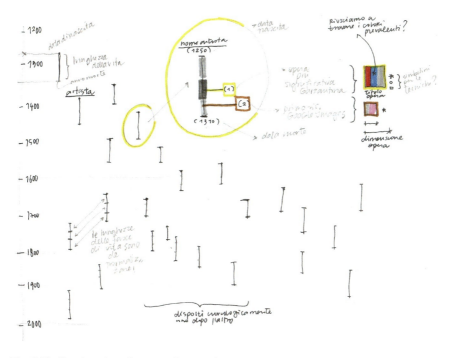

Fig. 3.12 Sketch, painters' ages, and masterpieces

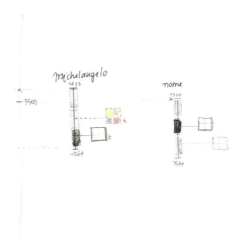

Fig. 3.13 Sketch, exploring singular elements

spectra within this slice of "the world of through time." Other pieces that followed this criterion are highlighted in Fig. 3.17.

While for the first group of visualization (visual-driven approach), it is the visual model we have in mind suggesting us which data we should look for; these latter

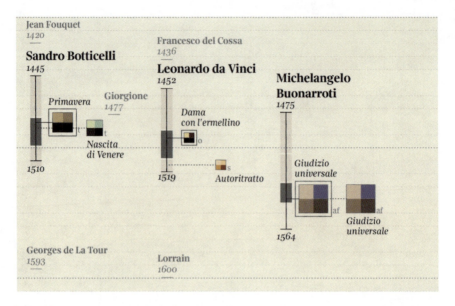

Fig. 3.14 Intermediate stage, digital version of former sketch

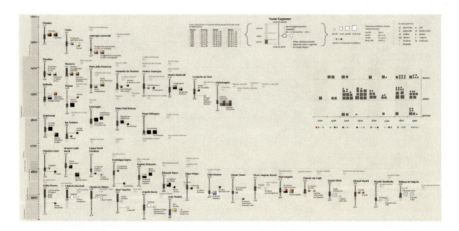

Fig. 3.15 Intermediate stage, adding legend, and reference table

kinds of pieces (data-driven approach) are visually developed in two different ways. Sometimes, the visual model would start from the data itself: Trying to sketch out an unusual but effective way to represent the overlapping information and pointing out the most interesting patterns we find. Other times we might be struck by a very big visual inspiration that could even not be related to the topic: We try to follow it and see where it leads us to. On top of that we figured out and we rarely got visual

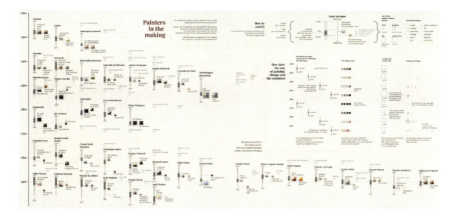

Fig. 3.16 *Painters in the making*, final spread version

inspiration from current data visualization; abstract art and paintings, musical scores, textiles, maps and urban forms, and architecture often drive our representations. We think this is a compelling approach to be pursued.

In fact, as the variety of data grows, as tools are very easy to be used to create "standard" data visualizations, the search for inspiration needs to extend outside the field. And this is important if we want to be innovative, if we want our projects to be remembered, or if we even aim higher at educating readers' eyes with new visuals able to display compound narratives, reflecting the complexity of the world of information out there.

We every time try to add something new picking visual inspirations from different fields but remembering how our brain (as readers) works toward finding meaning in things, which happens relying on something "familiar" even not manifest, something that pluck chords in us that sometimes we cannot easily name. Yet they resonate.

> Metaphor consists in giving the thing a name that belongs to something else.
>
> Aristotele

We then also every time try to build and frame our piece on something that has been experienced before (being a geographical map, a regular timeline or the disposition of elements in the piece according to unique intersection of parameters: Something able to transform the "strange" of the new visual experiments we include, into the known.

> In the end, creating is all about playing and innovating within familiar forms.
>
> Twyla Tharp

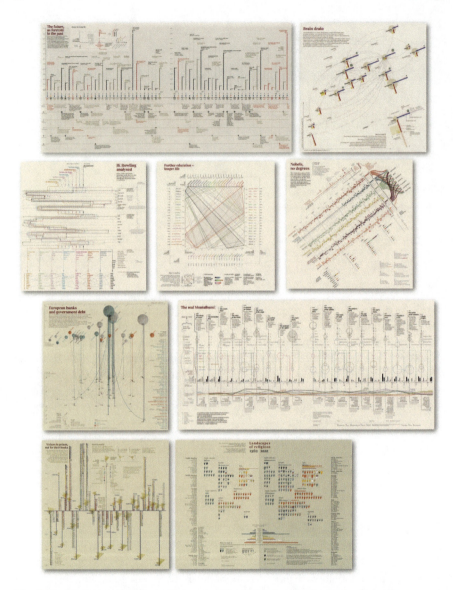

Fig. 3.17 *Landcape of religions; Writers in jail, not for their books; Brain drain, the more you study the longer you live; Nobel prizes and laureates; European banks and government debt; JK Rowling analyzed, visualizing the future*

3.4 The Importance of Inspiration from Different Contexts

As we see, the search for inspiration in unusual contexts is not a mere divertissement, but should be intended as an attempt to analyze the aesthetic qualities of things that are naturally pleasant to the eye, in order to understand how they can be eventually abstracted and reused as core principles and guidelines in building visual compositions.

This exercise can be an attempt to interpret both the single visual elements and the overall composition as construction blocks, iconic ingredients for other recipes, to be possibly reapplied to the things we are working on.

The hard question is indeed how to translate fascinations you have from art, architecture, music, and so on into actual clues for building something new in terms of data visualization.

> It is not what you look at that matters, it's what you see.
>
> Henri David Thoreau

Crucial here is training ourselves to look at things with a critical and curious approach, questioning ourselves about what we are seeing, what we are appreciating, and what is that attracts our attention, and why.

> Best question: Do you know how to see what's worthy of note? How do I see? When do I know what I'm seeing is either interesting or relevant? How do I mark this 'note-worthyness'? How do I share my 'things worthy of note'? Are there different ways to explain why I find things 'worthy'? Do I have hierarchies of 'worthy-ness'?
>
> Georges Perec

In "An Attempt at Exhausting a Place in Paris," Perec provides a "sociological survey" of the infraordinary: What happens when nothing happens? Over a period of 3 days, Perec watches a square in Paris, Place Saint-Sulpice, shifting locations periodically from café window to café window to park benches. By sitting (as opposed to walking or moving), he notes the birds, the weather, the passing of cars and buses, people walking and people sitting, people eating and how they eat, which hand they use or do not use.

His focus on detail, lists and lists, whether the items on his desk or the way of categorizing books (examples of his nonfiction from Species of Space), rather than exhausting the reader, excites him. For Perec, this is what is happening, even though nothing is happening, nothing noteworthy, and yet, this is life, this is happening. This is what happens when we stop moving and observe. It is possible to observe a lot, by watching, if paying attention.

But then, how to catch, capture, and translate all of this, being street scenes or more visual objects like paintings, architectures, or other images, into actual clues for adding novelty to our design?

3.4.1 Before I Think I Draw

Drawing plays an important role in the production and communication of knowledge, and in the genesis of new ideas. It has been largely described as a possible method of illustrating how instinctively our perception is directed toward finding meaning in things, toward recognizing things (Gansterer 2011). In addition, the act of drawing and the very fact we choose to stop and draw can be described as an act demanding the focusing of attention.

At Accurat, I am responsible for the design of the representation of information, my interest is in the graphics that respect the particular structure of a dataset, or of a complex system of information: Graphics that the form would not make sense of if you just plug in a different dataset, structures that suggest themselves the patterns to reveal the knowledge.

As a human being then, I draw a lot. When I am sketching the things that happen to attract my curiosity from abstract art, paintings or even architecture and landscapes, I always try to find a way to interpret both the single visual elements and the overall composition, the overall structure of what I am in front of. I always ask myself what I would like to read from the shapes, colors, and arrangement, trying to understand how their visual quality can be transferred to a different meaning and I truly understand why the things that inspired me work only when I am able to reapply the principles behind them to another context.

I am in fact mainly attracted by balance, repetition, and composition: I am getting used to always keeping a sketchbook with me, because I learned that I can really understand the patterns that I see in reality only when I try to reproduce them on paper. The very act of reproducing introduces a level of abstraction that helps focus on the aspects of the composition that caught my attention. Every time I am trying to copy the world I see I am surprised by how I end up losing myself in reproducing the structure of an architecture or the repetitive patterns in a collection of photographs that instantly evoke in me a sense of balance and order. I therefore always use drawing as my primary expression, as a sort of functional tool for capturing and exploring thoughts and exploring ideas. Of course, there is no fixed rule for that. But, I like here to encourage all to try it out, as Paul Auster did addressing writers.

Paul Auster, in the essay "Why Write," tells a story about growing up as an 8 year old in NY and being obsessed with baseball, particularly the New York Giants. The only thing he remembers about attending his first major league baseball game is that he saw his idol, Willie Mays, and immediately asked him for an autograph.

> "Sure kid, sure" Mays replied, "you got a pencil?" Auster didn't, neither his parents nor anyone else close by. Mays waited with patience, but when it was clear that nobody had anything to write with, he had to say "Sorry kid, ain't got no pencil, cant' give no autograph."

From that day on, Auster made it a habit to never leave the house without a pencil in his pocket: "It's not that I had any particular plans for that pencil, but I didn't want to be unprepared, I had been caught empty-handed once, and I wasn't about to let it happen again. If nothing else, the years have taught me this: if there is a pencil in your pocket, there's a good chance that one day you feel tempted to start using it. As I like to tell my children, that's how I became a writer."

It becomes even clearer when translated to visual design: You cannot just imagine the work; you have to put pencils to paper to generate and fix ideas; you have to sketch or draw things to produce and communicate knowledge.

> The act of drawing, the very production of a sketch or diagram in the context of scientific work is not free of cultural values and norms. It is a high-tension act, an act which demands the focusing of attention as well as a whole range of intellectual and practical techniques. Every stroke is the surprising result of an interested navigation within an area of possibilities
>
> Karim Harasser

I will now present some other pieces but this time focusing on how I literally sketched visual inspirations out.

3.4.2 Theory in Practice

3.4.2.1 The Brain Drain and a Visit to Moma's Inventing Abstraction Exhibition

This visualization (Fig. 3.18) explores the phenomenon of global "brain drain" in science, with an eye toward understanding the reasons why researchers might choose to leave their countries of origin and pursue careers elsewhere.

Combining three sets of data—a World Bank survey, results from a research paper titled Foreign Born Scientists: Mobility Patterns for 16 Countries, and The Times' ranking of the world's best universities—we contrasted the number of researchers per million people (y-axis) with the percentage of the country's GDP devoted to scientific R&D (x-axis). Also displayed are unemployment rate, female employment rate, percentages of foreigners and emigrants in population, emigrant researchers, and emigrant researchers returning to their country of origin. The background arcs map the principal relationships among the countries as regards migration flows.

As Maria Popova published on "Brainpickings"[2] the idea came to me after a recent visit to MoMA's Inventing Abstraction exhibition[3] (Fig. 3.19).

Abstract art and data visualization are related indeed in terms of visual languages, colors, and lines to create compositions, which can exist even

---

[2] See Brainpickings Web site, URL, December 28, 2013: http://www.brainpickings.org/index.php/2013/02/13/giorgia-lupi-brain-drain.

[3] "Inventing Abstraction" exhibition, MoMA, Leah Dickerman (Curator), December 23, 2012–April 15, 2013.

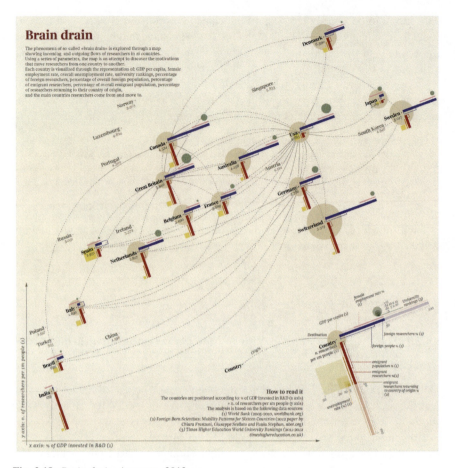

Fig. 3.18 *Brain drain*, Accurate, 2013

independently from visual references in the world. I wanted to come up with a data visualization able to replay that very geometric feeling, pleasant aesthetic and colors'-related flavor I had throughout my whole visit, passing by Mondrian's, Malevich's, and Kandinsky's pieces. It immediately occurred to me that each of the countries we were analyzing data on, should have been represented as a compound complex element, the parameters of which should have been visually related by the positioning, rotation, and spatial correlation of those geometrical shapes I was sketching down (Figs. 3.20 and 3.21).

As a conclusion, we cannot even talk about having big data here, obviously, but still, till the visualization was not complete, lots of interesting patterns and unexpected and possible correlations between elements of the countries would not be clear, we would not have noticed them, we would not understand and have the big picture at all.

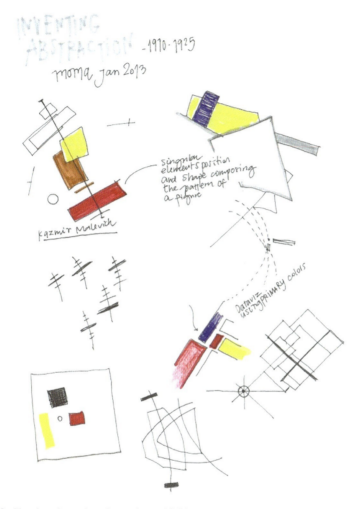

Fig. 3.19 Sketches, inventing abstraction exhibition

And this happens a lot if you are trying to deal with non-typical combination of information, when you are not dealing with a singular dataset you can simply draft a chart about: It is the visualization itself, the customized visual model we build (with a manifest inspiration from abstract art) that helps understand patterns and unexpected facts.

3.4.2.2 Nobel Prizes and Laureates and John Cage

This visualization (Fig. 3.22) explores the story of Nobel prizes through years. Visualized for each laureate are prize categories, year the prize was awarded, and age of the recipient at the time, as well as principal academic affiliations and

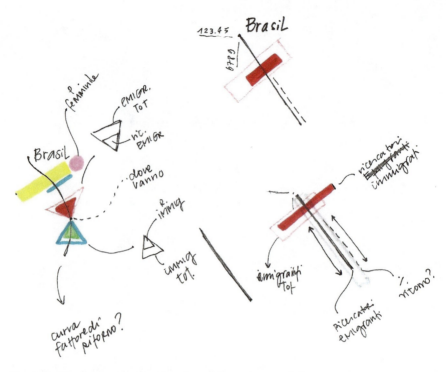

Fig. 3.20 Sketch, visual exploration of countries as compound elements

hometown. Each dot represents a Nobel laureate, and each recipient is positioned according to the year the prize was awarded (x axis) and his or her age at the time of the award (y axis).

Without going in depth into contents, I was thinking about one of John Cage's[4] most famous sheet, *Fontana Mix*[5], the first time I thought of representing ages of Nobel laureates evolving through time.

---

[4] John Cage was an American composer, music theorist, writer, and artist. A pioneer of indeterminacy in music, electroacoustic music, and nonstandard use of musical instruments; Cage is perhaps best known for his 1952 composition 4′33″, which is performed in the absence of deliberate sound; musicians who present the work do nothing aside from being present for the duration specified by the title. The work of John Cage and other contemporary composers is also called "graphic music notation": using non-traditional symbols and text to convey information about the performance of a piece of music.

[5] Composed by John Cage in 1958, *Fontana Mix* is a piece of tape music consisting of a score and four multi-channel tapes featuring sources derived from six different classes (city sounds, country sounds, wind-produced sounds, manually produced sounds, electronic sounds and smaller, amplified sounds). It has a graphic score consisting of 10 sheets of paper with curved lines and 12 transparencies, 10 of which contain a varied number of randomly distributed dots, 1 with a straight line, and the last with a grid pattern. See online, URL, October 31, 2014:http://www.solwaygallery.com/images/Exhibitions/THANKS/50th_BIG/8_Cage-FontanaMix-Ltgrey.JPG

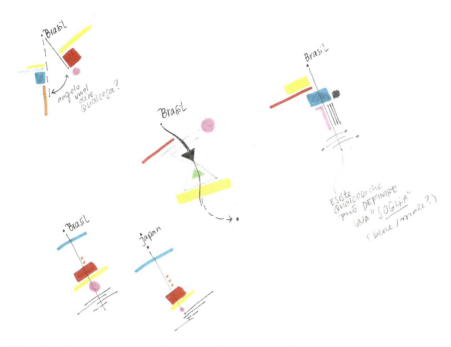

Fig. 3.21 Sketch, visual exploration of countries as compound elements

I love the way Cage composes the overall visual architecture of his pieces, of course, they are functional sheets to be played but they are also very graceful in terms of visual beauty.

In this sketch (Fig. 3.23), we shared with our publisher I was simply trying to follow this idea of building helping highlighting some differences we noticed within the data, and then, the visualization was pretty clear to me: A main compound musical partiture to display the punctual information of Nobel Laureates through times, categories, and ages; with the addition of side lines and charts to show aggregated data starting from the very frame of the main score.

Above (Fig. 3.24) are some intermediate stages as we started making the piece digitally. I have been asked a lot about "why would you tilt the visualization as it was on a slope?" Initially, the lack of space played its role on that choice, but this turn we tried while fitting everything into the art board, isn't it just incredibly more beautiful and elegant? (Cage was teaching again).

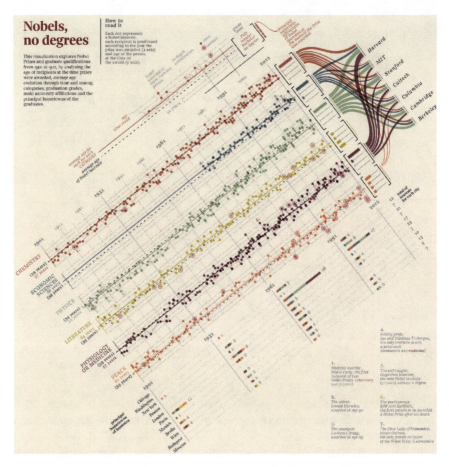

Fig. 3.22 *Nobel prizes and laureates*, Accurate, 2012

3.5 Conclusion

It is difficult indeed to conclude an article that was aiming at opening worlds and scenarios, rather than at answering questions or proposing methods and rules. As Moritz Stefaner highlighted in his speech "Finding Truth and Beauty in Data"[6]: "good visualizations not only answers questions, they generate new questions, they make you think and concern; good visualizations should tell stories, thousand stories able to provide multiple levels of exploration."

Whenever the main purpose of visualizations is to open readers' eyes to new knowledge and to reveal something new about the world, or to engage and entertain the audience about a topic, it is impractical to avoid a certain level of visual

---

[6] Keynote at European Communication Summit, Brussels, June 2013.

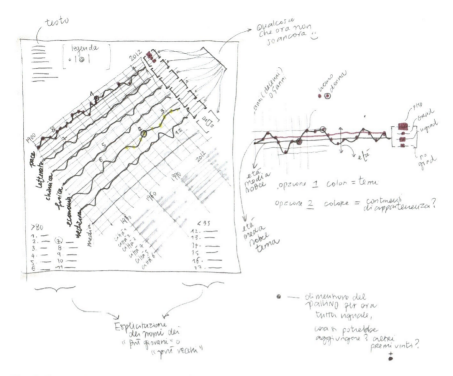

Fig. 3.23 *Nobel prizes and laureates*, sketch

complexity indeed. The world is complex, compound, rich in information that can be combined in endless ways, therefore catching new points of view or discovering something that you did not know before often cannot happen at a glance: This process of "revelation" often needs and require an in-depth investigation of the context. Consequently, we like to think at these kind of data visualizations we presented as visual ways to convey the richness, the involvement and feelings (being engagement or concern) that we experience in our everyday lives rather than simplifications of the world.

Thus, one of the important challenges for data-visualization design nowadays is to experiment on and find proper ways to express the data complexity, and more broadly the complexity and the multiple possible interpretations and contextualization of any phenomena in the contemporary world; which in opposition to reductions require the comprehension of the relations between the whole and the parts, at any time. And, as in the physical world, aesthetics plays an important role in shaping people's reactions and responses to any products, acting as the bridge between it and people's emotion and feeling. Therefore, experimental visualizations design should always aim at balancing conventions and familiar forms people are comfortable with, and novelty: Truly imaginative visuals able to attract individuals into the exploration, able to transform the strange of any visual experiment we

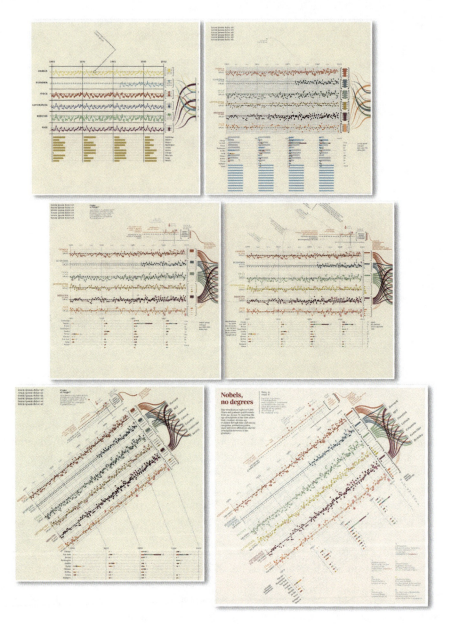

Fig. 3.24 *Nobel prizes and laureates*, intermediate stages

include into the known, and ultimately able to invite readers to explore the richness of the stories lying behind.

We are moving fast toward infinite possibilities for data analysis and display, with theoretical models, abstract formal languages, and open knowledge developed in other disciplines more than available to get inspirations from. Again, we think the question to set here is "how far can we imagine to go?" We do not know, but let us go there! It is in that spirit we share our exploratory work.

Credit and Acknowledgments The article takes its basis from former pieces we published on various online journals and magazines: the Parsons Journal for Information Mapping, Visualfitting everything into the art boardly, Visualizingdata.org, Brainpickings, Fast Company during years 2012 and 2013.

Accurat.it

Accurat is a design agency and consultancy based in Milan and New York transforming data into meaningful stories, and developing multimedia narratives and interactive applications. Accurat is directed by Giorgia Lupi, Simone Quadri and Gabriele Rossi. Visualizations presented in the chapter have been designed and produced by: Giorgia Lupi, Simone Quadri, Gabriele Rossi, Davide Ciuffi, Federica Fragapane, Francesco Majno, Stefania Guerra, Marco Bernardi, Matteo Riva, Pietro Guinea Montalvo and Elisa Raciti.

The full gallery of the visualization, December 28, 2013: http://www.flickr.com/photos/accurat/sets/72157632185046466

Corriere della sera, La Lettura, Serena Danna e Pierenrico Ratto (working team), December 28, 2013: http://lettura.corriere.it

The author wishes to thank Mike Dewar, Glauco Mantegari and Roberta Sferlazza for their precious help and contribution to this article.

References

Gansterer N (2011) Drawing a hypothesis, figures of thought. Springer, New York
Lupi G (2012) Non-linear storytelling: journalism through info-spatial compositions. Parsons J Inf Map IV. URL: http://pjim.newschool.edu/issues/2012/04/pdfs/ParsonsJournalForInformationMapping_Lupi_Giorgia.pdf

Author Biography

Giorgia Lupi (Co-Founder and Design Director at Accurat) is an information designer and researcher. Her work in information visualization frequently crosses the divide between digital and print, exploring visual models and metaphors to represent dense and rich data-driven stories.

She is co-founder and design director at Accurat, an information design company based in Milan and New York; Accurat analyzes data and contexts and designs analytical tools and visual narratives that provide awareness, comprehension and engagement. Since 2011 she is also a Ph.D. candidate in Design at Politecnico di Milano, within DensityDesign Lab.

During last year, she presented at several conferences: Cumulus Conference in Helsinki, The Eyeo Festival in Minneapolis, Data Visualization official meet up in New York, New York city Public Library public events' series, Media Evolution Conference in Malmoo, "Why Public Design?, Strata Conference in NY, Visualized Conference in NY, Tapestry Conference in Annapolis; she will be speaking at Resonate Festival in Belgrade, Eyeo 2014 Festival.

Her work has been featured in different magazines such as *The New York Times*, *Corriere della Sera*, *Fast Company*, *Slate*, *Forbes*, *the Atlantic Wire*, *Brain Pickings*, *Arcade Magazine*, *Courier International*, *Flash Art*, *Vogue Uomo*, *Popular Science* among all; and published in the recent *Infographic History of the World* Harper Collins, *Around the World* Gestalten books among all. Her work won important awards in 2013: Gold Medal for data-visualization and Special Mention for best studio at the Kantar Information is beautiful awards, Gold Medal for data journalism at O'Reilly Strata, Bronze Medal at Malofiej 2013, Bronze Lion at Cannes Festival 2013, Special Mention in Core77 Design Awards.

For more information and contact:

http://www.accurat.it
http://giorgialupi.net
https://twitter.com/giorgialupi

Chapter 4
A Process Dedicated to Cognition and Memory

Wesley Grubbs

Abstract The most effective form of communication is a visual one. Even when we speak we use visual metaphors to help our communication (i.e., to bark up the wrong tree or to push one's buttons). Focusing on a visual that helps bring attention to the story can also help establish a memory of the story, and this is a key aspect we take into consideration in every visualization we build.

4.1 Introduction

I have a confession to make. I am not a big fan of backyard barbecues. The reason is that the question of what I do is a nightmare to answer to anyone not directly in my field. Data visualization requires several far-reaching skills to be combined together. I am a programmer, statistician, graphic designer, artist, and creative director every day. The irony is that the work I do is geared at communicating and simplifying complex data to a broad audience. Maybe I need to make a visualization of what I do so I can enjoy the neighborhood block parties again.

If you are reading this book, I can sigh in relief because I can assume you know what data visualization is. Now that you know this, I wanted to share my thinking and process in the work we do at my studio, Pitch Interactive. But before that, I wanted to start addressing an important question: What makes a successful data visualization?

I am asked this question repeatedly, and it is never an easy one to answer. It is easy to say that the ultimate success is whether or not the visualization achieves its intended purpose. But the purpose is often fluid and changes as the project evolves.

For some of our work at Pitch Interactive, the successful part of a project is journey we take, the discoveries, the mistakes, even the dead ends that tell us to

---

W. Grubbs (✉)
Pitch Interactive, 2140 Shattuck Ave Ste 806, Berkeley, CA 94704, USA
e-mail: wes@pitchinteractive.com

© Springer-Verlag London 2015
D. Bihanic (ed.), *New Challenges for Data Design*,
DOI 10.1007/978-1-4471-6596-5_4

stop. Though we are typically the only immediate beneficiaries of such "success," we embrace all of these and this helps us continually evolve and innovate.

For other work, success was meeting or exceeding client expectations. These expectations could be whether or not a complex message was communicated in an easy-to-digest way or simply if the final result gave an impressive impact when the CEO revealed the company's annual figures at a trade show.

However, some of our most successful projects were ones we did internally to serve our own curiosity. Projects we did in between work when we needed to take a break from client needs and focus on our own. It is mostly these projects that were published in dozens of publications and showcased in museums like the MoMA in New York. And in some cases, such as our drones piece,[1] we were able to shed a clearer light on a politically sensitive topic and hopefully help better shape people's perspective on the issue. The success problem came from the fact that we were more personally invested in these projects. There were no committee or executives who were directly involved in the project we had to please. Also, the projects typically covered topics that were of public interest but not enough visible return on investment for someone to finance for us. We were working against our time and our own budget.

Regardless of whether a project was for ourselves or a client, the process is very similar and our approach follows a pattern. The process is as follows: to collect data, find questions to answer, refine data, reveal patterns, and communicate the story. And the order of this is shuffleable. The remainder of this chapter helps explain our approach and give light on how we accomplish the work we do.

4.2 Esthetics as a Tool

It never ceases to amaze me how frequent the importance of good design and esthetics is disregarded as unnecessary fluff or a distraction from the actual data. Design is critical in the work we do, and we spend a significant amount of time during the production and postproduction focusing on colors, type, alignment, and other visual elements because they are a crucial part to the communication aspect of the visualization. The human brain is wired to not only process but to attract to intricate imagery, for example, looking at a mountain range or watching waves on a beach. These are immensely complex images of terrain and motion, yet we often find ourselves mesmerized and amazed by them even if we cannot appreciate fine art. When intricate imagery has a story embedded within, it only stimulates our interest more.

What is even more interesting is how often we think in imagery. The next time you are discussing politics or an event to someone over lunch, try to pay attention to the images in your mind as you are discussing even numerical issues like money or

---

[1] See Out of Sight, Out of Mind, URL, February 6, 2014: http://drones.pitchinteractive.com.

the number of times something happened. Traditional diagrams like bar charts are excellent tools for conveying relatively simple relationships quickly, especially if you need to make a decision. However, they are not the best way to help us remember the story being told by the graph. When you discuss the latest sports or political news with a friend, notice that even with all the figures being discussed, your brain is not creating bar charts or scatter plots as you process that information. Your mind makes images of the topics you discuss. This is key in understanding our goal in the visualizations we create.

The purpose of esthetics in visualization is to help enhance cognition and memory. If you live in a city and go camping and look up to a clear sky full of stars, or when you listen to a favorite song or you look at a mountain and have this sudden sense of pleasure, something happens in your brain that is really magical. An experience is formed that your brain embraces. And if you are doing something else during that event, like eating breakfast or having a conversation with someone, there is a much higher chance you will remember what you ate or discussed for a long time. This is because different parts of your brain are forming connections about the experience and each part and each neurological connection better reinforces the other to help retain the memory you had. Our brains are highly visual and capable of processing very complex visual forms. Think about how complex looking into a forest is, or watching waves crash against a shore. Yet despite the complexities, these visuals have a tendency to actually relax and please us. Now, think of the experience of looking at a typical bar graph or the stock market. We feel tense. Our brains do not embrace these without effort. We actually have to focus on them to process the information they are providing. If you were to look at an eye of someone processing a scatter plot, you would see their pupils dilate because the brain has to switch to an analytical mode and this takes much more cognitive effort than sitting on a beach looking at waves and thinking about our existence. And if that effort is too much, our brains give up and the story is lost. Why is this? My personal theory is that waves, trees, and stars are all creations from natural causes, as is our brain. The structural, fractal-like makeup of all of these natural forms connects everything through familiar processes.

Whatever the reason, what I try to do is create a visual or esthetic that borrows forms or colors from shapes I find in nature whenever possible. For example, my color picking process come from photographs I will take of natural formations: sunsets, daisies, and clouds. Nature is far superior at picking and matching colors than I am.

When you expose someone to complex relationships form a database, you need to ease them into it so that their brain can accept if not embrace what they are looking at. Once you have that acceptance, you can ask the brain to analyze. You have just made analysis pleasing. Because of this, I think esthetics and decoration are crucial in aiding the cognition and memory that is needed to process a visual and, in most cases more importantly, to be able to retell the story. And with decoration, I do not mean to season a visualization with glitter (unless you can justify the use of glitter, which would honestly be pretty cool). The decoration I am referring to involves the framing of the visualization such as imagery that may

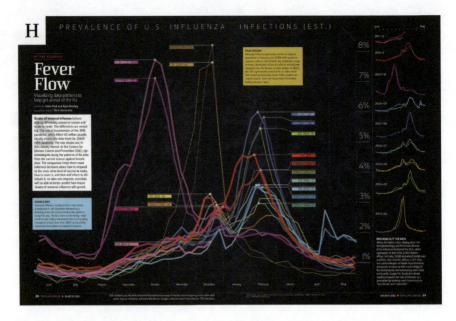

Fig. 4.1 *Fever Flow* (for Popular Science), Pitch interactive, 2012

reinforce the visual story, for example, the piece *Fever Flow* we made for Popular Science where we designed images of viruses to help further clarify the point of the story, which is about influenza trends and predicting them (Fig. 4.1).

We also need to consider and understand what it means to tell a story. Though our brains are capable of processing complex imagery, they are not very efficient in remembering exact figures, especially over a long period of time. The same goes for remembering exact words or phrases. And this is what I will address in the next section about cognition.

4.3 Imagery and Cognition

When I read a story to kids from a book, I have a manual to retell the exact same story every time if I read the book word-for-word. However, when I tell them the same story from memory, the versions of each story can vary and be retold in many forms. Take for example *Jack and the Beanstalk* (Lang 1965). Dozens of variations of this story exist. The reason there are so many different versions of this story is because this British fairy tale has been retold and redocumented by many people since it was first written in the eighteenth century. Jack's mother is either a widow or simple a single mother. The cow he trades for magic beans has different names. Sometimes, there are three magic beans, sometimes five. The giant's wife plays a larger role in some versions than others. This is all because when we retell stories, we stick to the main storyline, but we create different variations of the details that orbit the story.

The story is really quite simple: Jack lives with his mother, and they are poor. Jack trades his cow for magic beans. His mother is mad and throws the beans out the window. The magic beanstalk grows into the clouds. Jack climbs it to find a castle inhabited by a giant who eats just about anything. The giant has a lot of gold, and he is bad. Jack is good and poor so taking the gold is ok. He steals gold, a hen that lays golden eggs, and a golden harp. When the giant chases after Jack, he typically dies. Jack and his mother are now rich. The end. This is really the main story. The details we add are what make it interesting especially for a parent who is asked to repeat this story over and over again. It is the orbiting details that are really nothing more than decoration. But it is just that decoration that helps us visualize the story in our minds and remember it to retell it, in whatever variation we choose, at a later time. There is even a version where the giant killed Jack's father making Jack's killing of the giant an act of retribution (definitely not a child-friendly version by today's standards). Now bring this concept to the visual representation of data.

Looking back at Fig. 4.1, our Flu Trends graph, our goal was to help readers understand that the Center for Disease Control (CDC) has a difficult time predicting flu trends every year because the peaks and troughs do not always follow the same patterns. This is really all we are saying. And then, we add more interesting information on top of this, such as the peak numbers, and compare the CDC data with Google Flu trends to create a visually rich spread to help the readers take this image with them when they retell this story later. It is the image generated by the data that tells a much more powerful statement than just saying "flu is hard to predict." The visualization that we made reveals why.

Also, we did not want to add too much information here because it is important not to overwhelm the reader with too much information. We focus on labeling the peaks, but we keep the axis labels subtle. The information is there if you are looking for it, but it is not distracting you from the story being told. We intentionally kept labels to a minimum. Too many labels is like telling a reader "You're going to have to put some effort into reading me." It is like telling Jack and the beanstalk and saying exactly how high the stalk grew, the temperature that day, how long it took Jack to climb, how tall the giant was, how much gold he stole, etc. Sure a curious analyst would probably want to know all of this. But definitely not a typical five-year-old. They just want the story told in a way they can focus on the key parts.

4.4 Case Study: Invisible Montpellier

In 2010, we built a visualization for *Wired* Magazine depicting a day in the life of 311 calls in New York City for a spread in their October 2010 issue. In August of the same year, New York received its one hundred millionth 311 phone call. 311 is a number you call for non-emergency situations, such as if someone is parked in your driveway, a tree falls down (and nobody is hurt), a street light is left on, etc. Several

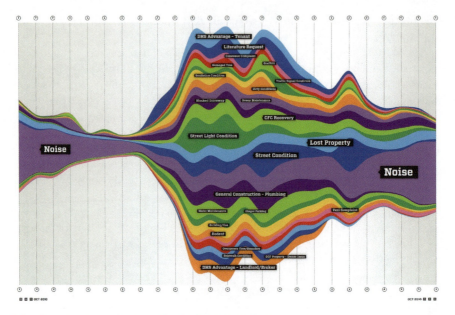

Fig. 4.2 *Invisible city*, Talk To Me (MoMA, NYC), Pitch interactive, 2011

metropolitans in the United States offer this service. Armed with two weeks' worth of data, we set out to create a visualization in less than a week (Fig. 4.2).

Not only did this piece make a full spread in *Wired*, but it was also showcased at the MoMA's Talk To Me exhibit in New York as well as several international publications and books. In early 2013, La Panacée, a new media art museum and institute in Montpellier, France, had a large-scale version of the 311 piece on display in their museum. La Panacée also commissioned us to build a visualization based on their own city complaint data, much like the data we used for *Wired's* 311 piece we did in 2010. This case study is about the exciting work we made.

4.4.1 Approach

There are many ways to go about visualizing data; however, there is one sure thing that must happen before you will have any idea of what you will visualize. You need to first have data. Sure, you can hypothesize about what you may want to show. I do this all the time. But a hypothesis is only just that if you do not have data to support your ideas. Once you have a dataset that you can work with, you can explore what you will want to do with it.

The dataset we receive from the city of Montpellier was even richer than the dataset we had from New York. However, we were missing one key data point that told us right away we could not do what we did for *Wired*: a time stamp. We were given a month's worth of data for Montpellier. Since the city does not have a 311 call service, they take complaints in different formats: letters, phone calls, etc. Below is a breakdown of several key attributes where the values were collected for each complaint submitted.

[ID][Type]
[Receive Date]
[Closed Date]
[Method]
[Agency Contacted]
[Address of Complaint Source]
[Quartiers]

4.4.2 Challenges

With a rich dataset in hand, we set to work on analyzing it to look for an interesting angle that we can focus on to communicate the data. The first concern was that we were asked to basically replicate the *Wired* 311 piece to represent the same information for Montpellier. Without the timestamp showing the time of day, we knew we could not convey a day-in-the-life visualization with a stream graph. Also the number of complaints was far less than what we had for New York. In two weeks' worth of 311 data from August of 2010, we had over 35,000 rows of data to parse through, plenty to extract patterns from. For the work we do, larger datasets allow us to not only look for patterns, but also verify those patterns. For example, the data were large enough to show a spike of complaints between noon and 2 pm, lunchtime, likely when people had a moment to make the call.

For Montpellier, we had approximately 1,000 complaints made during the month of January in 2010. Hardly enough to see whether there are any behavioral patterns in the complaints related to time. And when we tried to map out the complaints by week or across the month, the only things we saw right away were when there were weekends or holidays, day's complaints were not logged. And this was not interesting enough to build a story around.

However, the data did have one attribute we kept coming back to: an actual address of the source of the complaint. With an address, we could geolocate and get the latitude and longitude coordinates and map those to key areas around the city. This was something intriguing for us. With this, we could visualize concentrated points around the city where the complaints took place.

4.4.3 Process of Visualizing the Data

Knowing the geographic locations of the complaints was our focal point; we set to work at mapping out the points to start to study the results in a visual form. We built the hypothesis around the notion that geography would tell the story, and then, we built rough visuals to try and verify this hypothesis.

We started just plotting things out around the city. We also looked beyond the given dataset to see whether there were any interesting patterns or stories we could find among the geographical landscape of the city. With geolocation, there are dozens of APIs and resources to find the altitude of any given place around the world. In short, we were curious whether there was any correlation with altitude and complaints.

It was a fun experiment; however, it was unfruitful. When you are studying data and looking for patterns and stories, your best friend is curiosity. You should listen to it whenever it poses a question. You may not always be right, but the insights you get from it are invaluable.

Having pinpointed all the complaints around the map and looking at this from every interesting angle we could think, we decided to explore ideas around the quarters of the city, the primary regions. And then average the most frequent complaints for each quarter to basically try and label city sections with type of complaints.

Having these values plotted around the city, we wanted to create unique shapes for each type of complaint by drawing lines between all outer complain epicenters. We used an algorithm called convex hull which allows us to draw the shapes for each complaint type.

This was one of our early renderings that combined coded locations made with processing and some manual concepts with illustrator (Fig. 4.3).

But at this point, we were not completely certain this was the direction we wanted to take. Sometimes, to be more confident in the direction you are about to take, it is best to take a step back and try other ideas. At least form that, you can better understand what you do not want to. So we did something that we normally do not do so lightly. We decided to explore more abstract visual representations of the data that we had. After all, this piece was for an exhibit in an art museum (Fig. 4.4). We mixed statistical analysis with esthetic exploration.

This direction made sense from a data organizational standpoint, but it lacked a holistic story about the entire dataset unless we made this type of "blob" separately for each complaint. At that point, you overwhelm the viewer and we were not interested in doing this. We revisited our map and then dove deep into refining it and exploring more fine-tuned directions to take. In other words, we iterated version after version.

4.4.4 Iteration

Once we start to formulate a visual direction, then we start an iteration process for the designs. This process can sometimes be the most time-consuming part because it involves selecting the right typography, colors, finalizing shapes, and defining the overall look, feel, and narrative. It amazes me how often people seem to sidestep

4 A Process Dedicated to Cognition and Memory

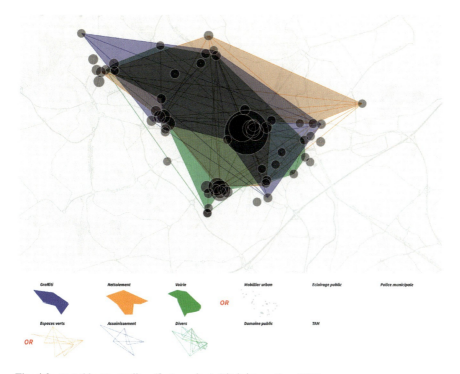

Fig. 4.3 *Invisible Montpellier* (first version), Pitch interactive, 2013

this process or stop short because it is so vital in ensuring that carefully thought out designs were made in this process.

Relying on code only, we replotted the lines and the hull (the filled areas) so that everything was coded directly from the data. To verify the placement, we overlaid a city public transport map to help the viewer more quickly identify that these shapes were driven by points in the city (Fig. 4.5).

But the transportation lines added too much clutter. We needed something else to anchor the geographic representation of the graph. Also, the colors were not working at all here.

Often, finding the right color can be one of the more difficult tasks in our work. Josef Albers, a Bauhaus-educated and world-renown color theorist, explains in his book *The Interaction of Color* (Albers 1975) that color is the most relative medium that exists. What this means is that if you have a red square on a white background, it will look very different if you change the background to black. Put a blue square next to it, and the red appears different still. Color is so relative that by just saying the color red to an audience with no image to show the red in question; it is likely that everyone in the audience will have a different shade of red in their mind. There are books written just on this topic alone.

The challenge is often knowing where to start, and from there, the process of color selection tends to evolve until you reach a point where you are happy with

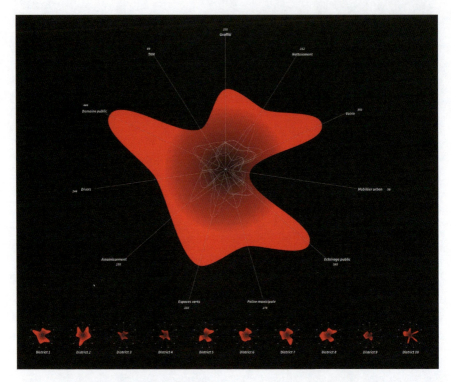

Fig. 4.4 *Invisible Montpellier* (third version)

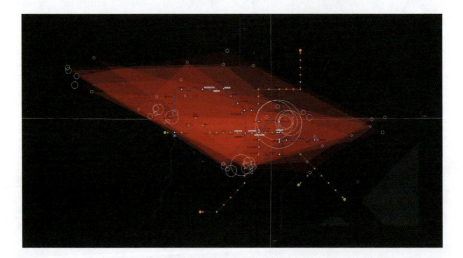

Fig. 4.5 *Invisible Montpellier* (study version)

4 A Process Dedicated to Cognition and Memory

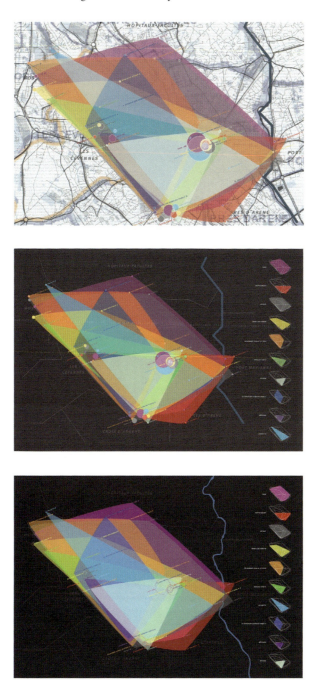

Fig. 4.6 *Invisible Montpellier*, (from top to bottom) fifth version, sixth version, final version

Fig. 4.7 *Invisible Montpellier* (printed version)

what you see. We had eleven layers of items stacked on top of each other. Finding the right mixture of colors that can visually stand out from one another and yet compliment one another was more of an art than a science in this case that required trial and error.

After many iterations and color interaction tests that we made, we finally came up with a set of colors we felt good with. Often, what we will do to get colors is take a photograph of something in nature, like a flower. We may pick one shade of yellow and then gather a list of other colors that compliment that yellow. Here is where we landed, this time with a map of the city (Fig. 4.6).

Once we had the map in the background again, the geographic anchor I mentioned above started revealing itself. The river Le Lez runs through the eastern edge of the city, a perfect visual anchor, not intrusive on the shapes but noticeable enough that you can quickly decipher what it is.

4.4.5 Final Result

Once we had our shapes and colors in place, we started dialing in on more details. Colors were tuned every so slightly, the background was worked on, typography was set, Le Lez was carefully illustrated, and we finally landed at a place we were satisfied with. We printed a test result in the studio, made further iterations, and then sent it to the client. They printed it on a 2.5 m × 1.5 m print (Fig. 4.7).

References

Albers J (1975) Interaction of color (Revised edn). Yale University Press, Connecticut
Lang A (ed) (1965) Jack and the Beanstalk. The Red Fairy book. Dover, New York (Original published 1890)

Author Biography

Wesley Grubbs (Artist, Data Visualizer and Founder of Studio Pitch Interactive) is an artist, data visualizer and provocateur based in Berkeley, CA. In 2007 he founded Pitch Interactive, a studio who's focus is weaving code with design with statistics to find versatile solutions to communicate complex data for clients such as Google, *Wired*, GE, *Esquire*, *Scientific American*, Popular Science and The Dow Jones.

Built upon his life experiences, degrees held in International Economics and Information Systems and his innate interest in the brain and cognition, Wes' work focuses on revealing patterns about human behavior and how our actions impact our surroundings.

Pitch Interactive's work spans illustrations, physical installations, console game user interfaces, software applications, websites and textiles. Their work has been showcased at the MoMA's Talk To Me exhibit in in New York, La Penacée's

"Conversations Électriques" in Montpellier, the McKnight Foundation's 30 year anniversary exhibit, the Foosaner Art Museum's The Art of Networks exhibit, The Max Planck Science Express Train, the Data Flow books and many other internationally acclaimed publications.

For more information and contact:

http://www.pitchinteractive.com
http://www.linkedin.com/in/wesgrubbs
https://twitter.com/pitchinc

Chapter 5
Graphics Lies, Misleading Visuals

Reflections on the Challenges and Pitfalls of Evidence-Driven Visual Communication

Alberto Cairo

Abstract The past two decades have witnessed an increased awareness of the power of information visualization to bring attention to relevant issues and to inform audiences. However, the mirror image of that awareness, the study of how graphs, charts, maps, and diagrams can be used to deceive, has remained within the boundaries of academic circles in statistics, cartography, and computer science. Visual journalists and information graphics designers—who we will call evidence-driven visual communicators—have been mostly absent of this debate. This has led to disastrous results in many cases, as those professions are—even in an era of shrinking news media companies—the main intermediaries between the complexity of the world and citizens of democratic nations. This present essay explains the scope of the problem and proposes tweaks in educational programs to overcome it.

5.1 Introduction

Can information graphics (infographics) and visualizations[1] lie? Most designers and journalists I know would yell a rotund "yes" and rush to present us with examples of outrageously misleading charts and maps. Watchdog organizations such as Media Matters for America have recently began collecting them (Groch-Begley and Shere 2012), and a few satirical Web sites have gained popularity criticizing them.[2] Needless to say, they are all great fun.

---

[1] I will be using the words "information graphics," "infographics," and "visualization" with the same meaning: Any visual representation based on graphs, charts, maps, diagrams, and pictorial illustrations designed to inform an audience, or to let that same audience explore data at will. I know that this decision will displease some scholars and practitioners, but I have my reasons. For more details (see Cairo 2012a, the Introduction in particular).

[2] See WTF Visualizations, URL, January 8, 2014: http://wtfviz.net/.

A. Cairo (✉)
University of Miami, Coral Gables, FL 33124, USA
e-mail: alberto.cairo@gmail.com

© Springer-Verlag London 2015
D. Bihanic (ed.), *New Challenges for Data Design*,
DOI 10.1007/978-1-4471-6596-5_5

The notion that graphics can indeed lie derives from Darrell Huff's bestseller *How to Lie With Statistics* that describes the most common kinds of visual sins, such as truncating the *Y*-axis of graphs. Edward Tufte, always keen on contriving catchy memes, tried to coat Huff's teachings with a scientific looking varnish and invented a unit of measurement called the "Lie Factor" (Tufte 1983). He even came up with a formula to calculate it. As it happens with much of Tufte's *œuvre*, it is impossible to know whether he was writing with tongue in cheek. I assume he did:

Lie Factor = Size of effect shown in the graphic/Size of effect in data
(The closer the Lie Factor is to 1.0, the more accurate the graphic is.)

Playful quantifying efforts aside, let me take the risk of sounding platitudinous: Charts, graphs, maps, and diagrams do not lie. People who design graphics do. This is a no-brainer if we stick to a well-known definition of the word "lie" in the literature about ethics: "An intentionally deceptive message in the form of a statement" (Bok 1999). The graphic is that statement, not the agent who creates it and delivers it.

A graphic can *mislead*, though. Misleading is not the same as lying because a graphic can lead readers astray without the conscious intervention of its designer. This distinction is not a technicality. According to professional ethics codes in journalism and graphic design,[3] knowing the truth and hiding it, or conveying it in a way that distorts it is simply unacceptable.

Codes of conduct are based on a priori rules, duties, and obligations. They are the embodiment of deontological ethics. Therefore, according to them, the intention of the agent is the key to analyzing if the actions she performs are right or wrong. On the other hand, designing a misleading graphic as a result of naive mistakes while analyzing or representing data is ethically neutral. I guess that we all can agree that lying is the worst action that any visual communication professional can perform, so lying will be the focus of the first part of this essay.

However, I find the alleged ethical distinction between lies and mistakes intriguing and worth playing with. Let's ask ourselves: Are the intentions of the designer really enough to evaluate the appropriateness of each graphic she creates? My hunch is that they aren't. Let me elaborate.

Professional codes of conduct implicitly differentiate between *truth* and *truth-telling*. In many circumstances, truth may be unattainable, fuzzy, or even unknowable, but that does not spare us of our obligation of being truthful. Someone who intends to communicate a message may not know all information needed or—more importantly for my argumentation—may not possess the skills to represent it correctly. But, if she strives to do her best, she will be acting ethically.

---

[3] For a sample of ethics codes: see, URLs, January 8, 2013:

http://www.journalism.org/resources/ethics-codes/
http://www.aiga.org/standards-professional-practice/
http://chrisalensula.org/the-ethics-of-visualization/
http://blog.visual.ly/a-code-of-ethics-for-data-visualization-professionals/.

So far, so good. This is simpler than the stuff discussed in any Ethics 101 course. As a designer or journalist, you are likely feeling comforted at this point: "Perhaps I make mistakes sometimes but I certainly don't lie," you may be thinking. Allow me to distress you a bit. Don't think just about *yourself*, the designer, when evaluating your own decisions while creating a graphic. Think of *who's going to decode it*, too. Read the following paragraph carefully:

When we see a chart or diagram, we generally interpret its appearance as a sincere desire on the part of the author to inform. In the face of this sincerity, the misuse of graphical material is a perversion of communication, equivalent to putting up a detour sign that leads to an abyss (Wainer 2000).

Notice that Wainer does not imply that the author is trying to actively deceive you. The "misuse of graphical material" could have been completely unintentional, the same way that putting the detour sign in the wrong side of the road could be the product of sloppiness or ignorance.

Switching the focus back and forth from the designer (the encoder) to the reader (the decoder) makes the distinction between lying and misleading much blurrier, for in the world out there the consequences of both lies and mistakes are equally grave: More noise and misinformed audiences. And so, paradoxically, I feel that what was going to be a short piece to help myself reflect about lying with information graphics cannot stick to lies alone. If we agree that infographics must represent a reality—data, information—with accuracy, we must not just obsess over the conscious actions of communicators. We should also point out the responsibility we have to educate ourselves to overcome our own biases, shortcomings, and knowledge gaps. We must work hard to eliminate or, at least, to minimize ambiguity, confusion, and potential errors of interpretation in our graphics. That will be the core idea of the second part and the conclusion of this essay.

5.2 Becoming a Good Liar

It would be preposterous to pretend that a single book chapter can outline the hundreds of ways information graphics can be used to deceive. Several books (Huff 1954; Jones 2006; Monmonier 1996), among others—have been written about the topic, so I won't repeat what they have already said. I'll just mention Howard Wainer's *Visual Revelations*, which explains that most graphic lies are based on three strategies:

1. Not showing much data
2. Showing the data inaccurately
3. Obfuscating the data

I have reorganized Wainer's list, a bit for the purposes of this essay. Here is my version:

1. Hiding relevant data to highlight what benefits us
2. Displaying too much data to obscure reality

3. Using graphic forms in inappropriate ways (distorting the data)

To become a good liar you will need to learn how to use these methods, so let's illustrate them with some examples.

5.2.1 Hiding Relevant Data to Highlight What Benefits Us

Any parameter can tell us something about a population, the same way that any properly obtained statistic can tell us something about a sample from that population. Most datasets are too complex to be scrutinized directly (try to extract meaning from a table with hundreds of numbers), so we use and manipulate statistics—mean, median, standard deviation, etc.—to analyze them. However, no parameter or statistic can reveal the entire truth of a dataset. Most visual liars are very aware of this fact, and they take advantage of it.

To give you a perfect example of this strategy, stick with me through a short detour which will become relevant in Part 2 of this chapter.

In the first decade of the Twenty-first century, the marketing and P.R industry hijacked the very word "infographics" (Cairo 2012b). This word has a long and noble tradition in the news, where it is used to define visuals that display rich data, maps to locate stories, and diagrams and illustrations to explain complex processes and procedures.

News infographics are, or should be, subjected to professional standards also applied to news stories. Marketing and promotional infographics are not, as you will notice if you do a little exercise: Go to your favorite search engine and type "best infographics" or "cool infographics." Go ahead, try it out. You will be shown piles of bad and uncool junk—and perhaps some gems; I'm willing to concede that it is possible to find shiny needles in any damp haystack, but hope weakens when one stumbles upon statements like this, common among the enthusiasts of these aberrations: "All online infographics are a form of content marketing link bait." (Krum 2013)

To what I say: Baloney. If producing "link bait," instead of something truthful, accurate, and deep is your goal, then you are not designing an infographic, but a poster, a flyer, or an ad. End of detour for now.

Lies in this kind of so-called "infographics" usually come in the form of cherry-picked statistics. Many of them are made of isolated figures surrounded by cutesy illustrations and colorful special effects. See the first man in Fig. 5.1, a graphic that mimics the style I'm describing. He proudly holds the average sales increase of his company over the last year: 12 %. Not bad, right?

But see what happens when we offer a more nuanced picture of what hides behind that figure (man on the right): Among the ten markets in which this company operates, five improved, but the other five shrank. Why would you hide this important bit of information from customers or investors? That was a rhetorical question, in case you did not notice. The first "infographic" is good "link bait." The second one is true.

Fig. 5.1 Two versions of a fictional graphic. The first one displays just the average change, hiding the fact that the ten markets in which this company operates behaved very differently between 2012 and 2013. That reality is shown on the second version of the graphic. The average of all the values encoded in the bar graph is indeed 12 %

5.2.2 Displaying Too Much Data to Obscure Reality

If extreme simplification and biased data selection are reliable strategies to obfuscate data, extreme and unnecessary detail may also be effective. Compare the two maps in Fig. 5.2 (data aren't real.) Neither of them is intrinsically wrong, but the first one is much better if our goal is to give readers a quick and clear overview of the data. The second one might be appropriate if the designer wants readers to explore poor areas in detail, perhaps with the help of interactive tools, such as zooms and filters. But, if the second one is reproduced at a small size on a printed page, it will become meaningless.

5.2.3 Using Graphic Forms in Inappropriate Ways (Distorting the Data)

This is my favorite strategy and, by far, the most common one. Many lies in this category are grossly conspicuous, but may go unnoticed if shown quickly on a screen, or if the viewer is distracted by visual bells and whistles.

State-owned media in countries ravaged by corruption and clientelism, like Venezuela or Spain, are reliable sources of joy for visualization professors who need to gather examples for class. A more colorful version of Fig. 5.3 was used by Venezolana de Televisión after the 2013 Presidential elections, in which President Hugo Chavez's successor, Nicolás Maduro, was challenged by an emerging leader of the coalition of opposing parties, Henrique Capriles. I redrew the graphic to be able to print it in B/W, so try to imagine it in all its original tacky tricolor magnificence (use the colors of the Venezuelan flag: Pure red, blue, and yellow).

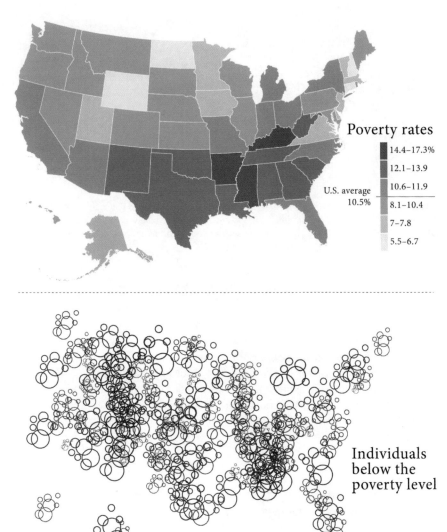

Fig. 5.2 Choropleth map and proportional symbol map. Data and locations are not real

When adding a zero-baseline, the picture becomes much more interesting—and discouraging if you are a fan of Maduro's (Fig. 5.4). I have heard designers defend graphics like that one arguing that viewers could just read the numbers. Right but, then, what are the damn bars for? Are they just ornaments? And I could add that most people won't have enough time to make up for the distortion if the chart stays on the TV screen for just a few seconds.

The next example is much more amusing. The original graphic was shown by TVE, the Spanish public broadcasting system, in September 2013. In this case, it's

PRESIDENTIAL ELECTIONS, 2013

Fig. 5.3 Presidential election results in Venezuela, based on a graphic by Venezonala de Televisión. Notice the truncated *Y*-axis which greatly distorts the difference between the percentages of vote

PRESIDENTIAL ELECTIONS, 2013

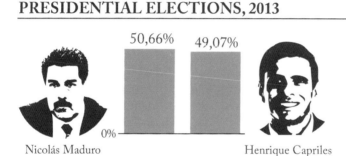

Fig. 5.4 An alternative version of the previous graphic in which a 0-baseline has been added, and the 3D effect has been removed

not just the *Y*-axis that was truncated; that's almost a trivial misdemeanor here. The problem is *where the X-axis begins*. See it in Fig. 5.5 (redrawn based on the original chart):

To understand why the designer is lying, you need to know that Spain's job market follows regular seasonal variations. A notable chunk of the country's wealth comes from the millions of tourists who visit it every year. Thus, unemployment tends to increase during the winter and falls sharply during the summer, something that becomes clear when 12-month data are displayed (Fig. 5.6). The most newsworthy bit of information revealed by this chart is that unemployment was *higher* in August 2013 than in July 2012, something that TVE did not mention. Maybe a smart government-appointed managing editor thought that she would not be able to use the right headline if data were correctly depicted.

Another very recent example was released by Partido Popular (PP), Spain's ruling party, in December 2013. Notice the inconsistent intervals on the horizontal axis: Until 2012, the graphic displays yearly data. Then, suddenly, it shows monthly

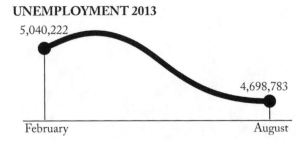

Fig. 5.5 The truncated Y-axis is not the only problem here. Based on a graph by RTVE (Spain)

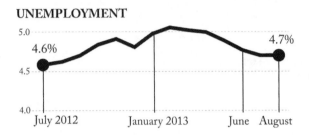

Fig. 5.6 Twelve months of unemployment rates, based on a graph by Eldiario.es. URL, January 8, 2014: http://www.eldiario.es/

change, but the choice of months seems arbitrary. The reason why the designer did this is not clear, but it may be related to the fact that President Mariano Rajoy approved a sharp increase in electricity prices for January 2014. Who knows? (Fig. 5.7)

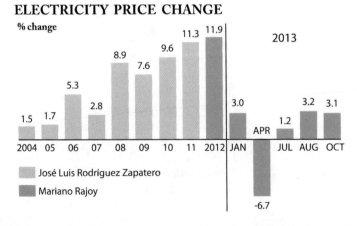

Fig. 5.7 Electricity price change, based on a graph by Spain's Partido Popular. Notice the inconsistent intervals on the X-axis

This bar graph is interesting not just because it is a good example of how sneaky politicians try to deceive citizens. It also leads us to the second part of this essay because it was uncritically reproduced by at least two of the main Spanish newspapers. The graphic passed through all the filters that journalists, designers, and the organizations they work for supposedly have to verify before they publish.

5.3 The Ignorance of Evidence-Driven Communicators

In section 1 of this essay, I explained that a single, isolated statistic can misrepresent reality. Nonetheless, allow me to cite this one, taken from a talk by Robert W. McChesney and John Nichols, authors of *The Death and Life of American Journalism* (McChesney and Nichols 2010): "The ratio of P.R professionals to journalists has climbed from 1.2-to-1 in 1980 to 4-to-1." If that fact doesn't worry you, it should.

It is true that many news organizations are not up to their own foundational ideals nowadays, but journalists, scientists, and information designers, people who I would propose to call evidence-driven communicators, are still the main line of defense against increasingly pervasive spin and bias in democratic societies:

> "Journalism is literally being rolled over by propaganda," said Nichols, who is a contributing writer for The Progressive and the associate editor of Capital Times, the daily newspaper in Madison, Wis. Eighty-six percent of all news stories that were printed or aired by Baltimore media in 2008 originated from what Nichols called "higher authorities," such as public relations firms or corporate press releases. That study, which was conducted by the Pew Research Center for the People and the Press, shows that traditional journalism has been reduced to "stenography."[4]
>
> John Nichols

The scarcity of honest communicators is not the only problem. Many journalists and designers are not prepared to identify sophisticated propaganda. This is a point made in the last few years in books like (Goldacre 2009) and (Patterson 2013), and in numerous media criticism Web sites,[5] although it is hardly a new discussion topic (Paulos 1988). On average, journalists and information designers are not seriously trained in the scientific method, research techniques, and data. And even those few who *do use* data regularly—computer-assisted reporters, particularly those outside of the USA and the UK—usually apply just techniques based on simple descriptive statistics (Defleur 1997).

Moreover, most journalists and designers lack a basic "bullshit detector," to use a term coined by Michael Shermer (Shermer 2011). To understand why this is problematic, remember that those professionals are responsible for holding the

---

[4] See Authors Advocate Government Subsidies for Journalism, URL, January 8, 2014: http://www.fordham.edu/Campus_Resources/eNewsroom/topstories_1771.asp.

[5] To cite just a couple or popular resources, see Lumley 2014; Malaprensa 2014.

powerful in check, and for filtering out noise so only signals will reach the public, at least in theory. If these evidence-driven communicators don't possess solid critical thinking skills, they won't be able to fight the kinds of lies outlined in Sect. 5.2.

Finally, not being versed in basic science and statistics may lead journalists and information designers to make mistakes that fall into categories that exactly match those of lies (hiding data, obfuscating data, and distorting data). The situation described in the previous paragraphs represents a huge challenge for modern democracies.

Let me exemplify my case with a story published by *Wired* magazine in August 2010. It was titled "The Web is Dead: Long Live the Internet" (Shermer 2011), and was written by Chris Anderson and Michael Wolff. The main point of the story was that the World Wide Web was in decline as a means to accessing content from the Internet: In 2010, according to the story, citizens were accessing digital content not just through their Web browsers, but mostly through smartphone apps and other services. The story was illustrated with a graphic similar to the one in Fig. 5.8, in which the decline of the WWW is striking: In 2000, it accounted for more than half of the traffic from the Internet. In 2010, it was barely 23 %.

Do you notice something fishy? Rob Beschizza, from the blog BoingBoing, did: "Without commenting on the article's argument, I nonetheless found this graph immediately suspect, because it doesn't account for the increase in internet traffic over the same period. The use of proportion of the total as the vertical axis instead of the actual total is an interesting editorial choice. You can probably guess that total use increases so rapidly that the web is not declining at all" (Anderson and Wolff 2010). It is not indeed, as Beschizza, proved with a graph in which the Y-axis corresponds to Internet traffic measured in exabytes (Fig. 5.9).

Now, ask yourself if you would call this a lie or a mistake. I don't really know what my answer would be. This could be the result of an overzealous editor trying to sell more copies of the magazine with an eye-catching and slightly controversial headline (*Wired* did receive a lot of attention thanks to this story). But were this a lie,

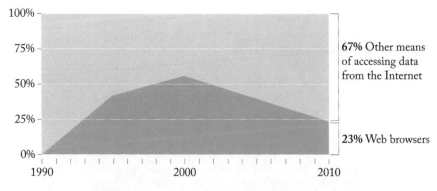

Fig. 5.8 The World Wide Web is dying, based on a graphic by *Wired* magazine

Internet Traffic in the U.S.

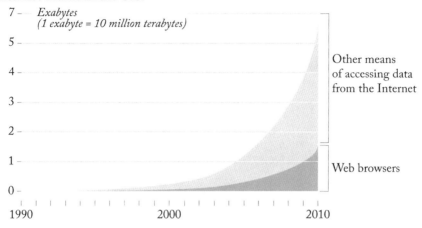

Fig. 5.9 The Web is hardly dying. Based on a graphic by Boing Boing (Beschizza 2010) BESCHIZZA Rob

a mistake, or a dubious editorial choice, does it really matter? The results are the same: Noise and a misled audience.

Another related challenge democratic societies face is that many designers of visualizations tend to yield to aesthetic preferences instead of striving for accuracy, precision, and depth first. Michael Babwahsingh, a designer based in New York City, has written:

> (The) growing popularization of information design techniques reduces the practice of information design to a "look," and may free creators of information design from any obligation to think deeply about the content and make sense of it first. What's more, the emphasis in many websites and publications has increasingly (and maddeningly) shifted from information that is logical and meaningful to information that is beautiful. (Babwahsingh 2013)
>
> Michael Babwahsingh

Visual appeal is undoubtedly a value worth pursuing, but not at the cost of hurting the integrity of the information, as this is one of the main sources for misleading infographics. The design of information graphics is as much an art as it is a form of engineering. There are certain guidelines and principles—I'm wary of calling them "rules"—grounded in visual perception and cognition that designers must learn an apply if they wish to communicate effectively; see (Few 2012; Kosslyn 2006; Ware 2012), among others. Unfortunately, judging by the work commonly seen in print publications, Web sites and, blogs that use infographics and data visualizations on a regular basis, many designers either are unaware of these principles or, worse, they willingly ignore them.

5.4 Conclusion: Fighting Noise with Knowledge

Visual lies are an inevitable side effect of freedom of expression. We'll never vanish them, no matter how aggressively we mock them or how thoroughly we analyze and denounce them. We will never get rid of mistakes, either. What we could do, though, is to better prepare future generations of evidence-driven visual communicators, individuals whose core goal is to inform audiences truthfully and accurately, not to sell stuff, to identify the former and avoid the latter.

Back in 1997, Jack Fuller, publisher of The Chicago Tribune, wrote: "We cannot accept the kind of ignorance of basic statistics that so often leads to preposterous reporting of scientific claims." Those words belong to a chapter titled "The challenge of complexity" (Fuller 1997), and they inspired Thomas E. Patterson to claim: "Journalists cannot meet democracy's needs unless they become 'knowledge professionals' who have 'mastery not only of technique but also of content'" (Patterson 2013).

Patterson proposes to develop a new kind of journalism education. He calls it "knowledge-based journalism." It combines deep subject-area expertise with a good understanding of how to acquire and evaluate information (research methods). I believe that the second component of Patterson's proposal—being able to gain knowledge in a systematic manner—is more important than the first one. Therefore, to conclude this essay, I'd like to argue that information design and visual journalism programs should incorporate the following items to their curricula, as a complement to the skills and principles they have traditionally taught:

1. A discussion on cognitive biases, which could be based on books such as (Chabris and Simons 2010; Kahneman 2011; Kurzban 2011).
2. An introduction to science as a method for inquiry; this would include lessons on how to read and interpret scientific claims.
3. As an extension of the previous point, an introduction to statistics which does not focus just on the mathematical minutiae—as fascinating as those can be—but on the conceptual side. It could be inspired by recent books such as Vickers (2009) and Wheelan (2012), and portions from older ones, such as Jaeger (1990, Weinberg and Shumaker 1974).
4. Foundations of computer programming.
5. Principles of cartography.
6. Visual perception and cognition applied to the design of information graphics. This part could be based on Few (2012), Kosslyn (2006), McEachren (2004, Ware (2012), etc.

Only by doing this, we will have a chance to minimize the harm that graphic lies and mistakes—being both inevitable—do to society.

References

Anderson C, Wolff M (2010) The web is dead: long live the internet. Wired. http://www.wired.com/magazine/2010/08/ff_webrip/. 8 Jan 2014

Babwahsingh M (2013) The real meaning of information design. Personal website. http://michaelbabwahsingh.com/2013/11/29/the-real-meaning-of-information-design/. 8 Jan 2014

Beschizza R (2010) Is the web really dead? Boing Boing. http://boingboing.net/2010/08/17/is-the-web-really-de.html. 8 Jan 2014

Bok S (1999) Lying: moral choice in public and private life. Vintage, New York

Cairo A (2012a) The functional art: an introduction to information graphics and visualization. New Riders, Berkeley

Cairo A (2012b) Reclaiming the word infographics. The Functional Art. http://www.thefunctionalart.com/2012/12/claiming-word-infographics-back.html. 8 Jan 2014

Chabris C, Simons D (2010) The invisible gorilla: how our intuitions deceive us. HarperCollins, New York

Cleveland WS (1993) Visualizing data. Hobart Press, New Jersey

Defleur MH (1997) Computer-assisted investigative reporting: development and methodology. Routledge, London

Few S (2012) Show me the numbers: designing tables and graphs to enlighten, 2nd edn. Analytics Press, Burlingame

Fuller J (1997) News values: ideas for an information age. University of Chicago Press, Chicago

Goldacre B (2009) Bad science: quacks, hacks, and big pharma flacks. Faber & Faber, London

Groch-Begley H, Shere D (2012) A history of dishonest fox charts. Media Matters. http://mediamatters.org/research/2012/10/01/a-history-of-dishonest-fox-charts/190225. 8 Jan 2014

Harris S (2013) Lying. Four Elephants Press, Vancouver

Huff D (1954) How to lie with statistics. Norton, W. W. & Company, Inc., New York

Jaeger RM (1990) Statistics: a spectator sport, 2nd edn. SAGE Publications Inc., Thousand Oaks

Jones GE (2006) How to lie with charts, 2nd edn. BookSurge Publishing, Charleston

Kahneman D (2011) Thinking, fast and slow. Farrar, Straus and Giroux, New York

Kosslyn SM (2006) Graph design for the eye and mind. Oxford University Press, United Kingdom

Krum R (2013) Infographics and relevance: 3 tips to improve SEO results. Vision interactive. http://www.vizioninteractive.com/blog/infographics-tips-improve-seo/. 8 Jan 2014

Kurzban R (2011) Why everyone (else) is a hypocrite: evolution and the modular mind. Princeton University Press, New Jersey

Lumley T (2014) NZ electoral visualizations. StatsChat. http://www.statschat.org.nz/. 8 Jan 2014

McEachren AM (2004) How maps work: representation, visualization, and design. The Guilford Press, New York

Malaprensa (2014). http://www.malaprensa.com. 8 Jan 2014 (in Spanish)

McChesney RW, Nichols J (2010) The death and life of American journalism: the media revolution that will begin the world again. Nation Books, New York

Monmonier MS (1996) How to lie with maps, 2nd edn. University of Chicago Press, Chicago

Patterson T (2013) Informing the news: the need for knowledge-based journalism. Vintage, New York

Paulos JA (1988) Innumeracy: mathematical illiteracy and its consequences, 5th edn. Hill and Wang, New York

Shermer M (2011) The believing brain: from ghosts and gods to politics and conspiracies. Times Books, London

Tufte EE (1983) The visual display of quantitative information. Graphics Press, Cheshire, p 57
Vickers AJ (2009) What is a p-value anyway? 34 stories to help you actually understand statistics. Pearson, London
Wainer H (2000) Visual revelations: graphical tales of fate and deception from Napoleon Bonaparte to Ross Perot. Psychology Press, London
Wheelan C (2012) Naked statistics: stripping the dread from the data. W. W. Norton & Company, New York
Ware C (2012) Information visualization: perception for design, 3rd edn. Morgan Kauffmann, Burlington
Weinberg GM, Shumaker JA (1974) Statistics: an intuitive approach, 3rd edn. Brooks/Cole Publishing, Three Lak

Author Biography

Alberto Cairo (Journalist, Information Designer and Researcher) teaches information graphics and visualization at the School of Communication of the University of Miami since January 2012. He is the author of 'The Functional Art: An Introduction to Information Graphics and Visualization' (Peachpit Press, 2012).

He has been a professor at the University of North Carolina-Chapel Hill, and director of infographics at El Mundo (Spain) and Época magazine (Brazil). In the past decade has consulted for media organizations and organized training programs in more than twenty countries.

For more information and contact:

http://www.thefunctionalart.com.
https://twitter.com/albertocairo.

Part III
Mapping and Visualizing Data

Chapter 6
atNight: Nocturnal Landscapes and Invisible Networks

Mar Santamaria-Varas and Pablo Martínez-Díez

Abstract Data visualisation has emerged as a key tool for urban design thinking that harnesses the immense power of information communication to illustrate the relationships of meaning, cause and dependency established between citizens and their environments. Technical advancements over the past decade have modified the way we sense, seize, use, plan and build present cities. Besides architecture of *stone and space*, we should recognise an expanding landscape of invisible networks. atNight project aims to explore, in a transversal manner, the potential of digital cartography to assign geometry and measure these intangible aspects of the reality. Using *nightscapes* as a paradigm, the research fosters urban planning from another perspective taking advantage of Big Data and its associated representation tools to interpret, intervene and rebuild contemporary and future metropolitan contexts.

6.1 An Approach to the Intangible Aspects of the City

For most of us, contemporary landscapes provoke visceral images of urban environments with streets and corners, gardens and walls, and trees and façades. Nowadays, multiple *urban things* (monuments, shops, apartments, vehicles and abandoned lots) build the backdrop of our social space.

M. Santamaria-Varas (✉)
School of Architecture of Barcelona (ETSAB),
Universitat Politècnica de Catalunya, Barcelona, Spain
e-mail: hi@atNight.ws

P. Martínez-Díez
Tech (UPC), Campus Sud—Edif. A, Avda. Diagonal 649,
08028 Barcelona, Spain
BAU Design College of Barcelona, Universitat de Vic (UVIC),
c/Pujades 118, 08005 Barcelona, Spain

© Springer-Verlag London 2015
D. Bihanic (ed.), *New Challenges for Data Design*,
DOI 10.1007/978-1-4471-6596-5_6

Both the urban designer Manuel de Solà-Morales (2008) and the sociologist Sennett (1994) define the urban experience as the contact produced between the body and the physical environment.[1] Inhabitants perceive cities in a tactile and visual manner when moving through private and public spaces. Interaction with the urban milieu happens on the *skin of the city,* whereas the collective identity is shaped by the multiple images collected.

Technical advancements over the past decade have completely changed the way we sense, seize, use, plan and build present and future cities. Besides architecture of *stone and space*, we should recognise an expanding landscape of invisible networks. While physically experiencing the city, inhabitants also generate a digital footprint, a generous amount of data which describes people needs, beliefs and reactions.

Mobile devices and the Internet have hybridised with social behaviours, enabling a more active role of citizenship in design processes. How can we use these new information and communication technologies to construct more intuitive and stimulating cities? As we face a crucial moment when traditional stereotypes are being questioned under economical, sustainable and social agendas, we need novel design strategies that unveil the connections between individuals, the collective identity and territorial realities. How can we promote meaningful innovation in terms of urban planning and management by determining the relationships between specific public places, social activities and digital behaviour? How might such an approach based on the representation of intangible aspects of the contemporary city help us to imagine new scenarios post-crisis?

6.1.1 The Agency of Mapping

Representation conditions the way one sees and acts within the man-made environment. A long history of carvings, engravings and paintings has demonstrated the importance of visual documentation in apprehending the city. According to de Azúa (2004), it was not until the early modern period that literature took an interest in geographical description, producing a gradual identification between urban and

---

[1] In the *Flesh and Stone*, Richard Sennett reviews history of people's bodily experience from ancient Greece nakedness and Roman geometry to the triumph of individualised movement in modern times. According to Sennett, "the spatial relations of human bodies obviously make a great deal of difference in how people react to each other, how they see and hear one another, whether they touch or are distant". On the other hand, Manuel de Solà Morales acknowledges "that in the surface of the city experienced in its tangible materiality, in its physical sensations, lie the origin and form of any kind of urbanity." He also introduces the idea of *urban things* to refer to common landscapes in metropolitan contexts: "a pavement, a glass façade, a wall, a ramp or a distant perspective interrupted by obstacles, a silhouette against the sky and a closed patio, bare, unfinished roads half occupied by provisional furniture [...] The city is the table that supports them and that presents them in their pure materiality, as realities identifiable in their differences, their relative position and their mutual reflections."

narrative space. Despite numerous examples (Dickens and London, Balzac and Paris, Galdós and Madrid), none of the major European cities could offer a coherent urban image during the nineteen century due to the industrial expansion. Azúa also points out, when quoting Benjamin (2003), that we need to adapt representation techniques to the changes experienced by cities. When referring to the modern age, Benjamin asserts that photography and cinema more closely embody citizen's perception following the new values and forms of production.

Especially after the urban explosion that took place after the Second World War, the city has become more difficult to envisage in its geographic and social dimension since metropolitan areas have metamorphosed into complex systems built upon invisible networks, *non-places*,[2] memories and changing landscapes. If painting and drawing portrayed the ancient settlements, literature recounted industrial urban areas and audio-visual techniques have been a compelling symbol of the twentieth-century metropolis, how can we grasp contemporary and future realities? And above all, how can new creative processes originate when framing, indexing and scaling existing contexts in consonance with citizen's perception of their actual urban experiences?

Maps, drawings and cartographies create and build the world as much as measure and describe it. Landscape architect Corner (1999) affirms that representation is particularly instrumental in the constructing and construing of lived space. In this active sense, "the function of *mapping* is less to mirror reality than to engender the re-shaping of the worlds in which people live. Thus, mapping *unfolds* potential; it re-makes territory over and over again, each time with new and diverse consequences".

Idelfons Cerdà 1855s topographical survey of Barcelona may serve as a case history. The engineer was commissioned to draw a plan of the flat territory extending from the city walls to the villages around the ancient centre. This document gave him a full control of land organisation (elevation, water streams, infrastructures, etc.). Five years later, he proposed a radical project, the Example, which has given Barcelona its current urban form. The accuracy of the previous drawing was essential to design the future metropolis. Cerda's example confirms that cartography, as creative practice, uncovers realities previously unseen or unimagined.

In recent years, a large body of research has investigated the potential of data visualisation applied to urban design. From academic experiences (Senseable City Lab, Centre for Advanced Spatial Analysis[3]) to independent scholars or design firm

---

[2] Augé (1995) coined the term *non-place* when referring to spaces of circulation, consumption and communication such as airports, railway stations, superstores, motorways and international hotel chains.

[3] Senseable City Lab activity at the Massachusetts Institute of Technology focuses on the representation of real-time city (Cambridge: MIT. URL, January 6, 2014: http://senseable.mit.edu/). CASA, University College of London, uses spatial analysis and GIS systems as basic forms of drawing space-time data (London: The Centre for Advanced Spatial Analysis. URL, January 6, 2014: http://www.bartlett.ucl.ac.uk/casa).

mapping projects (Stamen Design,[4] Eric Fisher,[5] Accurat[6]), such proposals radically create new ways to depict subjective and objective mechanisms taking place in cities through human–place transactions and collective dynamics alongside the tools required to synthesis them.

Castells (1989) redraws our attention to the evolution of the Internet and urban forms. Although it was thought that interactive communications could contribute to the demise of cities, the informational *space of flows* (a new type of space that allows distant synchronous and real-time interaction) interplays with the physical landscape. Assuming that the Internet and the mobile network virtually link particular places, we need to assess the implications and applications of these new technological paradigms to urban planning, possibly changing from a *top–down* to a *bottom–up* approach.

Historically, representation tools have empowered citizenship. From navigational charts to GPS, people have invented and used maps to help them define and order their world. Four hundred years ago in the Age of Exploration, cartographers employed compass lines to depict coastlines, rivers and harbours in the New World. Today in the Age of Participation, entities are re-envisioning mapping practices and adapting representation tools to the evolving need of personal registers and micro-geographies; at the same time, public bodies and private companies are opening Big Data for measuring the living condition of the city.

In the past, cartography has permitted a more conscious use of the territory, making citizens able to master space in their favour. However, mapping should go beyond geographical illustration to unmask invisible urban relationships. Our research team proposes to develop new instruments to capture the "ephemeral" besides the geometry of urban plots and facades, taking information from citizens' interactions (actions, activities, emotions) as the basis for a better planning of urban environments. In other words, we situate cartography as a collective enabling enterprise that both reveals and realises hidden potential.

---

[4] In the course of our research project, we have used graphical approaches and its associated tools provided by Stamen Design that guarantee the effectiveness and accuracy of results (San Francisco: Stamen Design. URL, January 6, 2014: http://stamen.com).

[5] As Nix (Nix 2013) states *Local and Tourist* project by Eric Fisher is an early example of the use of Social Data (Flickr). The cartographies represent on a single image both the perception of inhabitants and visitors within a city.

[6] As part of Urban Sensing European project, the research *Geographies of time* (Urban Sensing. URL, January 6, 2014: http://urban-sensing.eu), Accurat, erases the changing condition of city boundaries according to time and social streams (Accurat, Milan: Accurate. URL, January 6, 2014: http://www.accurat.it).

6.2 Nocturnal Landscapes: Uncharted Territories Approach

Cities are complex organisms with changing qualities that require variable interpretations and representations. Architecture creates an immutable scenario where dynamic networks of social and cultural relationships evolve following subtle and sometimes imperceptible patterns. In this context, the *urban night* has become a major challenge for urbanism and architecture,[7] as well as a representative archetype of intangible landscapes (Fig. 6.1).

As Kepes (1967) states, the city has a "double life": one under the sunlight and another under cars headlamps, neon advertising and streetlights. Nightscapes[8] offer an unusual and unexpected light show by elements such as sodium vapour lamps, which transform city colours and bright lights from store windows, which animate the streets sidewalks. The nocturnal metropolis is neither the complement nor the counter part of the daylight city but rather manifests special conditions.

In particular, night-time turns into a moment of revelation and radicalisation of certain phenomena, revealing nuances and differences with the day. For instance, whereas monuments shape the identity of many contemporary cites, at night, these main architectonic symbols fade into darkness. By contrast, entire buildings occupied by banks or hotels emerge as landmarks, uncovering a hidden cultural, economic and political order. These two cities project a completely disconnected image and follow a distinct hierarchy of values.

Nocturnal reference points depend on the presence and absence of light. Consequently, difficulties arise when attempting to describe these nightscapes based upon mutable fragments. While aware of night-time, there is an absence of a clear picture of it. One key to explaining this inadequacy resides in the same definition of the term landscape. Maderuelo (2005) and Berque (1994) argue that we only perceive what we can see: "landscape is a cultural construct founded on an aesthetic and visual relationship between man and environment".

In short, the nocturnal city is elusive. Despite the complexity and evident loss of orientation and legibility of the urban framework, when the night falls, the streets are still animated with life. The night-time can embody contrasting elements such as fear and danger, party and sin, dreams and excess. After the advent of electricity, night gradually shifted from this variety of contradictions to represent urban recreation and social life beyond the productive hours of day. At night, we join our

---

[7] Straw (2014) has brilliantly synthesised the "state of the art" of the urban night studies in relation to (1) the emergence of a new public culture of the night, (2) the night as a form and territory to be projected with light, (3) the status of urban night within the public policies and (4) the place of artistic practices challenging the day and night division.

[8] Throughout this essay, we use the term *nightscape* to refer to the nocturnal landscape of the city. We have borrowed this expression from Armengaud et al. (2009) who in their seminal book, define *nightscape* as (1) the mirror of human thought from rational thinking to poetic subject, (2) the infrastructural framework of territorial development and (3) a strategy to address contemporary architecture and landscape projects. However, we apply this concept in a broader sense.

Fig. 6.1 *Cartography Constellation Barcelona* from atNight project, Mar Santamaria, Pablo Martínez Díez and Jordi Bari Corberó, 2013

friends and we meet strangers (threatening or non-threatening) while invisibility and loneliness are present. In this regard, many experiences are still necessary to shine light over a significant night-time imagination despite meaningful attempts by the field of art, in the form of ephemeral lighting festivals, or the discipline of urban design.[9]

atNight involves ongoing research and inquiry focused on the (re)definition of the nocturnal landscape of contemporary cities and the role of night-time in the construction of "urbanity". Since nightscapes rely on intangible elements, this project handles data analysis and cartography as fundamental tools for a better understanding of artificial milieu. This investigation has resulted in a series of cartographies and registers of Barcelona nightscapes, as paradigmatic twentieth-century metropolis. The following section describes both the results of this analysis, the data visualisations and the methodological and technological procedure behind our research activities that vindicate the nightscape as a ground for reflection, awareness and action.

6.3 Urban Analysis

atNight evaluates the possibilities offered by digital technologies to propose new collaborative and efficient design scenarios. We aim at designing, testing and deploying strategies to collect, analyse and represent information. Specifically, we have used data visualisation to set up a possible interpretation of night values by harnessing the immense power of visual communication to explain the relationship of meaning, cause and dependency established between citizens and their environment.

6.3.1 Data and Workflow

The project addresses the mapping of intangible aspects for a better understanding of the city's visible surface and internal flows organisation beyond traditional topographic and zoning plans. The research team chose energy, visual structure, mobility and usage as the guideline topics that allow us to verify earlier hypothesis on nightscapes configuration.

We classified the data into three main categories according to the source. First, we used information from *Open Data* and public geographic services regarding

---

[9] Good example of Master Plans provides, at European level, urban lighting strategies both responding to technical and aesthetical complexity as Terzi (2001) in Rome, Antico (2011) in Antwerp and Gant and Narboni (2012) in Paris. However, the majority of European urban areas lack nocturnal planning schemes. Currently, nightscapes are being designed depending on electrical consumption, environmental impact and safety standards.

cartographic features and general statistics (demography, land use, streets layout, etc.). Second, we obtained mobility trends and energy consumption averages through agreements with public and private local agencies. Finally, we collected geo-social data sets by a systematic crawl of several location-based social networks API.

For example, we analysed geolocated Flickr, Panoramio and Instagram pictures, Twitter messages, Google Local markers or flow of movement across the city by taxi, bicycle or public transportation.[10] Availability of data collection in space provides an intensive and precise method of sensing citizen's activity in one to one scale. Hence, there is a need to design specific software and workflow to capture, integrate, visualise and share the collected data.[11]

The basic technological idea behind the project was to set up a platform that harvests data from geo-social streams and local *Open Data* providers, applies mining functionalities, extracts key elements and plots them on a series of maps. The research followed several phases—from the design of capture engines to the management of data via geographic information system—allowing parameters to be set and tuned in accordance with specific data sets.

Although numerous experiences in the field of data visualisation rely on clustering strategies to scrutinise and express information, we adopted graphical principles to envision interactions from a qualitative point of view.[12] We displayed the data individually depending on its exact location in time and space. No grouping strategies have been applied in order to maintain the detail of information while achieving an accurate level of abstraction to interpret its value.

We employed intuitive strategies to represent and communicate information from the simple existence of data and non-data—each value has been related to a geometrical feature—to the superimposition of several data sets. On the one hand, we assigned colour scales (similar or complementary tones) to easily distinguish dissimilar situations, mainly day and night perception. We established a hierarchy for certain values (e.g. the Klout score index that measures the importance of Twitter users) within the network by applying thickness and size to geometrical entities (Fig. 6.2). On the other hand, new information can be seen by modifying the opacity of data groups (overlapping interactions anticipate zones of higher urban intensity) or adopting multiplication or subtraction criteria of RGB values (Fig. 6.3).

---

[10] Social media streams and other local data providers have been analysed in the Fall/Winter 2012–2013. Flickr service has been collected since 2006.

[11] Link to the project repository code, URL, December 28, 2013: https://github.com/atNightmaps/atNight-py.

[12] Tufte (2001) examines in his pioneer works the general principles that have specific visual consequences governing the design, editing, analysis and critique of data representation to enhance dimensionality and density of information. As he suggests "to envision information is to work at the intersection of image, word, number, art. The instruments are those of writing and typography, of managing large data sets and statistical analysis, of line and layout and colour".

Fig. 6.2 Graphical strategies from *top* to *bottom*: data/non-data, colour and thickness

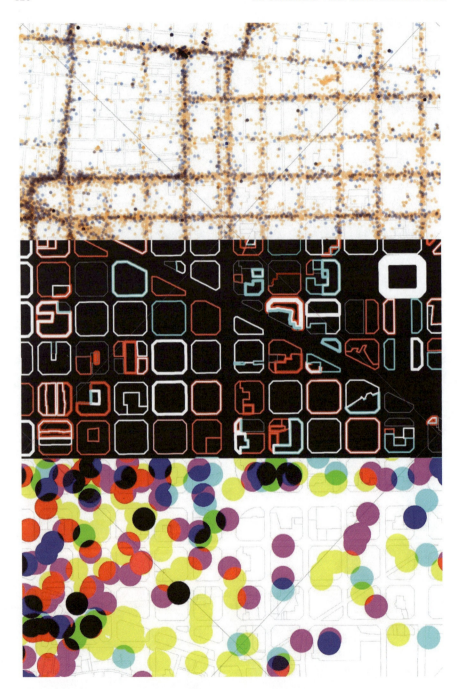

Fig. 6.3 Graphical strategies from *top* to *bottom*: opacity, subtraction and multiplication

We also combined the data with physical urban morphology (streets, blocks, plots, etc.) to discern whether the interactions belong to the public or the private domain. In some cases, we have also filtered and mapped information by keyword, i.e. showing only those data that contains a particular tag such as the name of the city. In any case, graphical functions (layering and separation, transparency and blending modes, micro- and macro-readings, compactness and distance) have given us the possibility to make visible both the complexity and the singularities of urban fabric—discovering elements, places and relationships that otherwise we cannot see.

6.3.2 Cartographies of Invisible Networks

From thematic cartographies, we have identified hierarchies and relevant values, patterns and symbols, traces and absences, transforming data sets into meaningful urban stories. To illustrate, we have highlighted the relationship between the busiest and populated streets, the most photographed sites in contrast to the most "experienced" areas or the sentiment that swings at different time of day. The cartographies that follow, classified in the three main guidelines of the project (visual structure, mobility patterns and activity), represent a first attempt to describe the uncharted territories of Barcelona.[13]

6.3.2.1 Visual Structure

Territorial Framework and Identity

Constellation Barcelona and *Barcelona is Barcelona* aim to be a germinal approach to verify the potential of geolocated data to reproduce territorial support. Both images portray the primary urban morphology as the mountain and coastal geography or the major territorial and civic axes (Diagonal, Gran Via, Rondes). Barcelona is immediately recognisable by means of Flickr, Instagram and Twitter geo-social interconnections. Nevertheless, when examined more closely, the urban plot is revealed not to be homogeneous. By filtering pictures and messages containing the term "Barcelona", we chart the places of identity from touristic sites (Sagrada Familia, Parc Güell or La Rambla) to spaces of proximity (the backstreet bar or the neighbourhood square). Monuments and large public spaces are popular among autochthons and foreigners; citizens also value anonymous sites which are the result of personal geographies.

---

[13] The complete set of cartographies can be consulted in the Cartographies section of the project website. URL, January 6, 2014: http://www.atNight.ws/cartographies.php.

Night-Time and Daytime

As indicated above, most visible areas predominantly design the shape of the city as perceived by their inhabitants and visitors. Geolocated Flickr and Instragram images illustrate the changing interdependence between urban form and citizen's impression during daytime or night-time (orange and blue, respectively). While diurnal and nocturnal cities mirror each other in the same physical support, at night, the periphery is more highlighted as can be seen in the cartography Barcelona Night and Day (Fig. 6.4). Similarly, uniqueness *versus* identity confirms from Instagram and Twitter data sets that day consciousness depends on the singularity of certain central areas and symbolic architectures. The nocturnal city takes the form of multiple focuses, which can seem anonymous places in a map, but actually uphold urban identity (Fig. 6.5).

Visible City, Living City

In addition, we compared visual structure, previously synthesized, with inhabitants' experience expressed via the Twitter platform. Do people feel different about their city at different time of the day? In the Barcelona Sentiment pair of maps, we can still make out the rough shapes of urban activity without a coastline, streets or reference points, just by visualising the change of positive or negative opinion according to site and time—we parsed the content of the Twitter message via semantic algorithms to detect the sentiment. Again, urbanity moves throughout the day from the centre to the periphery; the metropolitan boundaries emerge as the place for positive attitude as night represents the time of socialisation beyond productive hours. The visible city turns into the living city.

6.3.2.2 Mobility Patterns

Mobility cartographies unveil traffic flows and tempos in relation to the efficiency and density of the city plot. First, movement *versus* density examines how streets and private spaces relate to each other within the urban framework. Taxi rides from MyTaxi fleet are compared with population density per block via census data which erases the capacity of main streets and intersections to condense the inhabited areas. Second, Taxi Night and Day identifies certain interference between mobility performance from night to day (blue and orange, respectively, in Fig. 6.6). Daytime itineraries outline a city standing on homogeneity while a new hierarchical order appears within the road network during night-time. Third, Bicing Night and Day delineates the stop and go movement of the bike sharing system. Besides sketching the main routes connecting various parts of the city, we aim to represent the activity generated around Bicing stations in the lapse of exchange. These areas of influence, mapped from the availability of bikes per hour, design a pattern influenced by plain or steep topography (night and day overlap or alternate reciprocally).

6 atNight: Nocturnal Landscapes and Invisible Networks

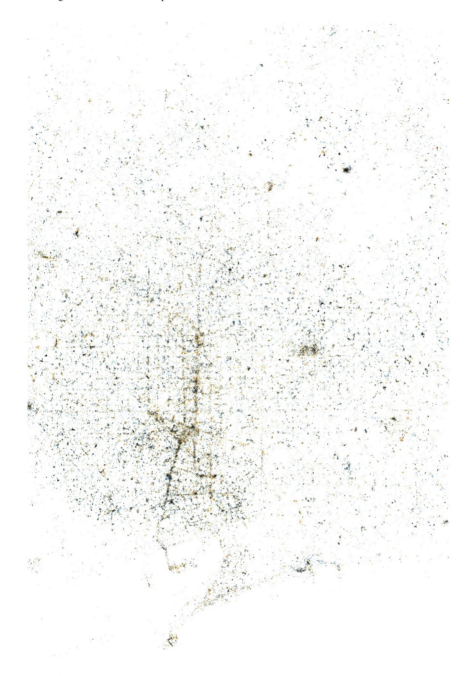

Fig. 6.4 Cartography uniqueness versus identity

Fig. 6.5 Cartography uniqueness versus identity

6 atNight: Nocturnal Landscapes and Invisible Networks

Fig. 6.6 Cartography taxi night and day

Finally, we evoke mobility spaces as the most relevant contemporary collective places. Like the arteries and veins of the city, the metropolitan transport system regulates Barcelona through input and output channels that dilate or contract according to passenger whims. Collective *versus* individual superimposes the number of users per line from Night Bus network to geolocated Twitter interactions to pinpoint contact areas in collective and individual behaviour as central transportation hubs.

Activity and Usage

The last couple of visualisations focuses on nightscapes from the urban activity viewpoint. On one hand, night-time reflects the city double life through Google Local markers. By overlapping daytime and night-time activities during regular business hours, we trace the borderline between everyday routine (residence and work) and exceptionality (leisure and social life). We detect a proliferation of nocturnal usages (eating, drinking, dancing) evenly dispersed across Barcelona which accentuate the permanent character of urban corners and public spaces within urban morphology.

Fiesta and Food Safety maps recreational areas (restaurants and nightclubs) and zones under surveillance (hospitals, ATM, police stations) by way of the same geolocated data. We can corroborate that nightscapes have turned into the time of urban free time and cultural life; at the same time, nocturnal hours embody transgression and crime. The distribution of this trio of concepts (fiesta, food and safety) divulges either a contiguity of homologous single-usage zones in the periphery of the city or the mixture of the three activities in the central areas. Even if night is still the context for forbidden feelings, we celebrate our nocturnal hours.

6.4 Conclusion

Over the next few years, we will reconfigure our cities. We should adapt urban contexts to new scenarios resulting from urban and technical advancements. The ability to transform society will be key to ensuring a prosperous future and a sustainable development. Recent technological breakthroughs have derived in the consolidation of a distinct urban model, the *Smart City*, which places sensors of all kinds to monitor life in real time (weather, traffic, flow of people, contamination, etc.).[14] What if these same digital achievements could work in reverse and could interact with urban data to raise a common imagination from the collective and collaborative contribution? In accordance with cities rearrangement, we should foster urban planning from another perspective taking advantage of novel representation tools to interpret, intervene and rebuild.

---

[14] Freire (2009) relates social technologies, collaborative empowerment and new Commons.

atNight aims to test precisely the potential of digital cartography to assign geometry and measure intangible conditions of the contemporary city. We have approached Big Data study and visualisation as a key method for urban design thinking. From the elaborated maps, we can extract general conclusions. The city can be depicted through the data collected by urban sensors or social networks. Such approach denotes a direct relationship between the urban space and the data it generates, allowing interpretations that speak not only about geometry but also about value structures. We could effectively recognise singularities within the urban fabric escaping from most canonical readings and take into account inhabitants subjective interpretation. In addition, we can demonstrate that the city presents a heterogeneous conduct. For instance, we obtained divergent densities of data at distinct points on the urban plot as well as diverse results when comparing the same values and locations according to the time of the day. In the light of these outcomes, we can prove that cities acquire fluctuating patterns depending on the area and the period of time. This hypothesis becomes measurable and therefore enables comparisons and objective evaluations.

Nonetheless, the project faces mandatory challenges and improvement objectives. For example, the duration of analysis cycles (from several months to entire seasons or years) and the amount of collected information should be increased. In this regard, any cartographical survey needs periodic iterations to redefine methodologies and determine variations and transformations of the case study (Barcelona). Additionally, we propose to adapt methodologies and tools to other significant metropolitan areas to test the capacity of intangible cartographies to address a variety of contexts.

In conclusion, it is crucial to delve into this line of research. Today, urban planning still relies upon traditional cartographic information (topography, plot division, usage) and neighbourhood-level statistics. These long-established practices are inadequate in comprehending the interaction of citizenship on the *urban skin* and the territory. The newly and accessible cartographic information will provide an inestimable tool for citizen empowerment, enabling individuals to take collective decisions about the intangible city we actually inhabit.

References

Antico S (2011) The light of Mechelen and Antwerpen. Luce, 1:20–25
Armengaud M, Armengaud M, Chianchetta A (2009) Nightscapes. Gustavo Gili, Barcelona
Augé M (1995) Non-places: introduction to an antropology of supermodernity. Verso, London
Azúa F (2004) La necesidad y el deseo. La arquitectura de la no-ciudad. Universidad Pública de Navarra, Pamplona
Berque A (1994) Paysage, milieu et histoire. Cinq propositions pour une théorie du paysage. Editions Champ Vallon, Paris
Benjamin W (2003) La obra de arte en la época de su reproductibilidad técnica. Itaca, México
Castells M (1989) The informational city: information technology, economic restructuring and the urban-regional process, Chap. 3. Blackwell, Oxford, pp 127–171

Corner J (1999) The agency of mapping: speculation, critique and invention. Mappings. Reakton Books, London, pp 214–252
Freire J (2009) Emerging urban planning: city, technology and social innovation. Domestic Landscapes, vol 4. Edge Networks, SEPES, Madrid, pp 18–27
Kepes G (1967) Notas sobre expresión y comunicación en el paisaje urbano. La metrópoli del futuro. Seix Barral, Barcelona
Maderuelo J (2005) El Paisaje. Génesis de un concepto. Abada, Madrid
Narboni R (2012) Les éclairages des villes: Vers un urbanisme nocturne. Infolio, Suisse
Nix M (2013) Visual simplexity. Entwickler, Frankfurt
Sennett R (1994) Flesh and stone: the body and the city in western civilization. Norton, New York
Solà-Morales M (2008) Matter of things. NAI, Rotterdam
Straw W (2014) The urban night. Cartographies of place. McGill-Queens University Press, Montreal
Terzi C (2001) Los planes de luz. Editoriale Domus, Milan
Tufte ER (2001) Envisioning information. Graphics Press, Chesire

Author Biographies

Mar Santamaria-Varas is an architect based in Barcelona who focuses her activity on research and teaching. She is graduated from the School of Architecture of Barcelona, where she teaches Urban Design and Planning. She has also lectured at ETSAV-UPC Vallès, ETSAR-URV Reus, IED Barcelona and École Spéciale d'Architecture-Paris as well as taken part in several juries and international workshops (as the first Architectural Association Visiting School Barcelona).

Her theoretical and educational activity investigates the understanding of new territorial and urban processes through landscape project—she has been scientific coordinator of Modeland INTEREG European Project. In recent years, she has engaged herself in the development of new cartography tools to represent urbanity by using new audio-visual media and data visualization. She has been awarded Grant for art and contemporary theory research by the Catalan Government (2013).

For more information and contact:

http://www.atnight.ws
http://es.linkedin.com/pub/mar-santamaria-varas/19/359/220
https://twitter.com/atnightmaps

Pablo Martínez-Díez is an architect graduated from the School of Architecture of Barcelona. He works in various architectural, curatorial and artistic projects, generating a critical dialogue on the role of ephemeral design through teaching and research.

He concentrates professional activity as a designer of light, objects and spaces (like Palau de la Música Catalana, Santo Domingo de la Calzada Cathedral or Ciudades del Agua Pavilion at the Zaragoza Expo). He has also participated in several architecture festivals as Racons Llum, Eme3, and Lumo in Barcelona and Lausanne Lumières in Switzerland.

His work has received several mentions and awards as IDDA Excellence Awards, ArtFad 2012, Mooi Award, Artsponsor, Injuve, Philips LUMEC. He lectures at several universities in Graduate and Master Degree (BAU Design College of Barcelona, BarcelonaTech, Istituto Europeo di Design).

For more information and contact:

http://www.atnight.ws
http://www.300000kms.net
http://es.linkedin.com/in/pablomartinezdiez
https://twitter.com/300000kms

Chapter 7
Visualizing Ambiguity in an Era of Data Abundance and Very Large Software Systems

Ali Almossawi

Abstract Data today is more accessible than it ever was, a development that makes it possible to tell stories that were previously beyond reach. Moreover, and thanks to advances in ETL and analysis tools, data is no longer restricted to the same traditional sources. This chapter walks the reader through the process of visualizing the ambiguous attribute of quality in two complex and mature software systems, namely, Firefox and Chromium, each of which has several million lines of code (LOC) and talks about the key issues that one might need to consider during such a process. This chapter presents the design structure matrix (DSM) as a highly effective and compact tool for visualizing quality in very large engineering systems and concludes with a discussion of a promising opportunity for this type of work as well as some new and forthcoming challenges.

7.1 Introduction

When Lord Reith ran the BBC back in 1922, he determined that the mission of the corporation was to "inform, educate and entertain." It is tremendously useful to think about the mission of data visualization in similar terms. In my own field of software engineering, data visualization, in its current incarnation, is proving an invaluable tool. Commercial off-the-shelf products are making it trivial to generate satisfactory transformations of raw data from an array of disparate formats, while a multitude of libraries, many of which are open-source, have helped reduce the gulf of knowledge for many academics and practitioners, a development which has helped make data visualization a more accessible field.

With paradigm shifts of this sort that afford convenience comes a change in standards and yardsticks, leading to temporal periods of what can best be described as a sort of *Wild Wild West* where pragmatism takes precedence over pedantry and

A. Almossawi (✉)
Mozilla Corporation, 2 Harrison Street, #175 San Francisco, CA 94105, USA
e-mail: almossawi@alum.mit.edu

where people feel at greater liberty to experiment and reinterpret conventional concepts, drawing in many cases from their own backgrounds and experiences. Hence, one observes a state where an artist's definition of data visualization is different than a statistician's. Having the good fortune to be part of a domain when it is at this phase is tremendously liberating, fun, and rewarding. Equally, having a clear mission in mind that succinctly defines the purpose of one's work is vital. I embarked on a project earlier this year whose aim was to find out something useful about the quality of a set of engineering systems. My arsenal of tools included static analysis, statistical analysis, network manipulation, and data visualization. The first iteration of the project was posted on Slashdot's front page[1] and generated a lot of discussion among developers, contributors, and readers. This chapter goes through the process that took this work to a satisfying conclusion and comments on the key issues that one may face during such a process, a process that was undeniably guided by those three key principles—to inform, educate, and entertain. Identifying these key issues will hopefully bring to light the challenges that one might face when visualizing ambiguous attributes of very large engineering systems.

7.2 Asking the Right Question Is Half the Answer

When starting on a new project, I rarely look for a story within a dataset, but rather start by posing a curious question or positing a claim with the intent of finding out later on whether the data affirms or falsifies that claim. This is typical practice in science. Visualization is hence only used if it is likely to serve a useful purpose rather than for its own sake. Not every story is worth telling, and so it easier to discard unsuitable stories when all one has is a question.

In this project, the question that I was interested in answering was "How good is the quality of the Firefox codebase?" This is quite evidently an ambiguous question, but it has the advantage of adequately framing the endeavor. We still do not know what *quality* means by asking that question, but we do know a few other things. We know that we are interested in a specific product, namely, Mozilla's Firefox Web browser. We also know that we are interested in measuring the goodness of said product, rather than its badness, which is a genuine difference rather than semantic gymnastics. Finally, we know that we are interested in measuring the quality of the codebase, rather than other factors that may contribute to the product's success, such as market share and user sentiment.

The issue of quality is an interesting one. It is a property that one typically aspires to maximize and is a critical contributing factor to the success of any engineering system. A fact of reality is that things tend to degrade with time due to the buildup of harmful elements. Within the context of software, degrading means

---

[1] URL, January 16, 2014: http://slashdot.org/.

that a system deviates in two specific ways: that it no longer functions as well and that it no longer satisfies its quality attributes. Quality attributes are those cross-cutting requirements such as maintainability and reliability, which one may argue are more important for a system's success than functional requirements.

A term that is sometimes used to describe a subset of these harmful elements is *technical debt*. It is a fitting metaphor, as it indicates that a software engineering malpractice here or a concession with respect to coding standards there can incur a debt that one must then pay off in the form of time, effort or defects. In fact, it is typical for teams working on large projects to pay off this debt every so often in the form of a formal exercise known as a refactoring effort.

To demonstrate how seriously technical debt can impact complexity, and hence quality, here is an example due to my colleague Dan Sturtevant (Sturtevant 2013). Imagine a simple system made up of 12 components, which is to say, 12 files. If we model that system using an adjacency matrix and wherever we find one or more dependencies between two components, we mark it with a dot, and we end up with a matrix such as the one in Fig. 7.1 that has a density of 47. If we then find the transitive closure for that matrix, a process by which we capture not only direct dependencies, but also indirect ones as well, we end up with a matrix that has a density of 81.

Now, if we add just two dependencies (Fig. 7.2) between two pairs of files, marked in red, in a way that breaks the design principle of modularity, and then find the transitive closure again, we end up with a matrix that has a density of 144. In other words, every component in the system can now potentially impact every other component were it be modified. By making only two changes, we have significantly increased the complexity of our system.

What have we seen so far? We have seen an adjacency matrix being used as an effective tool to tell a story about how dependencies in a system may impact complexity. Essentially, we have used it to provide evidence for our case that quality is an important topic to be concerned with and a worthwhile one to be asking questions about. The adjacency matrix's beauty is in its ability to effectively and compactly serve a utilitarian purpose. We will revisit it later on.

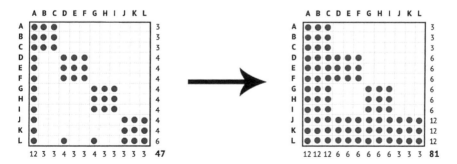

Fig. 7.1 The density of a simple system's direct (*left*) and indirect (*right*) dependencies

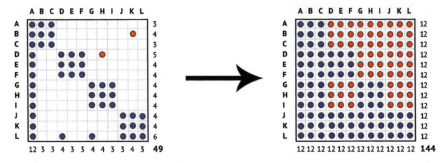

Fig. 7.2 The density of a simple system's direct and indirect dependencies after adding two new dependencies to it

7.3 Reducing Ambiguity

One of the key preliminary steps that a data visualizer has to perform is that of reducing ambiguity. Things in the real world are not always clear-cut and readily available, so many times, one has to deduce, infer, or assume things. The value of the end-result and its accordance with the truth are hence a function of the quality of those inferences and assumptions. When those assumptions are too far-fetched, one is left wondering how valuable the work really is.

Ambiguity is a challenge in a lot of engineering disciplines, which is why all of them, to my knowledge, prescribe methods for reducing ambiguity at the design stage, when it is cheapest. These methods include establishing a common vocabulary, documenting architecture using standard styles and prose and refining specifications so that they cover the entire gamut of granularity. The data visualizer walks a path that is not too different from that of engineers and so it is reasonable to expect him/her to think about what aspects of a project may be prone to misinterpretation. For instance, visualizing deaths due to gun violence may be easily defined with a footnote about what the category *deaths* constitutes. Visualizing potential years lost due to gun violence is a bit more challenging and without explicit caveats may be an all too liberal use of artistic license.

In our case, we have shown how one way of measuring technical debt is to look at dependencies between components. We can actually do better by looking at a set of measures instead, all of which have been shown in various studies to be strong predictors of quality. Hence, we use these concrete measures of quality to reduce ambiguity. This not only makes the final artifact more rigorous, but also more in line with the truth, which after all is what we are trying to convey to the reader. The following sections briefly describe these five measures.

Lines of code (LOC) measures the number of executable LOC in a system, ignoring comments and blank lines, and is widely used as a baseline measure of quality (Hatton 2008; Shore 2008). A system with more LOC is typically more difficult to maintain. LOC and defect density have an inverse relationship (Kan 1995; Basili and Perricone 1984; Shen et al. 1985), or according to some studies a curvilinear relationship (Withrow 1990). This may seem counterintuitive at first. The explanation is that firstly, architecture does not change at the same rate as LOC, and secondly, that a sizable number of defects occur within the interfaces that tie modules or components together. In this study, the set of analyzed files excludes unit tests and includes all files that meet our file-type filter.[2]

Cyclomatic complexity is a measure of the number of linearly independent paths within a software system and can be applied either to the entire system or to a constituent part (McCabe 1976). By viewing a block of code as a control graph, the nodes constitute indivisible LOC that execute in sequence, and the directed edges connect two nodes if one can occur after the other. So, for example, branching constructs like if-else statements would result in a node being connected to two output nodes, one for each branch.

Cyclomatic complexity is defined as:

$$v(G) = e - n + 2p \qquad (1.1)$$

The term $v(G)$ is the cyclomatic number of a graph G, e is the number of edges, n is the number of nodes, and p is the number of connected components, or exit nodes, in the graph. The additive nature of the metric means that the complexity of several graphs is equal to the sum of each graph. In our measure of cyclomatic complexity, we control for size and hence, the metric is per 1,000 LOC.

Direct dependencies measure the number of dependencies between components, where a dependency is a function call or a global variable, for instance. It is calculated by finding the density of an adjacency matrix that captures dependencies between components, as we saw earlier. Because only direct dependencies are considered in such a matrix, it is referred to as a first-order dependency matrix. The density of said matrix is hence the first-order density. Because of the sparseness of the first-order dependency matrices in this study, direct dependencies are per 10,000 file pairs.

Propagation cost measures direct as well as indirect dependencies between files in a codebase. In practical terms, it gives a sense of the proportion of files that *may* be impacted, on average, when a change is made to a randomly selected file (McCabe 1976; McCormack et al. 2006). The process of transforming a first-order dependency matrix that captures only direct dependencies between files to a visibility matrix that captures indirect dependencies as well is achieved through matrix multiplication, by raising the first-order dependency matrix to multiple successive powers until its transitive closure is achieved. Thus, a matrix raised to the power of two would show the indirect dependencies between elements that have a path length of two, i.e., calls from A to C, if A calls B and B calls C. Thereafter, by summing these matrices together, one gets the visibility matrix. For this ripple effect to be of use in analysis, the density of the visibility matrix is captured within a metric called propagation cost.

---

[2] c, .C, .cc, .cpp, .css, .cxx, .h, .H, .hh, .hpp, .htm, .html, .hxx, .inl, .java, .js, .jsm, .py, .s, .xml.

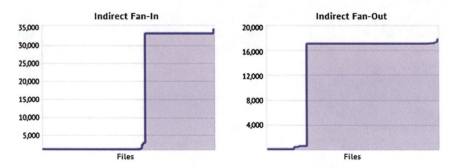

Fig. 7.3 Distribution of fan-in and fan-out for the 42,409 files in Chromium version 19.0.1084, taken from its visibility matrix, hence the qualifier "indirect"

Core files are files that are highly interconnected via a chain of cyclic dependencies. They have been shown in various studies to have a higher propensity for defects (McCormack and Sturtevant 2011; McCormack et al. 2010). A system's core is the percentage of files with one or more dependencies that have a high fan-in, which is the sum of incoming calls from other files, and a high fan-out, which is the sum of outgoing calls to other files. Here, by looking at the distribution of files' fan-in and fan-out values, it becomes apparent that neither is smooth, but rather exhibits at least one clear discontinuity point, as shown in Fig. 7.3. Since it makes practical sense to think of files beyond such a point as being part of a large cycle of calls, we use this point to calculate the core. A core file is therefore one whose fan-in is greater than the first discontinuity point in the distribution of fan-in values and whose fan-out is greater than the first discontinuity point in the distribution of fan-out values.

Let us take another break. I hope it is clear by now that the process of transforming an abstract set of concepts to concrete ones on is of paramount importance. In addition to the benefits mentioned earlier, notice that it also provides a great amount of visibility into the fundamental building blocks of the project, which in turn provides one with multiple opportunities for revision and retrospection. It remains that aesthetics play a key role in helping to draw a wondering eye to a particular piece of work, yet if the work does not lie on a solid bed of data and assumptions, its value remains superficial and its frailty may be exposed with the simplest of robustness tests.

7.4 Making the Takeaways More Meaningful

What we have done hitherto has been to pose a question and then reason about why it is a meaningful question, with the help of an effective visualization tool that we have so far been calling an adjacency matrix and then attempted to reduce ambiguity by defining quality using a set of concrete measures. We now have something

Table 7.1 Descriptive statistics for the set of Firefox releases ($n = 23$)

Metric	Mean	Median	Stdev	Min	Max
LOC	2,885,799	2,686,348	595,408	2,066,655	4,012,434
Files	13,266	11,907	2,844	9,478	18,460

Table 7.2 Descriptive statistics for the set of Chromium releases ($n = 23$)

Metric	Mean	Median	Stdev	Min	Max
LOC	4,767,974	5,157,958	1,951,894	896,204	7,331,584
Files	29,020	31,132	12,349	5,009	46,620

else to consider. Is the narrative that we have in hand adequate, which is to say, does it do a good enough job of informing the reader? In this project, a good way of conveying the value of these measures of quality that we have been discussing, which in a vacuum, may come across as bland, is to apply them to systems that people can relate to. Notice that I say *systems*. The problem with applying them to a single system is that the reader is left with the difficult task of trying to make sense of a set of measures that have no benchmark against which they may be compared.

For instance, if the cyclomatic complexity of a system is 180, what does that mean? Do we even care that it is 180 and not 179? Put differently, if it goes up from 179 to 180 in two consecutive releases, what conclusion are we to draw from that? Allowing the reader to reason about differences in terms of, say, orders of magnitude are more effective. For instance, if the cyclomatic complexity of system A is 120 and that of System B is 180, something meaningful may be determined from the fact System A's components are on average 50 % less complex. Hence, in this project, I analyze two popular, complex, and mature codebases, namely, Firefox and Chromium. They are products that are in the same domain, which introduces an entertaining element given that different people feel passionate about different browsers, but of course, this property is insignificant for analysis, which remains a domain-agnostic process.

The dataset for this study therefore consists of the codebases of 23 major releases of Firefox, versions 1–20, and the codebases of 23 major releases of Chromium, versions 1–23,[3] all of which are large enough for analysis, saving us from having to worry about adversely impacting our architectural measures as a result of having too few files. Our measures can become unreliable when codebases are too small, which makes intuitive sense. Furthermore, it is not very meaningful to talk about dependencies when the system size is small enough for one person to be knowledgeable about all of its modules. Tables 7.1 and 7.2 show some descriptive statistics for the 46 codebases. Firefox has around 2.9 million LOC on average per release and Chromium has around 4.8 million LOC on average per release.

---

[3] The Chromium major releases are per the builds specified in the release history section of the Google Chrome Wikipedia page.

7.5 Choosing the Right Metaphor

Unlike art, a metaphor that is too extravagant or forces one to stretch one's imagination leads to a piece that is at worst incomprehensible and at best deceiving. In both cases, it is of little value. A lot of the time, lines, bars, boxes, and shapes of that sort are more than adequate, particularly as a first step. In this project, given that we are comparing sets of time series data for multiple series, a line chart is the natural choice. I hope it is not too anti-climatic a development for the reader who might have been expecting something more elaborate.

In addition to line charts, we have another compact and very effective visualization tool at our disposal, namely the *Design Structure Matrix* (*DSM*), which until now we have been calling an adjacency matrix. DSMs are used in various engineering disciplines and in academia (Eppinger and Browning 2012), however, they remain an untapped resource in data visualization. A DSM proves useful not only for analysis, but also as a visualization tool, since it allows one to see various patterns of structural complexity within a system or module. The metaphors it affords are those of density and position.

By default, DSM sorts components, which in our case are source code files, by their hierarchical directory structure. In such a state, a square-shaped cluster indicates that there are many dependencies between files within a module. All the dots to the right and left of a cluster are files that the module depends on. All the files above and below it are files that depend on it. In many paradigms, modularity, which is to say, low coupling and high cohesion, is a desirable quality attribute. Hence, an ideal system is one that has modules that have more intra-module dependencies and fewer inter-module dependencies. By inspecting the various patterns in a DSM, it becomes possible to infer things about the system's quality.

In fact, we have seen in various studies that the different patterns of dependencies have varying effects on a system's quality. For instance, files that have a high fan-in appear in a DSM as vertical bands. These files are known as shared files, given that they perform common utilities such as printing, and have been shown to be a positive predictor of quality. Files that have a high fan-out are known as control files and have been shown to be a negative predictor of quality. Hence, one would prefer to see more of the former and less of the latter in a DSM.

Inspecting a DSM can also allow one to see indications of elevated levels of our three measures of inter-component complexity. Direct dependencies and propagation cost are easy to spot since they are represented by density: the denser a first-order DSM is, the higher the module's direct dependencies, and the denser a visibility DSM is, the higher the module's propagation cost. It is also possible to visualize a system or module's core. The process involves resorting a DSM's files using one of several algorithms, such as the ones mentioned in (Browning 2001), in a way that forces components toward one side of the diagonal. Because core files are cyclic, and therefore incapable of adhering to said constraint, they remain as the only cluster along the diagonal with files on either side of it. By observing the size of this cluster, one is therefore able to determine the size of the core.

7 Visualizing Ambiguity in an Era of Data ... 147

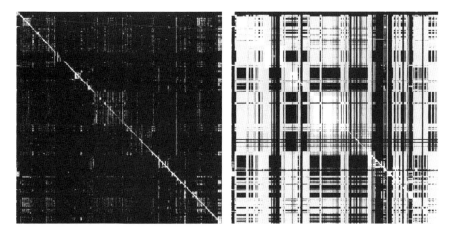

Fig. 7.4 Direct dependencies (*left*) and indirect dependencies (*right*) for Firefox version 16

Figure 7.4 shows an example of two DSMs for a system made up of 16,551 components. On the left is a DSM showing direct dependencies and on the right is one showing both direct and indirect dependencies. Not only are clusters in the latter matrix much more vivid compared to the former, but we also see that new clusters have emerged. The squares along the diagonals are modules; their intensity and size are an indication of a module's density of internal dependencies and size, respectively.

Figure 7.5 shows a set of DSMs for a particular module of a system. In this case, it is the *gfx* module in Firefox 20. Here, we can clearly see that the module is broken into identifiable sub-modules. We see a few components near the top that make a lot of calls to other components, though on the whole, components mostly depend on components that are within the same module.

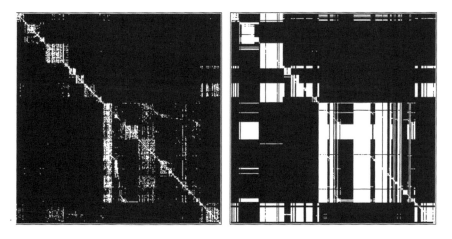

Fig. 7.5 Direct dependencies (*left*) and indirect dependencies (*right*) for the *gfx* module in Firefox version 20

7.6 Displaying the Results

When presenting data, two things are important to keep in mind. Firstly, that a lot of the time, less is more (Tufte 2001), and secondly, as the maxim in software engineering goes, pictures are of little value without prose. It is unfair to the reader to assume that they are able to decode the meaning of shapes and colors in a picture when they are not explicitly described in the form of a legend, captions, or text. When deciding how to present data, I almost always start with conventional shapes and only make the switch to something more imaginative if the switch is deemed worthwhile or if the context warrants such a switch. This is perhaps an engineering approach to data visualization that I suspect may not be shared by those coming from other disciplines. With such an approach, visualization is seen as an optimization step.

When presenting data that intends to tell a story, it is equally important to remain impartial. I have worked on projects in the past that have dealt with touchy subjects such as world politics, where the challenge is to present a story that is not intentionally divisive. Presenting facts from a neutral point of view can both tell the desired story and allow the piece to reach the very people who might otherwise have been turned away by it.

For instance, *Bahrain: Two Years On*[4] (Fig. 7.6) was a project that I published in 2012 to coincide with the second anniversary of the most recent civil unrest in Bahrain. It used interactive data visualization to highlight the tragedy of death, regardless of where it originates from and whom it impacts. Deaths are encoded using *position* and *color*, but with a twist: the bars extend to form the shape of a palm tree, which is a traditional symbol of the island and one that all factions of its society identify with. The objectivity of the piece coupled with its aesthetics appealed to both sides of the conflict, which was precisely the intention behind it.

Impartiality can easily be impacted by the way data is presented, albeit sometimes unintentionally. For instance, try modifying the aspect ratio of a line graph of two variables and observe the impact that has on the resulting slope and hence on the conclusion that a reader may draw from the graph (Cleveland et al. 1988). The same may be done with scales. In this project, impartiality is particularly important given that the topic is directly related to two products that different groups of people feel overly passionate about. Moreover, the author, in this case, is not an outside observer and so it may be perceived that I have a dog in the race and hence a personal interest in showing that one of the two systems is better than the other.

In the interest of brevity, what follows is a subset of findings for this project that show size, a measure of intra-component complexity (cyclomatic complexity) and a measure of inter-component complexity (core). With LOC, we see that Chromium is growing at a higher rate of 7.31 % compared to Firefox's 2.66 %. Chromium begins with a smaller codebase up until version 4 when it jumps by 52.61 % to 3.26 million LOC. In Firefox, the largest increase in LOC occurs in version 4 when

---

[4] Homepage project, URL, January 16, 2014: http://bahrainvisualized.com/.

7 Visualizing Ambiguity in an Era of Data ...

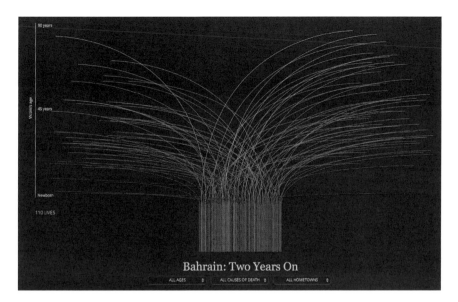

Fig. 7.6 *Bahrain: Two Years On*, Ali Almossawi, 2012

it goes up by 15.63 % to 2.64 million LOC. Recall that we are only looking at executable LOC (Fig. 7.7).

With cyclomatic complexity, we see that Chromium has a consistently lower cyclomatic complexity compared to Firefox, seeing as it is 122.26 on average compared to 188.37 on average in Firefox. In both systems, the general trend is a downward one. In Firefox, we see an increase in cyclomatic complexity between versions 1 and 3.5, following which the downward trend begins. In Chromium, we see a noticeable increase in version 2 and then a second noticeable increase in version 6. What the result tells us is that the files in Chromium are internally less

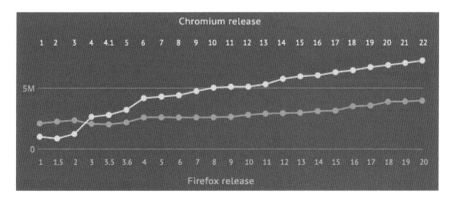

Fig. 7.7 A comparison of LOC

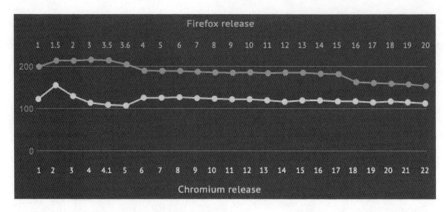

Fig. 7.8 A comparison of cyclomatic complexity

complex than in Firefox, meaning that the coding constructs in them result in fewer independent paths (Fig. 7.8).

With core, Chromium is consistently more complex than Firefox. The average core in Chromium is 33.22 % compared to Firefox's 21.39 %. Hence, the number of highly interconnected files in Chromium is noticeably higher than that of Firefox, which is to say that it is more likely for a new or modified file in Chromium to increase the system's complexity than in Firefox. As with propagation cost, we see that the mean growth of Firefox's core has decreased substantially following RRC, seeing as it dropped from 35.47 to 2.73 % and its standard deviation from 88.08 to 16.24 % (Fig. 7.9).

In order to explore these measures at the module level, we take our data and split it by module. Looking at the mean values for each module for each of the above metrics gives some insight into how the different modules fare. With Firefox, for instance, we find that a number of modules, such as *accessible* and *security*,

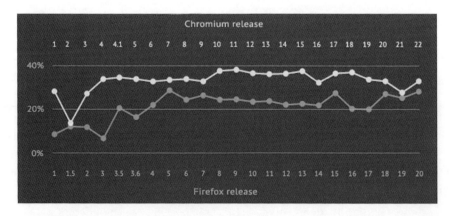

Fig. 7.9 A comparison of core

frequently appear among the most complex. The module, *accessible,* has the highest mean direct dependencies and is in fact almost double that of the next module. The module, *security*, has the highest core of any module, being as it is 25.91 %, meaning that nearly a quarter of its files are highly interconnected. By observing modules' visibility DSMs, we notice that with *accessible*, *security*, and *browser*, the number of non-zero elements is noticeably high, seeing as they constitute nearly 20 % of their respective matrices.[5]

7.7 An Opportunity Worth Capitalizing On

The project that was discussed in this chapter brought to light some interesting insights. However, it has a very serious limitation in that it is constrained by the paradigm that is presently dominant in data visualization. It is a paradigm that optimizes for singular insightful snapshots of phenomena, which means that as interesting as many of the present works may be, they remain bound by the number of dots in a spread or pixels on a screen.

One of the big opportunities moving forward is an idea of turning the artifact into a device so that the reader is no longer a spectator, but an explorer. Much like a telescope can be used to take one of a near-infinite number of snapshots of the night sky, so too should a visualization be able to do that. Several data visualizers have already started consciously capitalizing on this opportunity (MIT Media Lab professor Cesar A). Hidalgo has come up with a rather fitting metaphor for such visualizations, calling them *Datascopes*, a hat tip to the device that revolutionized our understanding of the planets and the universe.

How do we create these *Datascopes*? Simply put, we do so by rethinking the way we present data visualization, by moving the decisions about what dimensions to show from the data visualizer to the user. We provide dials that can manipulate a large dataset in real time and produce combinations of various fields that can then generate one of potentially millions of visualizations. This is a very basic concept that relies on a fundamental principle in UI design, which is to empower the user, while maintaining simplicity. In the context of data visualization, its effect is extraordinary.

With this project, how might we turn what we presently have into something that may be considered a *Datascope*? We would do that by allowing the user to determine (1) the level of granularity—component, module, or system, (2) the set of metrics to use for analysis, beyond the five that have been discussed here, seeing as there are tens of software metrics that are in use throughout the industry, (3) the set of systems, beyond the two that were analyzed here and the type of visualization,

---

[5] A spreadsheet containing the full results may be found at: *Firefox release and module data (URL, January 16, 2014: http://goo.gl/t6uq9p) and *Chromium release data (URL, January 16, 2014: http://goo.gl/tYmjl6).

beyond the two that were used here. These dials would provide the user with tens of thousands of unique visualizations to work with. The commercial static analysis tool that I use to generate some of the measures of quality from the two codebases does provide something similar to what I have described. However, it remains a siloed tool and does not quite afford exploration given that it is cumbersome, slow and of course closed. A Web-based *Datascope* of codebases does, I think, have a large potential to be immensely useful as a diagnostic tool as well a tool for predicting, whereas in a codebase, harmful elements are likely to build up.

7.8 Forthcoming Challenges

This chapter touched on one of the important challenges in data visualization, namely that of working with new types of datasets. As we have seen, in software systems, the codebase is itself the dataset and analyzing it requires a special set of tools, primary of which is static analysis. Tools such as the commercial package *Understand* can do much of the heavy lifting when it comes to static analysis by performing the necessary parsing and transformation of the source code so that concrete metrics may be generated from it. Working with source code may also require the need for network manipulation tools in order to aid in generating, analyzing, and repartitioning DSMs. Python and MATLAB are both viable options. As we begin to tap other sources for potentially useful data, we may need to adjust or update our armory of tools accordingly.

What we have seen in the past few years is a wave of new tools, most of which are open-source, that have made it unprecedentedly convenient for software engineers that are accustomed to a Web-based technology stack to get into visualization. The most well known of these tools is perhaps D3.js. Other tools, such as Tableau for instance, have done the same for business intelligence professionals. One of the big challenges going forward is creating new tools to attract experts in many of the other adjacent disciplines that would benefit from a data-driven approach. Three of many fields that come to mind are graphic design, product design, and industrial design.

A topic that is closest to my heart is the challenge of how one can use visualization to do good. Data visualization is a very appealing language, and as a result has the potential to improve people's lives by both conveying knowledge to a very wide cross section of people as well as eliciting it. In 2012, Cesar A. Hidalgo and I worked on a project called *Participie*[6] (Fig. 7.10) whose aim was to improve the way that people engage with policy-makers by creating a platform that affords direct participation. Rather than casting a vote once every 4 years, one would have the ability to directly participate on topics such as how the federal budget should be distributed. The platform helps one model such questions using some metaphor, say a pie, and

---

[6] Homepage project, URL, January 16, 2014: http://participie.com.

Fig. 7.10 *Participie*, Cesar A Hidalgo and Ali Almossawi (the Macro-connections group—MIT Media Lab), 2012

thereafter allows people to modify the various slices of that pie. A consolidated pie is updated in real time. The experiment allowed us to realize the value of using interactive data visualization to elicit answers to questions that were traditionally elicited using, say, multiple-choice questions in surveys. By letting people see in real time, the impact of their contribution to a discussion is very empowering to them on a personal level, insightful to policy-makers and of palpable value to society as a whole.

A sizable number of visualizations today live on the Web. As the Web's capabilities continue to improve and as it becomes an even more widely adopted platform for applications, 3D games, news, and general productivity and consumption, its effects will undoubtedly impact the types of visualizations that people create. At present, *client-side interaction* is perhaps one of the biggest sources of advantage that a Web-based visualization has over a print one. Interaction completely changes the way a story can be told as it affords abstraction in a way that is not possible with ink on paper. When visualizing multifaceted systems or stories, this attribute proves essential. It will be interesting to see how the field evolves and to what extent and how it adapts to improvements in technology.

As indicated in this chapter's title, we are presently in an era of data abundance. We find ourselves amidst acres and acres of land that has yet to be plowed and hence, we may have to go through a lot of dirt before we are able to reap much harvest. In the not too distant past, there was a wave of fascination with social media datasets such as those of Twitter. In the end, one wondered how much of that work lead to real insight, as opposed to trivial observations, and how much of it was just noise. One of the big forthcoming challenges is therefore to reach an elevated level of capability with the help of new tools, techniques, and taxonomies that would allow us to distinguish actual, useful data from noise. Advances on this front will do the field of data visualization a great deal of service.

References

Basili VR, Perricone BT (1984) Software errors and complexity: an empirical investigation. Commun ACM 42–52
Browning T (2001) Applying the design structure matrix to system decomposition and integration problems: a review and new directions. IEEE Trans Eng Manage 48(3):292–306
Cleveland W, McGill ME, McGill R (1988) The shape parameter of a two-variable graph. J Am Stat Assoc 83(402):289–300
Eppinger SD, Browning TR (2012) Design structure matrix methods and applications. MIT Press, Cambridge
Hatton L (2008) The role of empiricism in improving the reliability of future software. Keynote. URL: http://www.leshatton.org, 16 Jan 2014
Kan SH (1995) Metrics and models in software quality engineering. Addison Wesley, Boston
McCormack AD, Sturtevant D (2011) System design and the cost of complexity: putting a value on modularity. AoM 71st annual meeting
McCormack AD, Baldwin CY, Rusnak J (2010) The architecture of complex systems: do core-periphery structures dominate? MIT Sloan School of Management, Cambridge. Working paper No. 4770-10
Tufte E (2001) The visual display of quantitative information, 2nd edn. Graphics Press, Cheshire
McCormack AD, Rusnak J, Baldwin CY (2006) Exploring the structure of complex software designs: an empirical study of open source and proprietary code, vol 52(7). Institute for Operations Research and the Management Sciences (INFORMS)
McCabe TJ (1976) A complexity measure. IEEE Trans Software Eng 2(4):308–320
Shore J (2008) An approximate measure of technical debt. URL: http://jamesshore.com/Blog/An-Approximate-Measure-of-Technical-Debt.html, 16 Jan 2014
Shen VY, Yu T, Thebaut S et al (1985) Identifying error-prone software—An empirical study. IEEE Trans Software Eng 11(4):317–324
Sturtevant D (2013) Technical debt in large systems. Webinar. URL: http://sdm.mit.edu/news/news_articles/webinar_050613/sturtevant-webinar-technical-debt.html, 16 Jan 2014
Withrow C (1990) Error density and size in Ada Software. IEEE Software

Author Biography

Ali Almossawi (Data Visualizer, Software Engineer and Researcher) holds a Masters in Engineering Systems from the Massachusetts Institute of Technology (MIT) and a Masters in Software Engineering from Carnegie Mellon University. Ali currently resides in San Francisco with his wife and daughter, where he works for Mozilla's Metrics Team on data visualization projects, while continuing to collaborate with his colleagues at the MIT Media Lab.

Formerly, Ali spent time at both Harvard and the Software Engineering Institute (SEI) where his research involved creating predictive models of source-code quality as well as investigating architecture adaptability in software.

His work has appeared in *Wired*, *Scientific American*, *The New York Times*, *Fast Company*, *Slashdot*, *io9*, *The Dish*, *BuzzFeed* and others. Ali is also author of a book on critical thinking called *An Illustrated Book of Bad Arguments*, the 2nd edition of which is planned for release in August 2014.

For more information and contact:

http://almossawi.com
http://linkedin.com/in/almossawi
https://twitter.com/alialmossawi

Part IV
Interacting with Data

Chapter 8
Living Networks

Santiago Ortiz

Abstract Networks are extremely dynamic entities, yet the ways we conventionally visualize them seem to not being able to convey this dynamism. Bret Victor, in his remarkable conference entitled *Stop drawing dead fish* (Bret 2012), argues that the computer is a different creative medium, suited for dynamic representation and simulation, in which "[…] everything we draw should be alive by the default." In this text, I'll first explain why and how all networks are dynamic (Sect. 8.1) and then present different techniques I've been exploring in my own research and work to visualize networks as the dynamic and, to some extent, living organisms they are (Sect. 8.2). Paraphrasing Bret Victor: stop drawing dead networks!

8.1 Networks Are Dynamic

8.1.1 Dynamic Systems

Dynamic systems are the ones whose properties change over time, as opposed to static systems. Dynamism in systems could be driven by external causes (if the system is open), by internal ones or by a mix. In complex systems, for instance, dynamism is mainly produced by interaction among its components. The dynamic properties of a system (the ones that change over time) could be:

- The set of elements (constituents),
- Internal properties of the elements, such as state or stored information,
- Positions of elements (in the case the system has a spatial description),
- The configuration of interactions among elements (which elements interact with which ones),
- The properties and rules of interactions among elements.

S. Ortiz (✉)
Moebio, Buenos Aires, Argentina
e-mail: s@moebio.com

© Springer-Verlag London 2015
D. Bihanic (ed.), *New Challenges for Data Design*,
DOI 10.1007/978-1-4471-6596-5_8

Regarding the first property, the elements of the system, it makes sense to ask whether a dynamic system that changes its constituents continues being the same. Think of a human body that constantly renews its atoms, molecules and cells and nonetheless continues being the same system (what we call person).

Any system described as a set of elements and the relations among them is a network. If the elements or relations (their existence or properties) change over time, the network is dynamic.

8.1.2 Dynamic Networks

Networks found in nature, society, culture, and technology are dynamic in at least some of the following ways:

- Nodes and relations are created and destroyed, thus changing the structural properties of the network over time,
- Nodes, and eventually relations, have changing states (such as activated/unactivated or other more complex variables),
- In some physical networks, matter or energy flow through nodes and relations,
- Information flows through nodes and relations (either because of the mass/energy flow, or because nodes states are affected by the states of neighbor nodes, thus creating information patterns that travel through the network), and in some cases, information is also processed (modified within a single node).

Very often, networks match several or all the dynamic criteria. Take the human brain, and focus on the network of neurons: it is easy to see that all the dynamic properties are fulfilled:

- Neurons and synapses are created and destroyed throughout an individual's lifetime (in vertebrates, at some point, neurons stop being created, or almost, but new connections continue always appearing),
- Neurons are excitable cells, and at any given moment, they are active or inactive,
- Electricity and chemical components are transmitted from neurons through axons and synapses,
- Electricity and chemical components carry information, neurons store, and process information.

Other well-known networks, such as genetic expression networks or social networks, feature all the enlisted criteria. Notice that in the case of the genetic networks, one may study their structural changes throughout an organism's lifetime (changes due to epigenetics or contingencies such as a viral infection) or through millions of years (changes driven by evolution).

D'Arcy Thompson, a biologist expert in pattern formation and development (that is: an expert in biological dynamic systems), wrote in his book *On growth and form* (Thompson 1992) that "Everything is the way it is because it got that way."

This apparent truism has actually deep implications and offers new ways of research to understand complexity; it means that in order to deeply understand a system you should inquire about the formation processes that lead to the current system configuration. Sometimes, this could be the key to unstick a research that cannot explain why in a system things are the way they are. That is exactly what happened to Lazlo Barabasi and his team when studying structural properties of the networks, back in the 90s, specifically when they were trying to understand the law-power distribution of nodes degree, a property present in basically all interesting networks found in nature, society, culture, and technology. When they decided to understand networks as dynamic organisms, the questions changed from "why this" to "who it arrived to this," and their research accelerated, to the extent to identify some of the most important concepts existent in modern networks science, such as the preferential attachment.

> The scale-free model reflects our awakening to the reality that networks are dynamic systems that change constantly through the addition of new nodes and links.
>
> Lazlo Barabasi

Now take a network that is basically structurally stable: the gene regulatory network of an organism. Back in 2005, I visualized this network of the *E. coli* bacteria (Fig. 8.1). Blue relations indicate activation, red indicate inhibition, and green indicate influences that can change from one state to the other. In the last decades, biologists have learned that the genetic code alone does not contain the description of the organism in all its morphological stages. What drives this heterogeneous behavior over time is the complexity of interactions and the timing on which those occur. The first thing I thought when visualizing this data was that this network of activations and inhibitions describes an extremely unstable reality, with constant changes through time, exhibiting subtle and diverse patterns (I remember thinking on a symphony). The second thing I thought was how little my visualization helps to grasp the dynamical nature of the network. It is like taking picture of a football's team gathering before the match, showing it to a martian and explaining: "this is football."

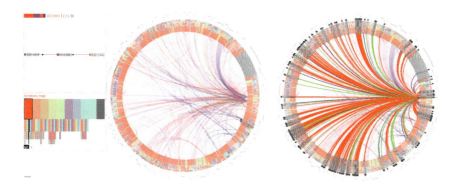

Fig. 8.1 *Gnom*: *E. coli* genes network visualized, Santiago Ortiz and his collaborators, 2005

There is another way networks are dynamic, yet this does not lie in their inner properties, but in the way, these networks can be captured by human conscience; in other words, in the way we read them. As it happens with a text, the ultimate process of reading is linear, even if the text describes a highly nonlinear reality, which is the usual case. We can describe a network enlisting all the nodes with their properties, followed by all the relations and their properties, but that way readers will hardly get a sense of the network structure (only trying to check if two nodes are connected could be a nightmare). In most cases, a network visualization, a classical dots, and lines diagram seem to be a good answer.

But when such a diagram is exposed to a reader, we can never tell what would be the order in which nodes and relations will be read. We can expect that the attention of the reader will be wandering through connected nodes (following paths) and sometimes jumping between non-connected nodes, etc., but also other complex attention/perceptual processes could happen. For instance, communities and clusters could be detected and isolated by attention (our brain is extremely good at this sort of tasks, to the extent of recognizing patterns that do not exist). On the other hand, some graphic properties such as the position of the nodes in the bi-dimensional space could affect dramatically the lecture and understanding of the network. The lecture of a network is then an extremely dynamic process, a pretty mysterious and an uncertain one. Two lectures of the same network visualized in a single way, and even performed by the same person, are never equal, as if networks whereas fluid and dynamic as rivers, evoking the famous Heraclitus quote. I reckon this different kind of dynamism as an extrinsic property of every network, a one that should be taken into account seriously when it takes to visualize networks. Networks are dynamic because they are narrative, and they are narrative in multiple ways: Each path or group of connected nodes tells a different story (Fig. 8.2) (Moretti 2013). When it comes to communicate networks, one has to take into account all these ways networks are dynamic.

8.2 Visualizing Networks

Communicate, write and read complex systems, and in particular networks, is a critical task, moreover when these systems become more and more used as explanations of reality. The visualization of networks is a very recent yet fast growing practice that faces huge challenges.

Many have suggested that what we call the emergent properties of a complex system are only illusions, probably necessary or at least unavoidable, due to our human limited capacity to comprehend the system by understanding their components and their huge amount of interactions. The complexity overwhelms us, and we have to content ourselves with descriptions of patterns that we call emergent and that often acquire a lead role: life, conscience, civilization, music, flavor, pleasure, culture, etc. Without denying the existence or importance of the emergency levels, a good visualization of a system should allow readers to see both the local properties

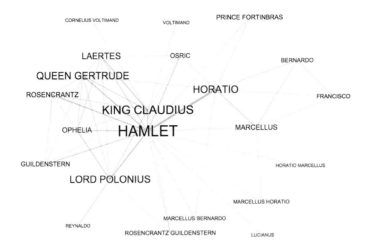

Fig. 8.2 *Hamnet: a network of characters interaction at Hamlet*, Santiago Ortiz, 2014

and interactions, and the whole pattern. Is that possible? The task seems to be, at least, extremely difficult and has subtle relations with a big issue physics face: the enormous difficulty to integrate theories that describe local particle systems (that happens to be time symmetrical) and the emergent statistical behaviors that are more close to our human scale (and that are time asymmetrical due to the entropy). "When looking at nature you must always consider the detail and the hole" says Goethe in the first verse of his poem Epirrhema (von Goethe 1994). Similar challenges are faced by psychologists and sociologist, biologist and ecologists, or even employees and managers in companies. A friend of mine that works in the problem of integrating the different complexity levels of the brain told me that neuron specialists, and brain researchers that take a more comprehensive organ approach, struggle a lot trying to understand each other; sometimes, it seems like both groups are studying a completely unrelated phenomena. It is difficult to create a narrative that encompasses the local and the global realities.

One of the most common network visualizations pattern is what is has been called—not without some sense of criticism and mockery—hairballs. These are diagrams with hundreds of nodes represented by dots or texts and thousands of lines or curves depicting relations. Some are quite attractive, probably not because of how well they convey information about the visualized networks, but because of the opposite, they make the case of how overwhelmingly complex these networks can be. It seems that this "unreadability" loose our brains and attract our minds as an abyss. In some cases, the visualization is not used as a way to convey information about the data, but rather to send a different kind of message: look how complex is my data. Better versions of these maps use colors to identify different communities (clusters) that in most of the cases have been algorithmically detected. If, on top of that, there is some basic interaction that allows the user to select a node, see its name, and the name of its neighbors (the local sub-network of a node), the visualization becomes much more informative. This is the case of the map provided by LinkedIn that users can generate with their own account information.

There are, nonetheless, many interesting contents and properties of the network that this image cannot convey. This image (Fig. 8.3) gives me an idea of the global structure of the network, but I get very little insight about the local properties. How to know, for instance, how many grades separate any two of my contacts? There are other network visualization methods that offer local points of view in which some of these problems are solved, but then the global view is missed, and to understand locality we often need to, at least, know the position of the locality in the whole. We want to see what is inside and outside the cavern.

In this section, I focus on the challenges associated with the dynamic properties of a network, including the last criteria I presented for dynamism: the dynamic way we read networks.

For that purpose, I feature examples out of my own research and production, in particular from three projects in which I systematically developed techniques to represent networks in a way they conveyed their dynamic properties.

For each project, I mention a specific challenge associated with the dynamism of the network and my approach to solve the problem.

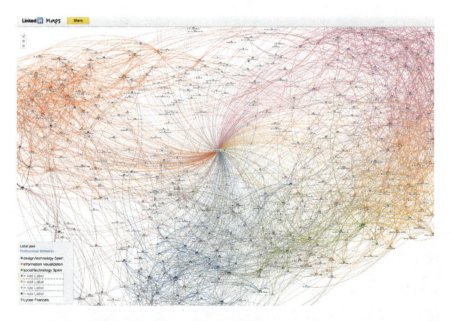

Fig. 8.3 My own Linkedin contacts network visualized. I can easily see which are the main clusters and recognize contacts that connect different clusters, thus constituting bridges for different contexts

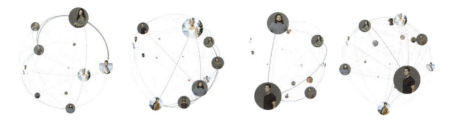

Fig. 8.4 *Lostalgic* (characters interaction at first 4 Lost episodes), Santiago Ortiz, 2012

8.2.1 Lostalgic

The data used in this project[1] (Figs. 8.4, 8.5, 8.6 and 8.7) are the corpus of scripts of all the episodes from the 6 seasons of the extremely successful American TV series *Lost*[2] (2004–2010): I grabbed all the data from the community-based wiki *Lostpedia*,[3] preprocessed it, and built a network in which the nodes are all the characters, and the weighed relations between characters indicate the intensity of the dialogues, if they exist.

Problem: Each episode is actually a different sub-network: Characters could appear or not, and the same is valid for connections. How to visualize this structural change through time?

Approach: My first approach is rather obvious: I visualize the sub-network of an episode and then remove and add nodes and connections whenever a different episode is selected. Instead of redrawing the network, nodes and relations appear and fade; there is no jump between episodes networks; instead, there is a continual structural transformation.

Then, I did the same but choose an adjacent matrix visualization method, instead of the classical graph. The result, in my opinion, is more interesting because it reveals more about patterns of characters and relations through different episodes.

Problem: The data of this project are a big script. I didn't want to just reveal patterns of the networks generated by characters interactions, but actually to allow readers to read the script. However, just making the texts accessible was not interesting. I wanted to use interactive visualization techniques to allow reading the scripts in a qualitative different way.

Approach: I developed two visualization views in which the script can be read. The first is quite linear and follows the hierarchical structure of seasons, episodes, acts, scenes, and dialogues. The linearity is partially broken with the option of zooming in and out, so the relation between global and local is not lost.

---

[1] Homepage project, URL, January 16, 2014: http://intuitionanalytics.com/other/lostalgic/.
[2] See Lost—Les disparus, URL, January 16, 2014: http://www.imdb.com/title/tt0411008/.
[3] URL, January 16, 2014: http://lostpedia.wikia.com/wiki/Main_Page.

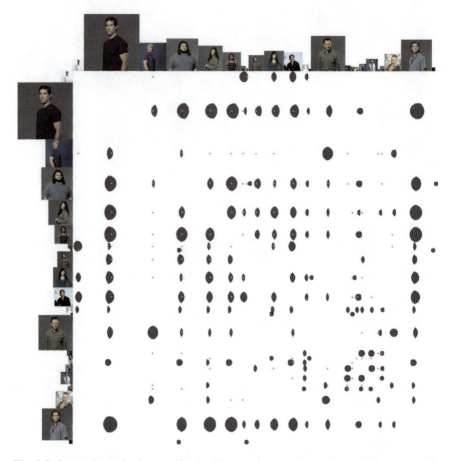

Fig. 8.5 Season last episodes are Choral: all the main characters appear and there are many interactions between them

The second way to read the script is through a technique I call "reenactment," in which texts and images are shown at certain speed, and timing takes into account texts length. This "simulation" imitates, as far as possible, the audiovisual experience of seeing an episode, taking into account the minimal resources at hand: texts and characters photos.

You can't replace the richer experience that an actual episode brings, but, what if this technique can be applied in other networks, not necessary based on scripts data?

8 Living Networks

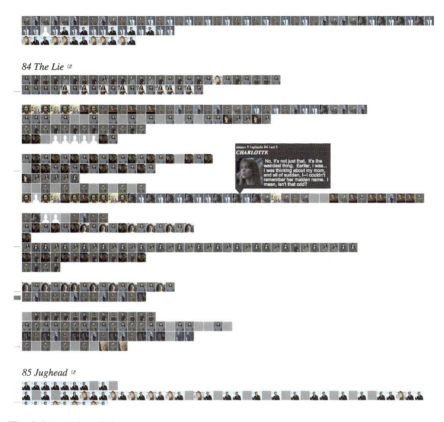

Fig. 8.6 *Lostalgic* (index)

Fig. 8.7 *Lostalgic* (reenactment)

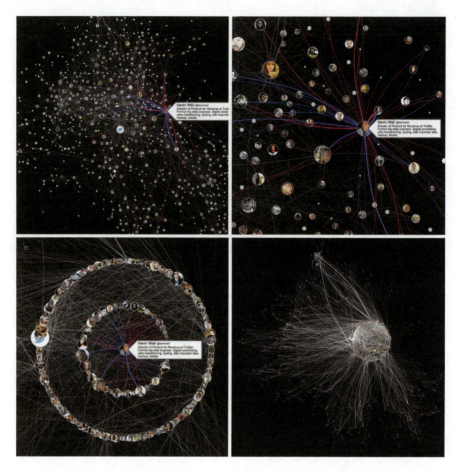

Fig. 8.8 *Newk*, Santiago Ortiz, 2012

8.2.2 Newk

Newk[4] (Figs. 8.8 and 8.9) visualizes 1 week of conversations of *Twitter* employees using Twitter. To gather the data, I used an official list of twitter accounts of employees of the company and the Twitter API to get all the tweets they wrote on the supervised week. Subsequently, I selected the tweets that where directed to (or mentioned) another employee account of. A directed relation is created if there is at least one tweet from an account to another. The number of nodes is round 1,000, and the number of relations is round 8,000.

---

[4] Homepage project, URL, January 16, 2014: http://moebio.com/newk/twitter/.

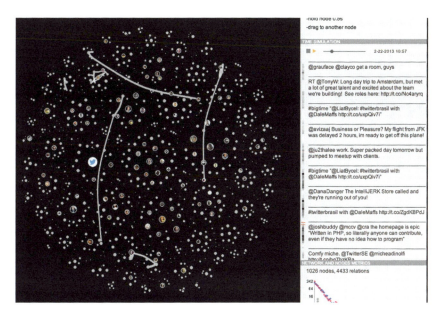

Fig. 8.9 *Newk*, Santiago Ortiz, 2012

Problem: The main challenge I assumed for this project was the continual transition of scales: how to allow both global and local views of the network, without having to shift from one visualization to another. Moreover, visualizing the local network of a node should be done in a way you keep the context: Not only the local network is visualized but also how this local group of nodes is related to the rest of the complete system.

Approach: To do this, there are several placement algorithms working at the same time. If no node is selected, the nodes are placed in a (pre-calculated) forces driven layout. Once a node is selected, its first degree connection nodes surround it in a small circle, and its second degree connection nodes surround it in a bigger circle. All other nodes in the network keep their normal position, except for a small radial distancing.

That way the local sub-network of the node, with one or two degrees distance, is featured without losing the view of the complete network context. By zooming in and out, one can appreciate the relative position and connections of the sub-network and the complete network.

Problem: A network of conversations is obviously dynamic, and each message takes place on a specific moment on time (is the consistence of messages through time that somehow define the more stable social links, also dynamic, yet in another timescale).

Approach: *Newk* contains a simulation mode (which echoes the previously commented reenactment mode in *Lostalgic*), when activated a timeline is displayed in the right panel, a time cursor moves throughout the week timeline, and messages

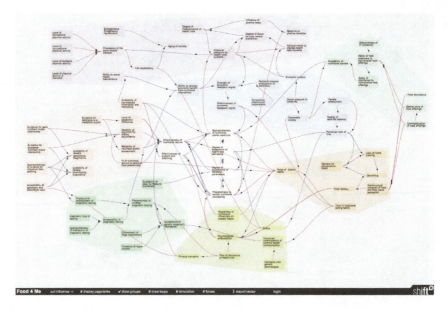

Fig. 8.10 *SYM*, Santiago Ortiz, 2013

are displayed in a stack, accordingly to their sending date. The timeline features the density of messages in all the 24 × 7 h, revealing a clear pattern of correlation with day/night in the USA. The rhythm at which new messages appear also conveys the change in time density. Meanwhile, in the network visualization side, the relations between two accounts are highlighted to indicate the nodes and direction of the message, and nodes move in the direction of the interlocutor, to better depict which nodes are active.

8.2.3 SYM

SYM[5] (Figs. 8.10 and 8.11) is a tool that allows the construction of networks, and, in particular, system maps. System maps are networks heavily loaded with narrative: Each node contains a concept or represents a variable present in certain reality and the same happens with relations that often represent an influence between two variables. *SYM* also visualizes these maps in different ways, harnessing the dynamism of the maps to the extent to visualize animated simulations.

Problem: System maps usually describe dynamic realities and moreover realities in which dynamism is central. For instance, the order in which a variable is affected by two others could be crucial, as well as the timings on the application of policies

---

[5] Homepage project, URL, January 16, 2014: http://vis.shiftn.com/sym/.

8 Living Networks

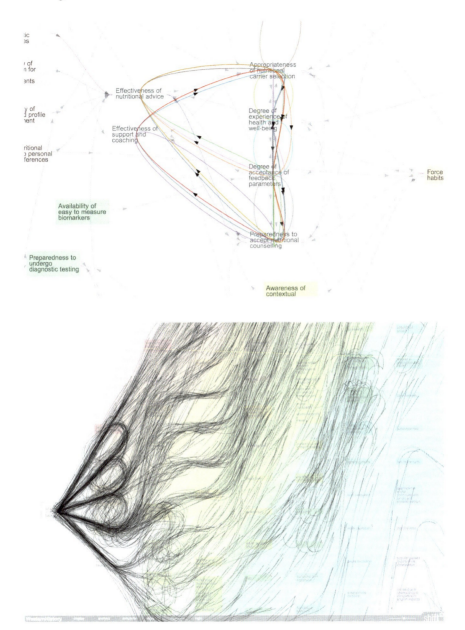

Fig. 8.11 Some of the loops detected in a network and flow simulation—particles flowing through network connections

that take into account these maps. Influences spread through the network in complex asynchronous ways, sometimes achieving unexpected regions of the system, or affecting nodes in loops with different time periods.

Approach: One of the *SYM* most interesting features is the detection and visualization of loops in the entire network or passing by a specific node

Problem: By visualizing the loops in the network, we can better understand the dynamics: Some of the patterns of the flow are revealed. But it continues being a static view, a photography, we cannot see or feel the actual temporality of the process.

Approach: *SYM* contains a set of tools to perform simulations that help understanding much better the behavior of the network, the order in which influences take place and the timings. Some of these simulations are helpful for understanding which nodes, en when, are affected by increasing the effect of a selected one. Other simulations allow the users to see the spread of information in the network topology throughout time. The most dramatic of these visual simulations is the particle-flow simulation, in which a beam of particles is launch from a selected node or, optionally, from all the nodes of the network.

8.3 Conclusion

Networks are dynamic and their dynamism is often more important than their structure. We cannot fully understand networks if we analyze and visualize them as static systems.

On the other hand, it is not easy to visualize dynamic properties of networks. One of the reasons of this is that they can be dynamic in many different ways: Networks change structure over time, its elements vary, information flows, and high-level properties emerge. Add to this instability that networks, being nonlinear by definition, are read in nonlinear and dynamic ways. We cannot understand a network at first sight (there is no gestalt principle for networks!), we can only navigate and explore them, and visualization should take this into account.

The visualization methodologies I propose in this article to communicate dynamic aspects of networks are just a few and maybe they are just early stages of solutions, but they help making the point that deep and effective networks visualization requires interaction and that challenges originated in dynamism should be addressed with dynamism. As in the zen *teisho* about the flag, the wind and the mind, the network moves, the visualization moves, and your mind moves.

Finally, although this article focuses on networks, one can make a similar case for any data structure and any reality information represents "everything is the way it is because it got that way." Think on any piece of valuable information: genomes, market data, health statistics, data produced by a particles collider, radio signals, development indexes, demographics statistics, social media data, etc. Nothing is static and nothing cannot be deeply understood without unveiling its dynamic patterns and properties.

Dynamic data visualization is itself a highly dynamic and very young field. It is not easy to produce (let alone produce fast) this kind of projects, for you need to be able to understand data in deeper ways (values, relations, interactions and changes), be a spatial–temporal designer (like an architect that designs mutable spaces), deal with graphical elements that change position, shape or that constantly appear and disappear, and whose behavior depends on the user or other elements behavior. But due to this complexity in the content and representation side, a deep understanding of human perception from a dynamic point of view becomes urgent. This is to say that DDV requires the assembling of multidisciplinary teams formed by data scientists, computer scientists, designers, and experts in perception, probably neuropsychologists.

Does it worth the effort? Well done DDV projects could describe and to some extent explain complex (sometime terribly complex) things such as an organism metabolism, a cell, an organ, the weather, the macroeconomy, a company, a community, cities, a machine, an ecosystem, cancer, etc., to name a few. Any of these dynamic systems cannot be well explained in a text or a static visualization, and DDV just opens a completely new window of communication opportunities. I go further and believe we might be in the eve of discovering a new form of writing.

References

Bret V (2012) Stop drawing dead fish. Talk originally presented to SF SIGGRAPH (on 16 May 2012), and recorded at the Exploratorium (on 27 Nov 2012). URL: http://vimeo.com/64895205. Accessed 16 Jan 2014
Moretti F (2013) Distant reading. Verso, London
Thompson D'AW (1992) On growth and form. Cambridge University Press, United Kingdom (New edn)
von Goethe JW (1994) Goethe, the selected works: selected poems, vol 1. Princeton University Press, New Jersey

Author Biography

Santiago Ortiz (Independent Data Visualizer, Visual Data Scientist and Researcher) (Bogotá, Colombia, 1975) studied Mathematics at the Universidad de los Andes in Bogotá, Colombia; he also studied music and literature. He has taught at the School of Fine Arts at the University of Porto in Portugal, and at the Departments of Arts and Mathematics of the Universidad de los Andes in Bogotá.

He has been a Professor of the Masters in Art and Technology at the European University of Madrid, and he teaches classes in Digital Design at the European Institute of Design in Madrid, among others.

In 2005 he co-founded Bestiario (Barcelona), the first company in Europe devoted to information visualization. Now he works as a consultant, helping organizations with the analysis and communication strategy of their information, and he specializes in networks, conversations and knowledge. He invents and develops algorithms, visualization methods, interactive narratives and new ideas for the internet.

For more information and contact:

http://moebio.com
https://twitter.com/moebio

Chapter 9
Epiphanies Through Interactions with Data

Dino Citraro and Kim Rees

Abstract By thoughtfully designing visualizations that engage and inspire people, we can help them reveal self-directed epiphanies. This develops a "pride of ownership" between the viewer and the data, thereby creating a deeper level of engagement, encouraging sharing information, and compelling action.

9.1 Imagining Data

Data are infinite. From the place that you are now sitting, data exist in every direction, at every scale, and in every configuration that you can imagine, and in any configuration that *anyone* can imagine. Data only exist because of our imagining. Data, unlike everything else in the universe, have no mass or energy and are completely the result of our observations.

We observe, and catalog, and reference. This is what we have done as a species since the long ago days when we decided that living together was better than living alone, and it is a result of our common observations that we realized the natural world is comprised of patterns. Depending on where you live, the sun rises and sets farther along the horizon during the year until one specific day in December when it starts moving backwards. And then come June, it stops moving backwards and starts moving forward again. And it does this every year in the very same way. The observance of this pattern, the cataloging of the natural environment, is data collection, and we have been good at it since before we were good at writing it down.

And now look at us. We have used our observations and our meticulous cataloging of nature to build entire libraries of observations that serve as the foundation of our beliefs.

D. Citraro (✉) · K. Rees
Periscopic, 235 NW Park Ave. Unit A, Portland, OR 97209, USA
e-mail: dino@periscopic.com

K. Rees
e-mail: kim@periscopic.com

© Springer-Verlag London 2015
D. Bihanic (ed.), *New Challenges for Data Design*,
DOI 10.1007/978-1-4471-6596-5_9

* **Influencers on the meaning of data**

The context through which we interact with data is essential to our understanding of the information it conveys. Our belief systems, our biases, the presentation methods used to reveal the data, and other factors influence our perception. As we interact with data, we must take care to understand these influencers to ensure the truths we glean from our interactions are genuine.

9.2 On Disbelief

Belief is a mind-set that cleaves to a worldview shaped by countless interactions, each building on the last—an unfathomable tower of bricks placed precariously one after another because our individual logic finds connections to create that order.

In fact, our brains function to reinforce this process. In a study using fMRI data to better understand conceptual change, it was shown that minor changes in a concept were easily integrated through normal learning processes. However, the experiment demonstrated that when people receive information that is inconsistent with their preferred theory, the absorption of that information is not easily integrated.

We found that when people were given data that were consistent with their preferred theories, areas thought to be involved with learning (i.e., caudate and parahippocampal gyrus) showed increased levels of activation relative to baseline. However, when participants were presented with data that were inconsistent with their preferred theory, the anterior cingulate, precuneus, and dorsolateral prefrontal cortex showed increased levels of activation. The anterior cingulate is thought to be a region of the brain associated with error detection and conflict monitoring, whereas the dorsolateral prefrontal cortex is thought to be one of the prime regions involved in effortful processing and working memory (Dunbar et al. 2007).

The integration of knowledge is difficult even on a physical level—it is not merely stubbornness or ignorance. It illustrates the quote:

> We don't see things as they are; we see them as we are.
>
> Anais Nin

We are the filter of information. It changes as it comes into contact with us. Werner Heisenberg was keenly aware of this, and his "Uncertainty Principle" (Cassidy 1992; Heisenberg 1927) has direct parallels when it comes to observing data.

9.2.1 Bias

Your experiences, your expectations, your desires, and every other factor that combines to make you unique from the next person you will see today are also the same factors that are making you uniquely biased.

Biases are beautiful colorations of the underlying data of the world and allow you to see a glass as half full, a rainbow in an oil slick, or goodness in an unexpected gesture. They also allow you to act and react in ways that others might not be able to, even if they had all the same information as you.

Even when viewing the same data, two observers can often form completely different stories, histories, and conclusions.

9.2.2 Skepticism

A prominent study shows that since the 1970s, conservatives' trust in the scientific community has been steadily declining.

> (…) The scientific community inevitably becomes entangled in polarized conflicts (e.g., economic growth versus environmental sustainability). As a result, science is "increasingly seen as being politicized and not disinterested." (Gauchat 2012)

Adding to that growing skepticism is the tendency for the progenitors of scientific research, namely academic institutions, to cloister their findings behind paywalls, enrollment fees, and other barriers. Further, publishers often hide this information from the public, leaving it to be delivered secondhand from mainstream media sources that have digested, summarized, and evolved the ideas into a whole variety of conveniently edited narratives. This unfortunate mutation of knowledge, while ostensibly done to disseminate information, creates the very latticework upon which distrust and ambiguity thrive.

9.3 Revealing Truth

Most people want to believe. When presented with information that challenges our belief systems, it is easy to understand why most of us will look for ways to support what we already believe, rather than find ways to discredit it. Belief is comforting, it is easy, it is self-affirming, and perhaps most important of all, it is normal.

The problem is, when it comes to data, this is an approach that can absolutely never be taken. Every individual that collects or visualizes data should follow a singular principle: *Any action that is made upon the retention and communication of data should further the revelation of truth.*

And belief is not truth. Belief is a subjective agreement made between yourself and the fuzzy set of lines you are drawing to lead yourself through an uncertain world. Truth is an empirical statement that can lead everyone to the same location, regardless of stating point or personal biases.

9.3.1 Truth in Numbers

Numbers are the currency of data, and numbers, like other currencies, can fluctuate depending on demand. When seen in the context of data, numbers often represent the synthesis of a concept or the tally of a set. A single observation no longer conveys the situational relationships of the setting from which it was gathered, but resolves itself as a member of a set. The set then also becomes a number and later evolves to be seated in a ratio between observation and conclusion.

This is convenient. It allows us to make inferences based on the translated data (numbers) of disparate sources and reach conclusions that are quite frequently accurate. However, as we trade complexity for convenience, we introduce a system of bartering that unintentionally gives rise to subjectivity and rhetoric.

Nick Diakopoulos, Tow Fellow at the Columbia University Journalism School, posits on The Rhetoric of Data:

> The Latin etymology of "data" means "something given," and though we've largely forgotten that original definition, it's helpful to think about data not as *facts* per se, but as "givens" that can be used to construct a variety of different arguments and conclusions; they act as a rhetorical basis, a premise. Data does not intrinsically imply truth. Yes we can find truth in data, through a process of honest inference. But we can also find and argue multiple truths or even outright falsehoods from data. (Diakopoulos 2013)

Our biases sprout these "multiple truths" when presented with conclusions that are at odds with our personal belief systems.

For instance, when considering neonaticide (the killing of a newborn), it is easy to reason that this behavior is an act of violence comparable with other types of homicide. However, if the data are scrutinized, it reveals a pattern that aligns the rates of neonaticide most closely with those of suicide. While each act is equally tragic, neonaticide is then seen as a *self-destructive* act, rather than one spurred by homicidal impulses (Meyer and Oberman 2001).

When visualizing data, it is imperative to foster an interface that allows access into the underlying, unbiased data. Inferences and summations are shortcuts that should be taken with extreme caution.

9.3.2 Respect

As we have seen, data are a representation of our desire to catalog and make sense of the world, and because we have become so good at retaining it, we have also become quite good at abstracting it.

Decoding our data abstractions can require a bit of work, and many who are new to the act of visualizing data incorrectly assume that their audience will be unable (or uninterested) in doing this work. This assumption leads to a dilution of the data into familiar metaphors and presentation methods, which it is hoped will increase comprehension and engagement.

While understandable, this approach is not only incorrect; it also does a disservice to all who strive to understand the underlying truth that the data represent.

Consider this example: If you have a baby at home who does not yet know how to talk, should you refrain from speaking to them until they do? Should you confine your communication to long glassy-eyed stares randomly punctuated by wildly exuberant tones? Of course not. Babies learn *through* being exposed to the communication they do not yet understand.

The essence of data visualization is knowledge. Embrace the act of educating your audience while they interact with data. Using visualization methods that are appropriate, though potentially unfamiliar, will not alienate your audience; they will further knowledge retention and the adoption of the underlying truths you hope to reveal.

9.3.3 Raw Data

Part of the problem with modern science is that the data and methods are held at the upper echelons. Government is beginning to open up their data and become more transparent, but academia has not made many efforts in opening its research. We are holding on to archaic beliefs that we can lead with blind faith.

As noted earlier, the public is becoming increasingly skeptical and distrustful, and one way to combat this is with open knowledge. By letting people look under the hood, so to speak, we give them an opportunity to approve or disapprove of the findings. Providing the raw data allows people to see the parts of the whole—in the same way, we look at the line items of a receipt to confirm the total purchase amount.

* From Influence, to Interaction, to Epiphany

While the factors just discussed have the potential to color our perception of data, there are ways we can help mitigate their influence. By creating interfaces that foster objective exploration of the underlying raw data, we can allow individuals to construct an understanding that is derived from self-directed epiphanies.

9.4 Interactive Data

A statistic is a tally. It is a collection of individual pieces of data all bound together and represented by a single value. For instance, 51 % of the US population is female. This statistic comes from the US census, a nationwide effort to count each citizen. We express these numbers in aggregate, as percentages, or fractions, or other tallies of the individual parts.

In Periscopic's interactive data visualization of gun violence in the United States,[1] we begin by showing all of the people who were killed by someone using a gun in 2010 (Figs. 9.1, 9.2, 9.3 and 9.4). This is another example of a statistic—a total of the number of people killed in a year.

While statistics can impart knowledge, they are commonplace, and as such can be overwhelming and forgettable. They are also impersonal—their aggregate nature

---

[1] URL, February 5, 2014: http://guns.periscopic.com.

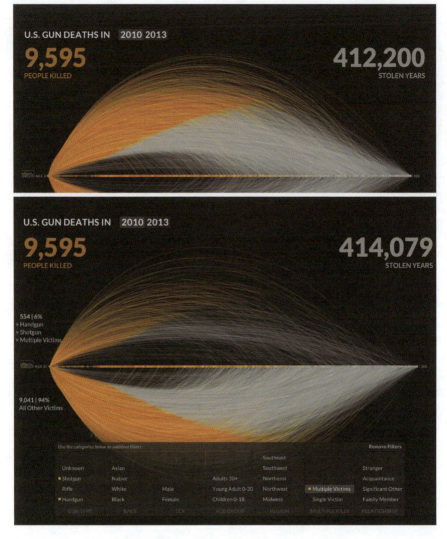

Fig. 9.1 *More than 400,000 stolen years: an examination of U.S. gun murders*, Periscopic, 2013, URL, February 5, 2014: http://guns.periscopic.com. This shows the mass of people who were killed in 2010 as reported by the FBI

lumps every available detail into a single entity, devoid of the multitude of nuance that comprises each datapoint.

Statistics are convenient, but should be seen as an entrance to a conversation, not an exit. By using statistics along with individual datapoints, a layered understanding emerges and invites the observer to travel deeper into the realm of inference and epiphany. A dissection begins that compels the observer to tease at the strands of the corpus, as if looking with a microscope toward an unseen foundation.

9 Epiphanies Through Interactions with Data

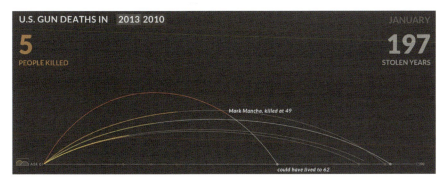

Fig. 9.2 This screenshot shows the beginning of the visualization—an animation of each person killed in the USA

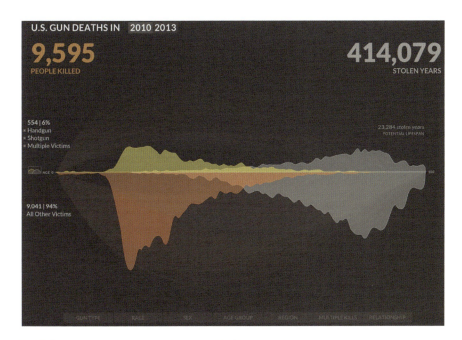

Fig. 9.3 The interface used to filter gun homicides

Fig. 9.4 Victim age distribution of those killed by gun violence

When visualizing the US gun deaths, we chose to reveal the data through its individual datapoints: each person—each casualty of gun violence. Upon launching the experience, visitors watch as five victims appear, each with three datapoints, their names, their age when they were shot, and a projection of the age they might have attained, had their life not been ended. The use of these datapoints establishes three essential aspects of the piece: the fact that each of these datapoints was a human, each had a life that was ended, and each left a void behind.

Presenting statistics at this human level fosters empathy and understanding. It allows visitors to walk hand in hand with the numbers—to navigate the "multiple truths" and begin to test which of them is truest of all.

9.5 Journeys, not Stories

The typical trajectory of a data story (or other communication) starts with an idea. Based on an idea, the author conducts research, identifies sources, gathers and parses the data, and summarizes the experience as a story of the findings.

Usually, the viewer is not privy to anything that happens prior to the very end result—the story. This is after the data have been gathered and analyzed and all the research is complete. The "story" is what the author wants the audience to know. As if to say "pay no attention to how I made this conclusion. Trust me!"

Data visualization is a vehicle for transporting the viewer from the story back to the original idea, allowing each viewer to follow the research, the sources, and scrutinize the data to see whether they arrive at the same conclusion, or story.

Whereas in other forms of communication, this approach might destroy the very messaging that is desired, in data visualization, the final message is less important than the journey of how the message was created. The data itself are the message. The relationships and conclusions that are drawn are by-products of the original need to document and retain an essential truth.

9.6 Epiphanies

Ownership is gained through investment. While it might seem counterintuitive to ask viewers of a data visualization to invest work into a system in order to receive ownership of its underlying truths, the act of interacting with data creates a series of connections that aid in memory retention (Metiri Group 2008).

Once a viewer begins to interact with the data, to manipulate it in a self-directed way, the visualization becomes a personal experience. Much the same as if thousands of people show up to a rally and create a common, group statement. Each will leave the rally having participated in a single event, but each will also have a unique experience. The stories of the event will be told from multiple perspectives, views, and reactions, and each will be personal.

When viewers are allowed to confirm the analysis and the statistics of a visualization, they become advocates for its conclusions. This act moves people beyond the passive absorption of statistics and propels them into the role of an ambassador of a self-discovered revelation (Fig. 9.3). This behavior can be seen in the gun violence data visualization.

Here, the viewer is provided with a set of access controls to all of the data and are allowed to configure it in any manner they find compelling. Viewers can filter by race, sex, region, and relationship to the killer, among other things (Fig. 9.4).

Selecting individual filters reveals only those victims that correspond to the selection. New statistics are generated with each refinement of the filters.

Through this behavior, viewers can easily form a hypothesis and test their assumptions. Each datapoint can be scrutinized and compared to any other to see what patterns might emerge.

This process of exploration, and the layered set of revelations that it produces, is the antitheses to the conundrum of the "multiple truths." Through this behavior, the act of interacting with data, viewers follow a self-directed path toward understanding and epiphany, and avoid the blind faith that is required in a traditional, statistics-first approach to data communication.

9.7 Summary

We live in the age of data exploration, an age that is creating entirely new perspectives of our world by using our collective observations to reveal patterns, correlations, and relationships: footfalls on the mental pathways of truth. No longer do we look to the horizon and wonder when the sun will return, we now conjure the sun in infinite permeations. We do this easily, earnestly, and sometimes, simply because we can. And just as explorers from past eras left trails for others to follow, as data visualizers, we must create the structures that will allow others to do the same.

References

Cassidy DC (1992) Uncertainty, the life and science of Werner Heisenberg. Freeman, New York

Dunbar KN, Fugelsang JA, Stein C (2007) Do naïve theories ever go away? Using brain and behavior to understand changes in concepts. In: Lovett MC, Shaha P (eds) Thinking with data. Lawrence Erlbaum Associates, New York, pp 193–206

Diakopoulos N (2013) The rhetoric of data. Nick Diakopoulos's website, 5 Feb 2014, http://www.nickdiakopoulos.com/2013/07/25/the-rhetoric-of-data

Gauchat G (2012) Politicization of science in the public sphere: a study of public trust in the United States, 1974 to 2010. Am Sociol Rev 77(2):167–187

Heisenberg W (1927) On the perceptual content of quantum theoretical kinematics and mechanics. Zeitschrift für Physik 43, p. 172–198. English edition: Heisenberg W (1983) Quantum theory and measurement (trans: Wheeler JA, Zurek W). Princeton University Press, New Jersey, pp 62–84

Metiri Group (2008) Multimodal learning through media: what the research says. report commissioned by cisco systems, pp 13–14. Available online, 5 Feb 2014, http://www.cisco.com/web/strategy/docs/education/Multimodal-Learning-Through-Media.pdf

Meyer CL, Oberman M (2001) Mothers who kill their children: understanding the acts of moms from Susan Smith to the "Prom Mom". New York University Press, New York, p 42

Author Biographies

Dino Citraro (Co-Founder, Managing Partner and Head of Strategy at Periscopic) is Head of Strategic Design and Operations at Periscopic, and has a strong background in problem solving, creative direction, and writing.

A twenty-year veteran of the multimedia industry, his work has spanned immersive online development, application design, interactive motion pictures, multi-player games, and in-teractive hardware installations.

He is the Visualization Editor of the Big Data journal, as well as a contributing blogger to several industry sites. He is also an accomplished photographer, a published poet, and has written and illustrated seven children's books.

For more information and contact:

http://www.periscopic.com
http://www.linkedin.com/pub/dino-citraro/7/38b/a65
https://twitter.com/dinocitraro

Kim Rees (Co-Founder, Managing Partner and Head of Information Visualization at Periscopic) is co-founder and Head of Information Visualization at Periscopic and a prominent figure in the data visualization community, actively speaking, writing, consulting, and otherwise participating in the field.

For more information and contact:

http://www.periscopic.com
http://www.linkedin.com/in/kimrees
http://www.twitter.com/krees

Part V
Exploring and Manipulating Data

Chapter 10
Sketching with Data

Fabien Girardin

Abstract The growing deployment of networked infrastructures has dramatically increased the amount of logs byproducts of people digital activities (i.e., digital footprints). That intangible material takes the form of cellular network activity, aggregated credit card transactions, real-time traffic information, user-generated content, or social network feeds. The capacity to transform this type of big data into insights, products, or services has called for new practices at the crossroad of design and "data science." This paper will discuss transversal incline of the innovation with digital footprints and will describe how sketching with data offers useful interfaces to the many stakeholders of innovative projects.

10.1 Introduction

The explosion in the use of mobile devices and social networks has generated large datasets of digital footprints. For instance, visitors to a city have many ways of leaving voluntary or involuntary electronic trails: prior to their visits tourists generate server log entries when they consult digital maps or travel Web sites; during their visit, they leave traces on wireless networks whenever they use their mobile phones; and after their visit, they may add online reviews and photographs. Broadly speaking then, there are two types of footprint: active and passive. Passive tracks are left through interaction with infrastructures, such as mobile phone, which produces entries in locational logs, while active prints come from the users themselves when they georeference tweets or their workouts with Nike+ enabled shoes.

We have been active observers and contributors since the dawn of that data deluge working on "making data talk" (Girardin 2009) and materializing services from multiple sources of digital footprints: cell phones, cars, shared bikes, digital cameras, and credit cards. For instance, our analysis on georeferenced photographs

---

F. Girardin (✉)
Near Future Laboratory, Av. Général Guisan 19 CP 242, Sierre, Switzerland
e-mail: fabien@nearfuturelaboratory.com

suggested that exploiting this dataset to know who visits different parts of the city at different times can lead to the provision of customized services (or advertising), the rescheduling of monuments opening times, and the reallocation of existing service infrastructures or the evaluation of specific urban strategies.

10.1.1 A Vision that Became Reality

The low cost and high availability of digital footprints now provide a new material to understand urban processes and design innovative services. Indeed, only a decade ago, the possibility of producing fully dynamic time–space diagrams from the fusion of human activities data and novel forms of analysis was only discussed in the conditional. For instance, Zook et al. envisioned in 2004: "When many individual diagrams are aggregated to the level of cities and regions, these visualizations may provide geographers, for the first time, with truly dynamic maps of dynamic human processes. One might imagine them as twenty-first century 'weather maps' of social processes" (Zook et al. 2004).

It is only very recent that the presence of digital footprints provides new means to analyze a city in real time and to replay its processes. These new potentials attracted the interest of large technological organizations, local authorities, and urban service providers in "big data." For instance, in a follow-up of our analysis on georeferenced photographs, we particularly showed the capacity to quantify the evolution of the attractiveness of urban space with a case study of the area of the New York City Waterfalls, a public art project of four man-made waterfalls rising from the New York Harbor (Girardin et al. 2009). Methods to study the impact of an event of this nature are traditionally based on the collection of static information such as surveys and ticket-based people counts, which allow to estimate visitors' presence in specific areas over time. In contrast, our contribution made use of the dynamic data that visitors generate, such as the density and distribution of digital footprints in different areas of interest and over time. Our analysis provided novel ways to quantify the impact of a public event on the distribution of visitors and on the evolution of the attractiveness of the points of interest in proximity. Local authorities used the results as part of their evaluation of the economic impact of the New York City Waterfalls.

In that specific project, we analyzed two types of digital footprints generated by phones or mobile devices that were in physical proximity to the Waterfalls: cellular network activity and photographing activity. Cellular network activity was measured by analyzing aggregate statistical data about number of calls, text messages, and overall amount of network traffic generated at each AT&T antenna every hour. Photographing activity was measured by adding up the number of photographers present in different areas of the city, and the number of photographs they took in each location. We acquired these data by analyzing photograph taken from the photograph-sharing Web site, Flickr. A major implication was to apply prior to any analysis conscientious, principled, and evident measures to protect people's privacy.

10.1.2 Societal Implications

Among the other pioneers in that type of investigations, we had to ensure that the social advantages of these applications were not in conflict with important privacy requirements. Digital footprints are both immensely empowering (for the people and places able to construct and consume them) and potentially overpowering as institutional and state forces are able to better harness information with growing personal and spatial specificity. In consequence, there are ethical and privacy implications to grapple with. In conjunction with people's own representation of traceability, there is a legitimate concern about the drift of research on digital footprints. Particularly, our work exemplifies the shift from large-scale top–down big-brother thread on privacy issues to more local bottom–up little-sister types of people monitoring, which makes the whole notion of opting out of technology adoption one of whether to opt out of society.

In fact, these digital footprints have become inevitable in contemporary society and also necessary if we wish to enjoy many modern conveniences; we can no more be separated from it than we could be separated from the physical shadow cast by our body on a sunny day (Zook et al. 2004). The growth of our data shadows is an ambiguous process, with varying levels of individual concern and the voluntarily trading of privacy for convenience in many cases.

In summary, at the same time, as digital footprints give us new means to model human dynamics and develop new services, they also challenge current notions of privacy. The works described in this article attempt to appreciate and use the complexity and richness of digital footprints without crystallizing into authoritarian structures.

10.1.3 Methodological Implications

The ability to replay the city shows that there are opportunities to propose novel ways to describe the urban environment and develop new solutions. However, there is a big assumption in seeing the world as consisting of bits of data that can be processed into information that then will naturally yield some value to people. Indeed, the understanding of a city and people goes beyond logging machine states and events. In consequence, let us not confuse the development of novel maps from previously uncollectable and inaccessible data with the possibility of producing *intelligent maps*. Our work precisely draws some critical considerations on the current state of the art. The first steps in our projects always aim to figure out: (1) What parts of reality the data reveal and (2) What we can do with them. For instance, not to confuse behaviors with endorsement that can be considered as a limitation of our New York Waterfalls case study that used the density of digital footprints as indicators of urban attractiveness. In similar studies, calibrations with ground truth information are necessary. Alternatively, some questions or problems can be answered mixing both quantitative and qualitative methods:

- The qualitative analysis to inform the quantitative queries: This approach first focuses on people and their practices, without the assumption that something computational or data process is meant to fall out from that. This qualitative angle can then inform a quantitative analysis to generate more empirical evidences of a specific human behavior or pattern.
- The quantitative data mining to inform the qualitative inquiries: In that approach, the quantitative data help to reveal the emerging and abnormal behaviors, mainly raising questions. The qualitative angle can then help explaining phenomenon in situation.

With complementary perspective on people behaviors or the actual use of the space, it becomes, for example, possible to develop new types of "post-occupancy evaluations" often overlooked in the practice of urban design and architecture or to design new services for and with local authorities, businesses, or individuals. Those approaches almost exclusively involve multidisciplinary teams.

10.1.4 A Multidisciplinary Process

Both societal and methodological implications require the involvement of many stakeholders often from different practices and objectives, from engineering to statistics, design, strategy planning, product management, and law. The process of innovating with digital footprints demands several steps, each with their own set of skills, knowledge, questions, and answers (Fig. 10.1): From the data access and collection techniques that feed data to obfuscations algorithms and big data management systems that are interrogated by basic data mining operation or advanced statistical inquiries. Often in parallel, information visualization techniques are used to build evidences and indicators. It is the engagement of multiple stakeholders of the project that provokes own questions, ideas, or scenarios and, therefore, new queries to the data.

Throughout our projects, we found the necessity to very quickly being able to visualize temporary results and share them with stakeholders of the project. We learned that this approach was useful to keep a proper momentum in projects that often need to fail, fork, or win within a few weeks or months. As a consequence, more and more results of our investigations became interfaces or objects with a means of input and control rather than only static reports. As it became prominent in our creative and innovation process with digital footprints, we called that practice "sketching with data." We will describe it in the rest of the paper with some examples of its virtues and the tool called Quadrigram[1] developed from our experiences.

---

[1] URL, February 7, 2014: http://www.quadrigram.com.

10 Sketching with Data

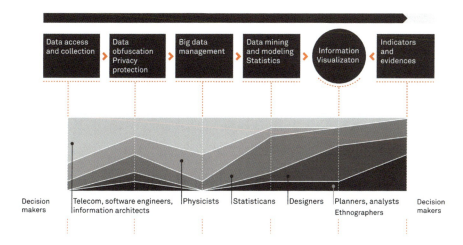

Fig. 10.1 Simplified overview of the multidisciplinary process of innovating with digital footprints. Each step demands specific skills, knowledge, questions, and answers

10.2 The Practice of Sketching

The practice of sketching is common in creative activities such as art or architecture often as a rough version of a work and preliminary attempts to complete something. For instance, in architecture (Schank-Smith 2005) mentions "Through visual artifacts, architects can transform, manipulate, and develop architectural concepts in anticipation of future construction. It may, in fact, be through this alteration that architectural ideas find form."

Sketches are meant to be fast; they are often produced with great speed to capture rapid cycles of ideation. The look of those attempts is not as important as the role they play in the creative process. Indeed, sketches are not precise or visually compelling but their expression is vague enough to allow illusions and analogies. It is through iterative attempts that they accompany brainstorming for the author or dialogues among members of a project. However, unlike prototypes, they do not aim for fidelity nor desirability or to prove a concept. Rather, they are used to discover a concept at the beginning of a project. In fact, they can be employed in all stages of the design process to form or deform ideas, even as an observational recording long after the project is completed. For instance, Leonardo da Vinci sketches have become art objects in their own right, with many pages showing finished studies as well as sketches.

Sketches do not have to be captured on paper. They can take the form of 3D models or interactive visualizations, and often the intention takes precedent over the media (Schank-Smith 2005). When translated to the world of data, sketching has a lot more in common with exploration in a workshop rather than it does working with Photoshop or other graphic design application. The process consists in generating early results and insights from manipulating data queries, establishing a blended approach with multiple sources, sculpting with algorithms, extracting with filters, and drawing results with libraries. Our use of sketches is meant as a medium for inspiration and transformation in multidisciplinary teams. In consequence, they need to be comprehended by a wide variety of professionals, from physicist and engineers to lawyers, decision makers, and strategists. They become *boundary objects* to explore ideas and solutions (Biddulph 2012). Borrowed from sociology, "boundary objects are objects that are both plastic enough to adapt to local needs and constraints of the several parties employing them, yet robust enough to maintain a common identity across sites" (Star and Griesemer 1989). Practically, they allow coordination without consensus as they can allow the specific understanding of each stakeholder to be reframed in the context of a common project. Instead of communicating across disciplines using vocabularies from different practices, a sketch reveals the data and its transformations in real running code and designs.

Sketches can be used to start a project with a multidisciplinary team that needs first to grasp the potentials of the data and their limitations. For instance, when an institution envisions the use of its own digital footprints to develop a real-time service, early sketches are needed to answer basic questions such as what is the expected amount of data, what are the expected signals in the data, or more precisely what time frequency can be associated to the notion of *real time*. Practically, a time series visualization can help each team member understand whether their service can deliver new information every minute, every hour, every day, or every week.

Besides the practice of prototyping often employed in the domains of engineering and experience design, there exist other creative forms to express ideas and concepts with programming and data. Under the umbrella term of *creative coding,* there is a growing community of professionals who use the language of code and data as their medium. Their work, which often evolves through iterations of sketches, includes everything from digital art to elaborate interactive installations, all with the goal of expanding our sense of what is possible with data and software. In those approaches, the necessity to program in machine language cuts the direct relation between the physical action of the hand and the result. The tool becomes an abstract and is not more a direct extension of the body and the mind.

10.3 The Tools

Based on our first experiences in sketching with data, we detected two increasing demands within innovative institutions. First, there is a necessity of multidisciplinary groups to think with liberty with data, outside of coding, scripting, wizard-based, or blackbox solutions. Then, we perceived the demand to diffuse the power of information visualization within organizations to reach the hands of people with knowledge and ideas of what data mean. We were struggling to combine the tools

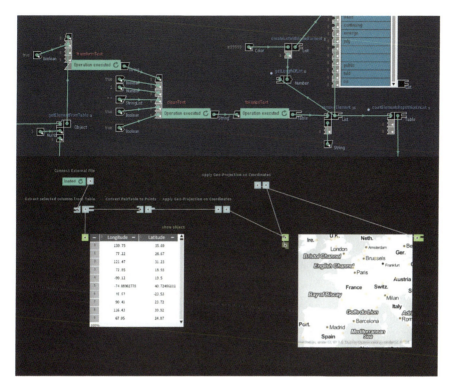

Fig. 10.2 Quadrigram is a Visual Programming Environment composed by pre-programmed modules that perform operations. The modules are linked between themselves to produce a visual workflow in a boundless canvas

that could allow people with the knowledge and data but not the technical skills to develop their own ideas and scenarios. We needed them to manipulate their data as artists or architects that explore ideas with freedom with an easy to erase pencil or chalk stick. The response came with the development in collaboration with the design studio Bestiario[2] of a Visual Programming Environment called Quadrigram (Fig. 10.2). We wanted the tool to be particularly designed for iterative data exploration and explanation.

With Quadrigram, the abstract notions of programming became transparent as it offers the opportunity of manipulating data as a living material that can be shaped in real time. By living data, we mean data that are constantly changing and accumulating. They can come from social network, sensor feeds, human activity, surveys, or any kind of operation that produces digital footprints. The tool was meant not only for *data scientists* but rather everybody with knowledge and ideas as an expansion of their mind. We designed it with the principles described in Table 10.1.

---

[2] URL, February 7, 2014: http://www.bestiario.org.

Table 10.1 Summary of the key principles used to design Quadrigram

Principle	Description
Non-linearity	The infinite nature of a boundless workspace provides a natural way of developing ideas without the need of keeping linearity. Therefore, users can sketch and arrange their ideas on demand, increasing naturalness
Iteration	Quality is a consequence (among others) of understanding, the more you iterate the more you master, and one of the key techniques in data visualization is iteration
Data as a living material	Data are constantly changing. The ability to work with real-time data empowers to detect changes fast and act, consequently, enhancing efficiency and control capacity

To create Quadrigram as a tool to sketch with data, we had to rethink the approach people take to work and create with data. We describe the four main challenges in the following sections.

10.3.1 Redefining Work with Data

The design of Quadrigram lays on this very idea of *feedback loop*. It is designed for iterative exploration and explanation. Each iteration or *sketch* is an opportunity to find new questions and provide answers with data. Data mutate, take different structure in order to unveil their multiple perspectives. We like to think that Quadrigram offers this unique ability to manipulate data as a living material that can be shaped in real time or as Mike Kuniavsky describes in Smart Things: Ubiquitous Computing User Experience Design: "Information is an agile material that needs a medium" (Kuniavsky 2010).

With the diffusion of access to data (e.g., the open data movement), investigation with data has become utterly multidisciplinary. Projects embark teams with fast prototyped tools that promote the processing, interpretation, and reinterpretation of insights. For instance, our experience shows that the multiple perspectives extracted from the use of exploratory data visualizations is crucial to quickly answer some basic questions and provoke many better ones. With Quadrigram, we suggest a novel approach to work data in which analysis and visualizations are not the unique results, but rather the supporting elements of a co-creation process to extract value from data. In Quadrigram, the tools to sketch and prototype took the form of a Visual Programming Environment.

Table 10.2 Summary of the qualities of a tool that supports sketching with data

Quality	Description
Power	Puts power in the hands of more people that have the knowledge, though not necessarily the technical ability to program solutions from scratch. Everyone from project managers to interns can now play a part
Speed	Reduces development time and effectively the cost of implementation, by handling complex, programmatic patterns
Creativity	Inspires solution-finding by exposing the full flow of data, whereas standard programming languages abstract and compartmentalize the process
Flexibility	Ensures flexibility by allowing for multiple pathways toward a solution, working in the freedom of two dimensions. By contrast, wizards and standard tools assume user-intent and impose fixed sequences that are rigid and one dimensional, limiting the potential for discovery along the way

10.3.2 Reducing the Barriers of Data Manipulation

Visual Programming Environments have flourished in the domain of information technologies, starting with LabVIEW[3] in the 1980s and then spreading to the emerging fields mixing data with creativity such as architecture, motion graphics, and music. In these domains, they have demonstrated virtues in reducing the barrier of entry for non-experts. In the Visual Programming Environment we developed, users manipulate in an interactive way pre-programmed modules represented as graphical elements. When connected, these modules form a "data flow" (also called dataflow programming) that provide a constant *What You See Is What You Get* (WYSIWYG) view of the result of the program ideal for quick *trial and error* explorations. This way the tool allows for the evaluation of multiple pathways toward the correct solution or desired result. It inspires solution-finding for non-technical professionals by exposing the full flow of data.

Visual programming language (VPL), as per its nature, facilitates the learning process, not only for those who are neophytes in programming, but also for those who know how to use non-visual programming languages. In comparison with standard programming languages and wizards, we wanted Quadrigram to lower the barrier of data manipulation, reducing time to produce sketches, inspire creativity, and ensure flexibility (Table 10.2).

10.3.3 Creating a Coherent Language

A major challenge when grouping tools to work with data within a common Visual Programming Environment has been to define basic building blocks of a language. We started with an exploratory phase that led to the release of an experimental

---

[3] LabVIEW, February 7, 2014: http://www.ni.com/labview.

environment called Impure and its large sets (500) of diverse pre-programmed modules. This free solution generated a decent community of valorous 5,000 users. We used Impure as testbed for our ideas and performed the necessary user studies to come up with a coherent basic language. We particularly focused on specific action verbs that enclose the most common operations on data (e.g., *sort, search, insert, merge, count, compare, replace, save*). These actions are performed on Data Structures (e.g., *create List, sort Table, replace String, cluster Network, compare Date, resize Rectangle, load Image*) within specific domains (e.g., Math, Geography, Statistics). The language is complemented with a set of visualizers categorized according to their objectives to reveal aspects about the data (e.g., compare, contextualize, relate). Through this axiom of *actions—structure—domain,* user can find the appropriate module within a diverse toolset. However, that language did not eliminate all the barriers to manipulate data. For example, we found out that users needed some conceptual knowledge of data structures prior to sketching their own solutions.

10.3.4 Taking Advantage of an Ecosystem of Great Tools

Quadrigram was not meant to replace an existing tool or practice but rather to complement other more sophisticated tools of data analysis in order to embellish the last stretch of the run. Indeed, *big data* have been often understood in a vertical manner—terabytes of information. In Quadrigram, we approach *big data* in a horizontal manner—through the multi-diversity of datasets. Like many other platforms, Quadrigram connects to various types of data sources (e.g., databases, APIs, files) to load data within a workspace. But, we also wanted users with detailed needs to take advantage of R scripting[4] to perform advanced statistical method or Gephi[5] to layout large networks. The main challenge was to find and implement a protocol to communicate Quadrigram data structure back and forth with these great tools. In other words, we wanted users to perform analysis in R as part of their data flow. Similar to the architecture of distributed systems, the solution was to pass around serialized Quadrigram data structures, which offers a pretty unique mechanism to store and share results of data manipulations that we call "memories." Consequently, Quadrigram is a *sponge* capable of absorbing information from many diverse sources and permits its users to visualize it as part of the activity of sketching.

When sketching with data and with Quadrigram, we aim at bringing data closer to existing investigation and innovation processes. In the following section, we highlight qualities of this practice in our experiences of answering questions and designing new services with a wide diversity of digital footprints.

---

[4] The R Project for Statistical Computing, URL, February 7, 2014: http://www.r-project.org.

[5] Gephi, an open source graph visualization and manipulation software, February 7, 2014: https://gephi.org.

10.4 The Virtues of Sketching with Data

Originated within the posture of the investigators, our work has gradually evolved into helping institutions and companies in transforming digital footprints into insights, products or services. This practice requires the basic skills of *data science* (i.e., data analysis, information architecture, software engineering, and creativity) with a capacity to play the interface with wide variety of professionals from physicists and engineers to lawyers, strategists, and designers. This transversal incline of the investigations and innovation with digital footprints requires the knowledge of the different languages that shape technologies, report on the geography of their use, and describe people practices. The models of inquiries blend the qualitative field evidences with quantitative observations from logs with the use of sketches to share a common language.

10.4.1 Share a Common Language

As a first practical example, we would like to describe the capacity of sketches to bring diverse practices with their own specialisms and specific understanding of an idea to be reframed in the context of a common project. Indeed, many of our projects required the joint understanding of space (e.g., a territory, its rules, cultures, history), of the networks that compose the space (both physical infrastructures and digital activities) and the human behaviors manifested in that space. In a project called *Footoscope*,[6] we employed these prisms to explore new ways to analyze and experience football (i.e., soccer) with the increasing presence of data in the game. Sports have always kept a tight relationship with data to measure performances. It has been particularly the case to improve athletes' capabilities with motion analysis or objectify team sports that are easily fragmented into single events (e.g., Sabermetics). With new means of producing statistics via video and sensor technologies, other sports have started the search for objective knowledge.

In the domain of football, companies such as Prozone and Opta Sports have led the innovation in data collection. In parallel, some academics have been exploring this new terrain to apply their statistics-led methodology (López Peña and Touchette 2012). Similarly, designers have also started to transform these new measures (often in real time) into sophisticated visualization to augment the spectator's experience. In *Footoscope*, we explored new possibilities of sports *datatainment*. Through multiple simple sketches, we described visually the morphology and tactics of a football team according to raw data on its passing game (e.g., passes between players, positions of the players when receiving the ball, playing time) transformed into indicators and visualizations (Fig. 10.3). We sketched the results with Quadrigram with amateurs to help us decipher data of teams they know or want to

---

[6] URL, February 7, 2014: http://www.footoscope.com/.

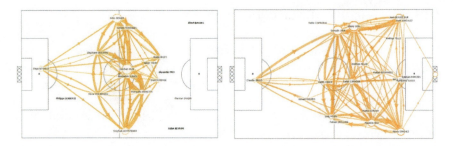

Fig. 10.3 *Footoscope*, Near Future Laboratory, 2010. *Footoscope* helped football amateurs decipher data of teams they know. Here, two sketches produced during the investigation of the World Cup 2010 with the incapacity for Switzerland (on the left) to manage the distances between its lines, with its defense and strikers compacted at a short distance. This contrasts with a more balanced team that takes a greater advantage of spaces, such as Chile (on the right)

explore. Sharing those attempts to understand the game differently allowed amateurs to extract and focus on key information that is otherwise hidden within football data. In a form of rapid visual business intelligence, this analysis and its visualization became the supporting elements of a co-creation process to extract value from data. For instance, in a team of amateurs, we deciphered the World Cup 2010 statistics to reveal the network, spatial and behavioral elements of each team. That anecdotic work generated dialogues between statistics and people with knowledge of the terrain to produce a new apprehension of the game. When shared online, the sketches provided a common language for different practitioners to grasp the potential of data in football.

10.4.2 Qualify the Results

In this second practical example on the virtues of sketching with data, we showcase our project at the Louvre Museum. Not only because we have kept fantastic memories of the breathtaking context, but also because we learned a lot from the analysis of digital footprints to provoke qualitative knowledge. The Louvre is, by far, the most visited museum in the world with 8.5 million visitors and more than 40,000 visitors at peak days. In Paris, it is one of the main drivers of "cultural enthusiasm" that is an inherent feature of the city. In consequence, the museum witnesses levels of congestion, which, beyond a certain threshold, can be described as *hyper-congestion*. This phenomenon has some direct negative consequences on the quality of the visitor experience as well as on the organization and management of the Museum (e.g., increased stress level of the surveillance staff).

The Study, Evaluation, and Foresight Department of the Museum performs extensive surveys, audience analysis, and on-site observations to ensure a good visiting experience. However, the information they collect is punctual and only partially feeds the visitors flow models necessary to setup and evaluate some of the museum strategies. In an exploratory project, they wanted to investigate new solutions to help answer their concerns with *hyper-congestion*. In response, we first investigated the collection of new empirical data on the flows and occupancy levels of visitors in key areas of the Louvre and the developed diagnosis indicators that capture the changes of visitor behaviors according to the congestion in the museum.

In collaboration with a real-time traffic information provider, we specifically designed sensors that audited during short periods of 2 weeks the presence of Bluetooth-enabled mobile phones on a key trail that leads to the Venus de Milo. The analysis of the collected longitudinal measures of presence and flows of visitors quickly led to the development of an indicator that unveiled areas in which the congestion of a room changes the visitor's presence times and flows. While unprecedented in the history of the Louvre, some results produced more questions than answers. We faced a new set of inquiries that quantitative evidences from sensors could not answer but field observations could. For instance, what event provoked the congestion, what aspects of the visiting experience were affected, or why some rooms do not show symptoms of *hyper-congestion*?

In response to these interrogations, we returned to the sketches produced as part of our data analysis. Yet this time, we did not complete them for the decision makers but for the security staff. Indeed, on-side personnel offered untapped knowledge on visitor practices and flow management strategies. So we setup different meetings at the museum and used our sketches to have the staff qualify the results of the audits. Their evidences from the field explained some irregularities and completed the understanding of visitor behaviors. For instance, a simple decision to close a door provoked changes in the measures of visitor flows.

In that experience, we learned the types of questions the analysis of digital footprints can answer. For instance, "how many observations can we produce?" "what do the data tell about a population?" "what evolutions can we measure over time?" "can we categorize these evolutions?" "what are the trends and the outliers?" or "what are the flows that connect different places?" (Yoshimura et al. 2012). Yet, the understanding of an environment such as the Louvre goes beyond logging machine states and events. This project showed that the qualitative view from the staff reinforced the quantitative measures and consolidated the overall knowledge on *hyper-congestion*.

10.4.3 Innovate with Data

This final practical example highlights the virtues of sketching when conceptualizing and building services based on digital footprints. With data, there is always the risk that teams jump to technical solutions before evaluating whether solutions will work. Our approach in that domain often focuses on building the simplest possible thing. Indeed, once we can prove something is working and we can prove that users want it, the next step is to improve the service. So, we start considering if each task can be divided into small actions that end-users or experts can perform. That way the design and development of a data product or service starts with something simple that lets a team determine whether there is an interest to go further. Sometimes, the idea behind these sketches will survive into the finished version because they represent some fundamentally good concepts that the team might not have seen otherwise; sometimes, they will be replaced by a different approach or technique.

We applied this approach in several data product projects including steering the development of mobile recommender systems or interactive dashboards of commercial activities for municipalities and retailers. Following the spirit of sketching, the results can be produced quickly (e.g., in a few days, if not a few hours), they are never complete or precise but they are vague and good enough to let you know whether it is worth going further. This approach has some similarity with prototyping that embraces the notion of the minimum viable product and the simplest thing that could possibly work. However, sketching with data does not aim for fidelity nor desirability or to prove a concept. Rather, we use them to form and deform ideas and concepts within a multidisciplinary team. They can be employed in all stages of the design process, for instance to check the sanity of the data or evaluate some aspects of the user experience.

10.4.3.1 Checking the Sanity of the Data

One of the biggest challenges of working with data is getting the data in a useful form. As teams want to jump to trying to reach a common language and design the product, this critical task of cleaning the data is often overlooked. In fact, in the data science community, there is a rule that says that 80 % of the work in any data project is cleaning the data. For instance, the geocoding process of transforming database of postal addresses into geographic coordinates might lead to erroneous information due to interpolation or ambiguous street names. Those common issues can be evaluated with mapping a sample of the dataset and verifying visually the accuracy ratings of each address (Fig. 10.4). This type of sketch is meant to quickly check the sanity of the raw data and whether the errors or imprecisions can be assumed over the course of the project.

10.4.3.2 Defining a Clear Focus

The practice of building products with digital footprints and its integration of user experience design processes is still relatively young. In consequence, many solutions have a tendency to engage users with the many, often irrelevant, insights that can emerge from data analysis and visualization. For instance, a wide variety of indicators can be developed when designing a dashboard for municipalities to measure the commercial impacts of events. Indicators of data scientists might fail to understand because local authorities and city planners simply do not grasp or communicate their potential. The quick production of interactive sketches with real data is often a good approach to evaluate the potential of different indicators with

Fig. 10.4 Sketch of the result of transforming database of thousands of postal addresses in Madrid into geographic coordinates with the potentially erroneous results highlighted in *red*

their views (e.g., distances to purchase of local citizens, commercial routes among the main areas of commercial influence, balance of trade with neighboring cities). These early results are material to collect feedback and iteratively engage users in considering the evolution of their work using new metrics and indicators. Consequently, each insight is designed according to the specific action the user can take leading to a product with a clear focus.

10.5 Conclusions

The growing quantity of digital footprints gives us new means to model human dynamics and develop new services. However, evolution presents both societal and methodological implications that require the involvement of many stakeholders often from different practices and objectives from engineering to statistics, design, strategy planning, product management, and law. Throughout our projects, we found the necessity to very quickly being able to visualize temporary results and share them among different specialisms. As it became prominent in our creative and innovation process with digital footprints, we called that practice sketching with data. We participated in the development of a tool that treats data as an agile material. In Quadrigram, a sketch is an opportunity to find new questions and provide answers with data. A sketch becomes *boundary objects* for inspiration and dialogue in multidisciplinary teams.

Our experience shows that digital footprints are not sufficient to give full answers and solutions about people, their behaviors, and usage of technology. Yet the world of "data science" and computer science still lacks in sensitiveness to the limitations of quantitative evidences and the models we can build on them. We have often been confronted to these limitations. Several of our projects with digital footprints taught us that there are insights that only the articulation of digital footprints and contextual observations can provide.

On a more general picture, sketching with data is an approach of the investigation and design with digital footprints, which is close to academic research. Sketching implies taking time to think, open, and contrast ideas, and staying humble, not being afraid of expressing doubts. When it comes to mixing practices, this practice of the researcher driven by doubt but confident in its methods is what we believe drives to relevant insights and solutions.

References

Biddulph M (2012) Prototyping apps using real data. Presentation at the EAN's World of Data event in London. Available online: http://www.slideshare.net/mattb/eans-world-of-data-prototyping-apps-using-real-data, 7 Feb 2014

Girardin F (2009) Aspects of implicit and explicit human interactions with ubiquitous geographic information. Ph.D. thesis, Universitat Pompeu Fabra

Girardin F, Vaccari A, Gerber A et al (2009) Quantifying urban attractiveness from the distribution and density of digital footprints. Int J Spat Data Infrastruct Res 4:175–200

Kuniavsky M (2010) Smart things: ubiquitous computing user experience design. Morgan Kaufmann, Burlington

López Peña J, Touchette H (2012) A network theory analysis of football strategies. In: Clanet C (ed) Sports physics: proceedings of 2012 Euromech physics of sports conference. Éditions de L'École Polytechnique, Palaiseau, pp 517–528

Schank-Smith K (2005) Architect's drawings: a selection of sketches by worlds famous architects through history. Routledge, London

Star SL, Griesemer J (1989) Institutional ecology, translations and boundary objects: amateurs and professionals in Berkeley's Museum of Vertebrate Zoology, 1907-39. Soc Stud Sci 19(3):387–420

Yoshimura Y, Girardin F, Carrascal JP, Ratti C, Blat J (2012) New tools for studying visitor behaviours in museums: a case study at the Louvre. In: Proceedings of the international conference in Helsingborg (ENTER 2012)

Zook M, Dodge M, Aoyama Y, Townsend A (2004) New digital geographies: information, communication, and place. Geogr Technol 155–176

Author Biography

Fabien Girardin (Data Visualizer, Software Engineer and Researcher) is a co-founder of the Near Future Laboratory a thinking, making, design, development and research organization speculating on the near future possibilities for digital worlds. He is active in the domains of user experience and data science.

In his work he mixes qualitative observations with quantitative data analysis to gain insights from the integration and appropriation of technologies. Subsequently, he exploits the gained knowledge with engineering techniques to prototype and evaluates concepts and solutions for service providers and decision makers.

In response to the increasing demand of clients to think and sketch freely with data, he co-created the visual programming environment Quadrigram.

Fabien holds a Ph.D. degree in Computer Science and Digital Communications from Universitat Pompeu Fabra in Barcelona, Spain. Along his academic journey, he was also affiliated with the Senseable City Lab at the Massachusetts Institute of Technology (MIT) and worked at the Swiss Federal Institute of Technologies Lausanne (EPFL).

For more information and contact:

http://nearfuturelaboratory.com
http://nearfuturelaboratory.com/seventh-and-half
http://www.girardin.org/fabien
http://www.linkedin.com/in/fabiengirardin
https://twitter.com/fabiengirardin

Chapter 11
Information Visualizations and Interfaces in the Humanities

Giorgio Uboldi and Giorgio Caviglia

Abstract For the last few years, computers and the Internet have been changing the way research is conceived, conducted, and communicated, transforming scholarly publication and collaboration, and supporting the creation, the storage, the analysis, and the dissemination of data and information. While natural, medical, and social sciences have a long and established tradition with these technologies, most of the humanities disciplines have found it difficult if not impossible to integrate computational tools, based mostly on quantitative approaches, with their research methods. In the last 20 years however, new research areas and activities have emerged from the intersection between humanities and computing. Today, what is known as *digital humanities* represents a heterogeneous set of studies and practices that aims at understanding the implications and the opportunities that digital technologies can provide as media, tools, or objects of study in the humanities (Schreibman et al. in A companion to digital humanities. Blackwell, Oxford, 2004; Gold in Debates in the digital humanities. University of Minnesota Press, Minneapolis, 2012; Berry in Understanding digital humanities. Palgrave Macmillan, New York, 2012). These new relationships between the digital and the humanities are rapidly demanding for new modes of observation and interpretation. Information visualizations and interfaces appear as essential tools to explore and make sense out of big and heterogeneous amounts of data (Manovich 2013). But, in a context where most of the methods and the technologies are still adopted from other disciplines, the biggest challenge seems to be imagining new genuine research tools capable of embedding and valorizing the humanities endeavor (Drucker in Culture Machine 12:1–20, 2011). The work presented here aims at deepening the relationships between designers, humanities scholars, and computer scientists through the outlining of new research tools and processes based on

---

G. Uboldi (✉)
Politecnico di Milano, Piazza Leonardo Da Vinci, 32, 20133 Milan, Italy
e-mail: giorgio.uboldi@mail.polimi.it

G. Caviglia
Stanford University, 450 Serra Mall, Stanford, CA 94305, USA
e-mail: caviglia@stanford.edu

© Springer-Verlag London 2015
D. Bihanic (ed.), *New Challenges for Data Design*,
DOI 10.1007/978-1-4471-6596-5_11

humanistic data and digital environments. Furthermore, it explores the possibilities and challenges set forth by information and data visualizations as tools to support scholarly activities.

11.1 The Digital and the Humanities

Digital technologies have radically changed the way scholars work, collaborate, and publish their research by supporting the creation, the storage, the analysis, and the dissemination of data and information. While many areas of study within the natural, medical, and social sciences have a long tradition with these technologies, most of the humanities disciplines have found it difficult or inappropriate to integrate computational tools and methods, usually conceived to perform quantitative analysis, in their studies. However, in recent years, new research activities and opportunities have emerged from the intersection of humanities and digital disciplines. Today, terms such as "digital humanities" or "cultural analytics" describe an heterogeneous set of studies and practices that aims at understanding the implications and the opportunities that digital technologies can provide as media, tools, or objects of study in the humanities (Schreibman et al. 2004; Gold 2012; Berry 2012; Hayles 2012).

The enormous quantity of information that emerges from digital collections and, in a more general sense, forms the digital world offers different opportunities to rethink traditional research activities and tasks in the humanities (Moretti 2005; Manovich 2007). These new objects of study are rapidly demanding new modes of observation, exploration, and interpretation: the biggest challenge in regard to re-shaping the relation between the digital and the humanities is to imagine and build new research tools able to embrace and emphasize the endeavor of humanities (Drucker 2009). In this perspective, information visualizations and interfaces have increasingly emerged as essential tools to explore and make sense out of the growing quantity and variety of available data. The design of visualizations and interfaces to access, observe, and explore these new cultural data emerges as a central issue in current research initiatives and agendas. Furthermore, humanities provide an almost-perfect context for design and design thinking (especially, communication and information design) to apply and study their situated, interpretative, and user-centered approaches.

11.2 Designing (in) the Digital Humanities

Like, and perhaps, more, than other disciplinary contexts, the digital humanities are increasingly looking at design as an indispensable set of practices and knowledge to be integrated in their activities (Balsamo 2009; Burdick and Willis 2011; Lunenfeld et al. 2012). The question about the design of digital tools to support humanities inquiry is still central in current research activities and agendas (McCarty 2003; Drucker 2011). A look at the "Digital Humanities 2.0 Manifesto" by Schnapp and Presner (2009) suggests that design can be considered not only as a set of necessary notions and expertize but a possible way to (re)define the field itself, focusing on the "attention to complexity, medium specificity, historical context, analytical depth, critique, and interpretation."

According to Anne Burdick, the perspectives for design in the digital humanities appear, at the same time, challenging and dramatic. Challenging because design thinking, as situated, interpretative, and user-oriented, is well suited to the kind of initiatives carried out in the field (Burdick and Willis 2011). Dramatic because while design is already practiced and sometimes even "theorized," in English, Literature, or History departments, it is hardly recognized as such, and "designers are nowhere to be found" (Burdick 2009). As chair of a graduate media design program and as a designer in computational linguistics, literary science, electronic literature, and media theory, and criticism, Burdick notes that design "is not the discipline that we know and love—that is, it's not the province of design practitioners, researchers, and educators. Instead, 'design' is variably a value-add, an everyday event, a working method, a byproduct, a literacy, and a complete abstraction" (Burdick and Willis 2011). Within the digital humanities, most projects are still designed by scholars, research assistants, or IT staff. The lack of designers and, especially, design researchers leads often to an appropriation of the theoretical aspects of the discipline by humanities scholars.

This can be partially explained by the fact that one of the most interesting and, perhaps, most significant changes that digital technologies have brought into humanities research is a reconsideration of the relation between practice and theory, making and thinking. Some digital humanities scholars (Schnapp and Presner 2009; Ramsay 2011a; Lunenfeld et al. 2012) consider theory and knowledge production as deeply and inevitably anchored in the action of making, building, modeling, and creating. Ramsay (Ramsay 2011a) argues that "there's always been a profound—and profoundly exciting and enabling—commonality to everyone who finds their way to digital humanities. And that commonality [...] involves moving from reading and critiquing to building and making." According to Ramsay, the capacity to build and make—in other words, to design—becomes not only an important skill that can facilitate the work of a scholar, but it represents a distinctive element of a digital humanist, something so peculiar that it can be used—in an extreme way—to establish "who is in and who is out" the field (Ramsay 2011b). The centrality of the project in the research process appears as a fundamental aspect in the theorization of digital humanities and, at the same time, represents a strong element of connection with design and design research, for so long focused on the epistemological implications of practice (Schon 1983; Buchanan 2001; Friedman 2003).

11.3 The Nature of Data

The technologies and tools conceived in recent years to collect, encode, and store information in digital formats, have made it possible for humanities scholars to access, explore and analyze an incredible amount of data coming from digitized analog artifacts as well as born-digital artifacts. However, the concept of data has a short history when compared with the sciences and their long tradition of quantitative and qualitative approaches. Trying to grasp what data are in the humanities, —how it is captured, and how it is curated and manipulated—are important steps in the direction of designing visualizations tools capable to embed and express the nature of humanistic inquiry. As Borgman (2007) suggests, when looking for an official definition of data, the one associated with archival information systems can work as a useful starting point: "a reinterpretable representation of information in a formalized manner suitable for communication, interpretation, or processing. Examples of data include a sequence of bits, a table of numbers, the characters on a page, the recording of sounds made by a person speaking, or a moon rock specimen" (Reference Model for an Open Archival Information System, 2002). In the realm of the humanities, however, it is challenging to find a common point of view on how to think about data.

As an attempt to better clarify the different perspectives on data in the humanities, it could be useful to look at the relationship that some disciplines, especially arts and literature studies, have with their data sources and compare them with data sources in other disciplines. For scientists, the creation and collection of data are usually conceived for specific scientific purposes. Every step of the process, in most cases, follows formal research protocols and research questions. Although data sources in the field of science may be very heterogeneous depending on the discipline, the control over the data is crucial and plays a key role in research activities.

On a continuum of data source and control, social sciences occupy the middle ground between sciences and humanities: "those at the scientific end of the scale gather their own observations, whether opinion polls, surveys, interviews, or field studies; build models of human behavior; and conduct experiments in the laboratory or field" (Borgman 2009). Other social scientists rely on born-digital data coming from the Internet, not only to study online culture, but in a more general sense, as a unique data source to analyze society and culture (Rogers 2009).

The situation in the arts and traditional humanities is quite different. In most cases, they do not create their own data but rather rely on records, whether newspapers, photographs, letters, diaries, books, articles; records of birth, death, marriage; records found in churches, courts, schools, and colleges; or maps. Basically, "any record of human experience can be a data source to a humanities scholar" (Borgman 2009). That is why data in the humanities are slightly peculiar: Firstly, it deals with analog, non-discrete data that need to be translated in digital formats to be used and processed; and secondly, the data at hand are usually part of semiotic systems that transgress physical measures. Data extracted from language, texts, paintings, and music have dimensions that depend on semantics and pragmatics and find meaning in context (Schöch 2013). As Lynch (2002) explains, it is important to distinguish between raw materials and interpretations in the digital humanities even if the distinction is not easily perceptible: "it's hard to completely isolate interpretation from raw materials; interpretation creeps in everywhere, for example in descriptive metadata that is part of the digital collection" (Lynch 2002).

It is for these reasons that speaking of data within the field of humanities is problematic and has proven to be controversial. Disapproval has come also from conventional scholars who approach data and quantitative methods of analyzing them warily. The obviously empirical data-driven research in the humanities appears to be at odds with principles of humanistic research, such as context-dependent interpretation and the aims of the researcher.

Another important aspect about data and data sources in the humanities is its implicit uncertain, incomplete, and often ambiguous nature. Despite of huge efforts of digitization by private and public actors, we are in fact still far from the ultimate goal of creating a complete record of human culture and creativity. In addition to this, we cannot forget that during the centuries, many sources have been irremediably lost and many others are located in different places and collections. The results are incomplete datasets, incoherent metadata categories, uncertain provenance of records, and all the implications of separately collected and sourced. These are huge drawbacks but are also conventional questions of source interpretation that need to be adjusted rather than abandoned to empower new techniques of reading.

The vision about what data are in the digital humanities is so complex that some scholars, particularly Johanna Drucker, have argued the adequacy of the term "data" itself. According to Drucker, the etymology of the term, which comes from the Latin "that which is given," is in fact quite problematic in the humanities. Because it is difficult to think about a fact or information completely independent from its observer, Drucker (2011) prefers to use the term "capta" literally "that which has been captured." The underlying assumption is the idea that capturing data is an act of constructing and it is the result of a series of specific decisions taken by the observer.

Similarly, Owens (2011) argues that data are not a given, but it is always manufactured and created. Furthermore, Owens goes on to explain that data are approachable from different perspectives: It can be regarded as an artifact intentionally created by people, as a text that is subject to interpretation, and as computer-processable information that is to be analyzed through quantitative methods. This means that data are not set in stone. It is "a multifaceted object which can be mobilized as evidence in support of an argument" (Owens 2011).

This idea of data suggests closer attention to the process and technologies used to obtain and store information. In this sense, the idea of materiality, often considered foreign to the realm of the digital, assumes new meanings and relevance. Materiality is an important aspect that contributes to influence the choice behind the generation, the study, and the interpretation of the data. No procedure or technology is transparent: Each of them forces to take decisions, according to the specific needs of the curator as well as the technological peculiarities of each step within the workflow. The digitization of data is not merely a passive technical process but it requires active decision making on how and what to digitize but also on what not to digitize.

11.4 Visualizations and Interfaces

In the last years, new interpretative models, capable to overcome the limits of standard and consolidated techniques to study human culture (e.g., close reading, deconstructionism), have been required and explored by humanities scholars, especially from History and Literary studies (Pope 1995; McGann and Samuels 2004; Moretti 2005). While digital technologies are not always directly involved in the definition of these new forms of "reading" and "writing," the possibilities that they provide to work with data and information have been immediately exploited to apply and test new theoretical and methodological research practices. Information visualizations and dynamic interfaces, especially in the form of web applications, have been rapidly incorporated within research projects in the digital humanities (Jessop 2008).

The need to make sense of data from digital collections has urged scholars to adopt tools that were actually developed and designed for other disciplines such as natural and social sciences (e.g., Gephi, NodeXL, ManyEyes, Excel, Tableau and many others). These tools offer new perspectives on the study of large volumes of data now available to humanities research. They have been adopted in the early stages of the digital humanities, especially in the field of literary studies, characterized by an abundance of material resulting from the digitization of books and volumes. Many digital humanities projects have taken advantage of visualizations to highlight underlying patterns, trends, and structures out of large datasets. Abstract models, drawn from disciplines external to literary studies—such as graphs, maps, and charts,—have been developed to observe literature and history from different points of view. In opposition to the well-established practice of close reading, which places great emphasis on the single particular over the general, the concept of distant reading, introduced by Moretti (2005), proposes distance as a "condition of knowledge" that favors reduction and abstraction in order to make sense of more general phenomena. For instance, a look at the number of published novels over time may shed light on the evolution of literary genres in relation to political and historical occurrences and in different circumstances. By utilizing quantitative methods and information visualization, Moretti moves toward the analysis and the renewal of the "very nature" of literary history, how it was devised and managed, changing it into "a patchwork of other people's research, without a single direct textual reading" (Moretti 2005).

Another example of this kind of approach to visualization is the idea of cultural analytics proposed by Manovich (2007). Stemming from the idea of culture as data that can be mined and visualized, Manovich proposes to apply the same methods and approaches adopted by scientists, businesses, and government agencies to

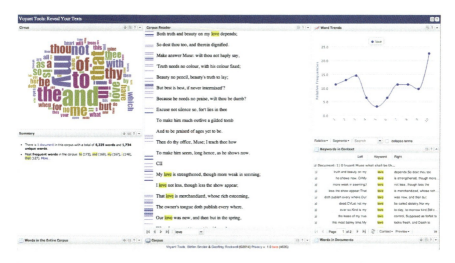

Fig. 11.1 *Voyant Tools*, a dashboard interface by Stéfan Sinclair and Geoffrey Rockwell that allows to perform lexical analysis through several algorithms, metrics, and visualizations, 2014

generate knowledge and understand culture. Therefore, the role played by visualization, especially that of Visual Analytics, becomes central "to really make sense of rapidly changing global cultural patterns and trends today" (Manovich 2007).

By looking at the digital tools currently used in support of humanities research, many of them can be described as dashboards. By dashboard, we mean here those interfaces usually constituted by a set of juxtaposed modules, individually focused on a specific aspect of a predetermined analysis carried out on a data set or a text. This model comes from disciplinary contexts outside the humanities, such as statistics, business management, and economics, where quantitative analysis is one of the predominant investigative tools. These interfaces make use of standard visual representations (e.g., tag clouds, charts, bars, scatterplots) and are frequently adopted in literary studies to perform text analysis or other kinds of text and data mining. The underlying idea behind a dashboard is to provide the user with a tool capable to perform analysis and show the results, employing different views at once. One of the most interesting (and popular) projects in this sense is *Voyant Tools*[1] (Fig. 11.1). As part of a collaborative project aimed at developing and theorizing text analysis tools and text analysis rhetoric, *Voyant* allows to perform lexical analysis (e.g., study of frequency or distribution data) through several algorithms, metrics, and visualizations. The available visualizations, based both on quantitative and qualitative data, come mostly from the tradition of visual analytics, and the user experience is shaped using standard design patterns (e.g., linking and brushing techniques). Each panel/module offers also some basic parameters to be changed by the user (e.g., using a stop words list, the number of segments to be

---

[1] URL, March 23, 2014: http://voyant-tools.org.

Fig. 11.2 *Kindred Britain*, Nicholas Jenkins, Elijah Meeks and Scott Murray, published by Stanford University Libraries, 2013

visualized). *Voyant*, like many other text analysis tools, is a perfect example of a tool designed for—and used by—humanities, which has imported logics and methods from other disciplines.

Somehow in contraposition to the concept of "black box" provided by dashboards, where most of the analysis happens behind the screen and the users can only wait for the result, we find browsers. Often used with entire digital collections, browsers[2] allow users to observe and explore the data, in terms of general (macro) overviews as well as specific (micro) focuses on single items or metadata. Browsers usually present isomorphic representations of the data collection, where each item has a correspondent symbol or image in the browser. Sorting and filtering operations can be performed in order to focus the attention on particular subsets. The design of these tools is often based on the concepts of transparency and flexibility: The interface is conceived in order to "disappear," leaving the user free to explore the data and minimizing the interference.

Kindred Britain[3] (Fig. 11.2), a project developed at Stanford University, represents an interesting example of this kind of interface. The tool, based on a database of nearly 30,000 individual records, allows the users to visually explore Britain as a large, extended family, in which any individual can be connected to any other through relationships. One of the explicit goals of the project is to create an interface that is both explanatory, highlighting some of the interesting stories discovered, and exploratory, encouraging the users to explore the data through the use of the network, the timeline, and the map. Focused on the exploration of the data,

---

[2] See Mandala Browser, URL, March 23, 2014: http://mandala.humviz.org.
[3] URL, March 23, 2014: http://kindred.stanford.edu/.

the interface elements and the visualizations are designed in a way to allow intuitive interactions (like the drag and drop of a node over another to explore the path between them) and clear visual insights, emphasizing the idea of "connectedness" offered by the database.

11.5 Bringing Design Practices in the Humanities

While many projects in the digital humanities have been oriented toward the adaptation of existing methodologies, others have developed a different approach to data visualization. In fact, the will to conceive genuine ways to explore the possibilities given by digital technologies has urged scholars to reconsider the use of certain approaches in order to preserve the interpretative nature of humanistic modes of inquiry. Visualization, rather than a tool to answer a priori-defined research questions, can become the starting point for open-ended explorations that can continue also outside the tool itself, by further readings and/or reinterpretations of the sources.

Within this perspective, visualizations, and interfaces are not conceived as tools, but rather, as steps of an understanding and "transformation process" (Masud et al. 2010) involving new ways of thinking and making with and through digital technologies. The interaction with the interface and the visual representation of the data can suggest new research questions, confirm or infirm methodological claims or reveal something previously unnoticed in the data. Such an approach can be found in the speculative tools developed by Drucker and Nowviskie (Nowviskie 2004; Drucker 2009). Here, interfaces are seen as environments where to perform subjective readings of texts (or other cultural objects). Embracing the idea that the best way to understand a work is actually to intervene on it (McGann and Samuels 2004), modifying it and "playing" with it, this approach aims at building digital tools capable to host such a game (Ivanhoe,[4] Temporal Modeling[5]). Goal of the game is to construct an interpretation of the text by modifying it or by adding comments, annotations, or relationships between its elements. Similar to a role-playing game, a set of rules (more or less implicit) usually define what and how the players can intervene on the text.

From a design perspective, what seems very interesting in this idea of "interpretative game" is the affinity with a design activity itself. In this sense also, the interaction in terms of representation and modeling the material makes us think about the interface as a design tool.

In this sense, more than in building the tools, design participates as the heuristic practice that implies the interpretation of the data. This perspective introduces the idea of a different kind of relationship between the humanities and design, where

---

[4] URL, March 23, 2014: http://www2.iath.virginia.edu/jjm2f/old/IGamehtm.html.
[5] Temporal Modeling. URL, March 23, 2014: http://www2.iath.virginia.edu/time/time.html.

designers can bring into humanities research their experience and sensitiveness in defining the research process itself, as practice-based, situated, subjective, and performative.

11.6 Conclusions

As we have seen, design can play a key role in defining new research and epistemological processes within the digital humanities. If on one hand, the role of designer seems to be associated, and limited to, the technical aspects of visualization and development of the interfaces, on the other hand, it appears clear that the nature of the humanities inquiry provides an almost-perfect context for design, especially communication and interface design.

In this sense, design not only represents a suitable and important partner in the definition of digital humanities projects, but the humanities inquiries themselves can be conceived as design processes.

The introduction of digital technologies has driven the attention toward the relationship between practice and theory as a significant aspect of the modern research activity of humanities scholars.

The centrality of the project and the development of tools designed to make sense of the abundance of digital data transformed the scholars in professional practitioners focused on the making as an indispensable condition for theory production.

Encompassing the long tradition in design research and design theory that recognizes the activity of design as a form of knowing-in-action and designers as reflective practitioners (Schon 1983), we argue that the modern form of collaboration between scholars and designers should be reconsidered toward new models of interdisciplinary cooperation where the ability of making and modeling of both designers and developers is combined with the interpretative work of scholars for the generation of new and hybridized ways of thinking. In such a perspective, the core of the cooperation is based on a continuous interaction between scholars, communication designers, and developers in order to establish new form of modeling and interpretative activities.

In the professional world, designers have been taking on more strategic roles in interdisciplinary fields. Communication designers have developed to be key players between the information at hand and the research to the communication of the message across different disciplines, be it humanities or any other subject. Furthermore, due to the fact that communication designers have experience with digital tools and interfaces, they are vital in developing strong relationships with programmers and computer scientists.

References

Balsamo A (2009) Design. Int J Learn Media 1(4):1–10
Berry DM (2012) Understanding digital humanities. Palgrave Macmillan, New York
Borgman CL (2007) Scholarship in the digital age: information, infrastructure, and the internet. MIT Press, Cambridge
Borgman CL (2009) The digital future is now: a call to action for the humanities. Digital Humanit Q 3(4)
Buchanan R (2001) Design research and the new learning. Des Issues 17(4):3–23
Burdick A (2009) Design without designers. Keynote for a conference on the future of art and design education in the 21st century. University of Brighton, England
Burdick A, Willis H (2011) Digital learning, digital scholarship and design thinking. Des Stud 32(6):546–556
Drucker J (2009) SpecLab—digital aesthetics and projects in speculative computing. University of Chicago Press, Chicago
Drucker J (2011) Humanities approach to interface theory. Cult Mach 12:1–20
Friedman K (2003) Theory construction in design research: criteria, approaches, and methods. Des Stud 24(6):16
Fuller M (2008) Software studies. MIT Press, Cambridge, pp 334
Gold MK (ed) (2012) Debates in the digital humanities. University of Minnesota Press, Minneapolis
Hayles KN (2012) How we think: digital media and contemporary technogenesis. University of Chicago Press, Chicago
Jessop M (2008) Digital visualization as a scholarly activity. Literary Linguist Comput 23(3):281–293
Lunenfeld P, Burdick A, Drucker J et al (2012) Digital_Humanities. MIT Press, Cambridge
Lynch C (2002) Digital collections, digital libraries, and the digitization of cultural heritage information. 1st Monday 7, p 8–9
Manovich L (2007) Cultural analytics: analysis and visualization of large cultural data sets. URL: http://www.manovich.net/cultural_analytics.pdf, 23 Mar 2014
Manovich L (2013) Software takes command (International texts in critical media aesthetics). Bloomsbury Academic, INT edition
Masud L, Valsecchi F, Ciuccarelli P et al (2010) From data to knowledge. Visualizations as transformation processes within the data-information-knowledge continuum. In: Proceedings of IV'2010, pp 445–449
McCarty W (2003) Encyclopedia of library and information science, 2nd edn. Dekker, New York, pp 1224–1235
McGann J, Samuels L (2004) Deformance and interpretation. Radiant textuality. Palgrave Macmillan, New York
Moretti F (2005) Graphs, maps, trees. Verso Books, New York
Nowviskie B (2004) Speculative computing: instruments for interpretative scholarship. Ph.D. thesis, University of Virginia, Virginia
Owens T (2011) Defining data for humanists: text, artifact, information or evidence? J Digital Humanit 1(1). URL: http://journalofdigitalhumanities.org/1-1/definingdata-for-humanists-by-trevor-owens, 23 Mar 2014
Pope R (1995) Textual intervention: critical and creative strategies for literary studies. Routledge, London
Ramsay S (2011a) On building. URL: http://stephenramsay.us/text/2011/01/11/on-building.html, 23 Mar 2014
Ramsay S (2011b) Who's in and who's out. URL: http://stephenramsay.us/text/2011/01/08/whos-in-and-whos-out.html, 23 Mar 2014
Rogers R (2009) The end of the virtual: digital methods. Amsterdam University Press, Amsterdam
Ruecker S, Radzikowska M, Sinclair S (2011) Visual interface design for digital cultural heritage. Ashgate Publishing, Hardcover
Schon AD (1983) The reflective practitioner: how professionals think in action, 1st edn. Basic Books, New York
Schnapp J, Presner T (2009) The digital humanities manifesto 2.0. Humanities Blast. URL: http://www.humanitiesblast.com/manifesto/Manifesto_V2.pdf, 23 Mar 2014

Schöch C (2013) Big? Smart? Clean? Messy? Data in the humanities. J Digital Humanit 2(3)
Schreibman S, Siemens RG, Unsworth J (2004) A companion to digital humanities. Blackwell, Oxford

Author Biographies

Giorgio Uboldi is a Designer and Research assistant at DensityDesign Research Lab (Politecnico di Milano) where he handles different design projects and carries out research in the field of information visualization and communication design. During his Master studies he developed a specific interest for Information Design and Digital Humanities. Giorgio obtained his Master degree in Communication Design at Politecnico di Milano in 2013. For his thesis project, he worked on an interface for the study of social networks in the Humanities in collaboration with DensityDesign and Stanford University.

In the last years he has also been involved in different research and teaching activities at Politecnico di Milano and multiple freelance collaborations.

For more information and contact:

http://www.giorgiouboldi.com
https://twitter.com/giorgiouboldi

Giorgio Caviglia is a Designer and Researcher working at the intersection of Information Design and Digital Humanities. His work investigates new research practices for the study of cultural and historical data, bringing together scholars, designers and developers in the definition of digital and visual methods. From 2009 to 2013 he has been part of the DensityDesign Research Lab (Politecnico di Milano) as senior designer and researcher, working on visualizations and data-driven tools for the Humanities and Social Sciences. After receiving his Ph.D. degree in Design at the Politecnico di Milano in 2013, he is currently a Postdoctoral Scholar at Stanford University, within the Humanities + Design lab at the Center for Spatial and Textual Analysis (CESTA).

He has been involved in research and teaching activities at Politecnico di Milano, Stanford University, ISIA Urbino. His work has been featured in numerous exhibitions (SIGGRAPH, MIT humanities + digital, MediaLAB Prado Visualizar, MMLA, EXPO2010 Shanghai, Triennale di Milano), publications and showcases (Visual Complexity, Malofiej, Data Flow, Design for Information, Corriere della Sera).

For more information and contact:

http://it.linkedin.com/in/giorgiocaviglia
https://twitter.com/giorgiocaviglia.

Part VI
Translating and Handling Large Datasets

Chapter 12
Big Data, Big Stories

Richard Vijgen

Abstract In the past 30 years, the world has become ever more computerized. Advancements in technology and the dropping costs of digital storage meant that corporations, governments, and academics alike started to collect and record information on an unprecedented scale. The rate at which new information is stored is growing to such an extent that today, 90 % of all the worlds' data have been created in the last 2 years (IBM Big Data, December 28, 2013: http://www-01.ibm.com/software/data/bigdata). Contained within these ever-expanding data structures are the stories of our time. As an increasing share of cultural exchange becomes "born digital," the question becomes how to represent all this information. While managing and reading Big Data requires technical skills, finding the narrative in the database is as much about digital culture as it is about technology. Can data design extract the big stories from Big Data? I will explore this question using a case study from my own practice: *The Deleted City*, A Digital Archeology. I will discuss how data technology invented and distributed by the software industry can, in the hands of designers and creative coders, amount to a data culture.

12.1 Big Data

Big Data is everywhere, an academic subject just a few years ago; it is now part of popular culture and debate. With the ever-dropping cost of storing digital information and the increasing power of databases like NoSQL to manage enormous amounts of data, collecting and archiving information has become easier than ever before. The possibility that something useful can be found in these ever-expanding streams of information has led many organizations to adopt data collection as a *de facto* strategy. At the forefront of these developments are companies like IBM,

R. Vijgen (✉)
Richard Vijgen Studio, Prinsessestraat 43,
6828, GT Arnhem, The Netherlands
e-mail: mail@richardvijgen.nl

evangelizing the use of Big Data management software such as Hadoop to allow clients to extract meaningful signals from previously discarded data noise. The sales pitch is often about efficiency, finding patterns, discovering niches, and targeting consumers with ever more precision. Businesses and governments alike find opportunities in analyzing these growing datasets that are increasingly available, compatible, and accessible.

As with any new technology, Big Data does not just bring increased economic efficiency and productivity; it changes the way we exchange and think about information. You can think about Big Data as a collection of technologies and methods; the result is a new way of exchanging ideas, a new rhythm, a new scale, and a new way of perceiving information. Today, Big Data has moved out of the laboratory and into the newspaper. This new face of information has become part of our digital culture, and its properties are beginning to feel more and more familiar.

With digital data as the dominant structure of information in today's society, collection information has become easier. Keeping it to yourself, however, has become increasingly difficult. Digital information is fluid by nature; it has the intrinsic "whish" to be multiplied. Unlike a traditional library or archive, the digital equivalent can be copied to a $20 USB stick in less than 60 s and be set free by uploading it to the Internet. Even though file sharing is often seen from a perspective of piracy, multiplying and moving digital information is the equivalent of water flowing to the lowest point.

The lowest point for digital information is the Internet. At the off-ramps of the digital highway are the FTP servers, the Torrent trackers, UseNet groups, and whistle-blower communities. The datasets that are found here are the products of corporate, academic, or governmental data mining, scraping, and archiving. Some are published, and some are leaked or even stolen. For who is prepared and able to investigate, these archives contain the stories of our time. Finding these stories allows data journalists and data designers to find the big story in Big Data, break the news, shape our worldview, and in the process contribute to our contemporary data culture, where digital data are increasingly part of public life.

Big Media events like Wikileaks or PRISM show what happens when journalists investigate Big Data to report the stories from the data frontier. To do this, journalists must apply the same techniques used by the engineers at IBM to look for patterns, anomalies, and niches as meaningful signals amidst data noise. As the Big Data consultants at IBM and other companies offer significantly more support to those collecting the data than to investigative journalists, this kind of research often requires a kind of DIY approach. As (big) digital data are increasingly part of public life, their representation is a challenge for designers, journalists, and storytellers of all kinds. The native structure of large digital data structures is usually not suited for storytelling. Unlike digital video and audio, which can be broadcast, digital texts which can be printed, large data structures such as databases need to be translated into a more suitable medium for storytelling.

12.2 Representing Big Data

Translating digital information to the medium best suited for the story is the essence of my practice. Digital information is a description of reality; it is the residue or imprint of a human activity, a natural phenomenon or a registration of a mechanical or digital process. Sometimes, it is all of the above. Sometimes, a dataset describes a phenomenon that we are familiar with, that we recognize and form a mental image of, helping you to relate to the subject. In other cases, a dataset might represent something abstract such as metadata. This is especially the case with "born digital" information, information collected by software describing the behavior of systems or people. This information describes an abstract structure that only exists in digital space. As such, it does not have a natural appearance that we can form a mental image of. As a designer, these sources are particularly interesting because representing them visually requires interpretation and its appearance is derived from the structure of the information.

Each project starts with a conceptual analysis of the dataset, does it tell an apparent story or is the story still concealed in the data structure? When there is a clear angle, the research can be targeted to the representation information. Often, the perspective is still ambiguous; a dataset might contain a story, but the story still needs to be constructed from the many bits and pieces of information that are contained within the dataset. In order to find a story or visually represent a complex dataset, you must try to understand the dataset, and get an idea of its size and its structure. Once you understand the formal properties of the dataset and familiar with its subject and context, you can try to "read" it.

12.3 Case Study: *The Deleted City*

My practice revolves around finding the stories in the dataset. A project that illustrates all the different stages involved with this process is *The Deleted City*. As a design studio, I initiate research projects with the goal of identifying and stretching the limits of information design. *The Deleted City* started out as a research project when I read about the *Archive Team* distributing a very large BitTorrent file of the defunct online community Geocities. As a former participant in this network, the 650 GB file captured my attention. The file contained an important piece of digital history and the collaborative work of 35 million people who created a homepage within this network. The problem was that the file itself was very hermetic. Although it contained readable information, the structure of the information was shapeless. Without being able to visualize, it is very difficult to understand a complex information structure of this size. Being convinced that this very large file was worth being saved for posterity, I felt it would be a perfect test case for information design to translate the dataset's digital structure to a readable visual system, revealing its cultural and historic value. *The Deleted City* illustrates the process of finding, analyzing, framing, and visualizing Big Data. From it emerges a contemporary perspective on the first decade of the Internet as a popular medium.

12.3.1 The Deleted City, a Digital Archeology

The Deleted City is a digital archeology of the World Wide Web as it exploded into the twenty-first century. At that time, the Web was often described as an enormous digital library that you could visit or contribute to by building a *homepage*. The early citizens of the net (or *netizens*) took their *netizenship* serious and built homepages about themselves and subjects they were experts in. These pioneers found their brave new world at Geocities, a free Web-hosting provider that was modeled after a city and where you could get a free "piece of land" to build your digital home in a certain neighborhood based on the subject of your homepage. *Heartland* was—as a neighborhood for all things rural—by far the largest, but there were neighborhoods for fashion, arts, and *far east-related topics* to name just a few.

Around the turn of the century, Geocities had tens of millions of *homesteaders* as the digital tenants were called and was bought by Yahoo! for three and a half billion dollars. Ten years later in 2009, as other metaphors of the Internet (such as the social network) had taken over, and the homesteaders had left their properties vacant after migrating to Facebook, Geocities was shutdown and deleted. In a heroic effort to preserve 10 years of collaborative work by 35 million people, the *Archive Team* made a backup of the site just before it shut down. The resulting 650-GB BitTorrent file is the digital Pompeii that is the subject of an interactive excavation that allows you to wander through an episode of recent online history.

12.3.1.1 Geocities

In the fall of 2009, a message appeared on the homepage of www.geocities.com. It stated "On October 26, 2009, your Geocities site will no longer appear on the Web, and you will no longer be able to access your Geocities account and files." With this message, Yahoo! informed the world that it was about to shut down and delete the popular online community it had acquired 10 years earlier for $3.57 billion.

Geocities, founded by David Bohnett and John Rezner in 1994, was an online community modeled after a city, which would become one of the first large online communities on the Internet. For many who were new to the Web (including myself), Geocities provided a very attractive model to get involved with this new medium. In the early and mid-1990s, it was not uncommon to describe the Internet using metaphors like "digital city" or "global library"[1]. Geocities was in a way a little of both; it followed the analogy of a city by arranging users' homepages in thematic neighborhoods, while at the same time providing (in a playful manner) a framework for contributing to an ever-expanding body of knowledge.

---

[1] E.g., *eWorld*, an early popular online service operated by Apple and placed on the market between June 1994 and March 1996. See on Wikipedia, URL, October 31, 2014: http://en.wikipedia.org/wiki/EWorld#mediaviewer/File:EWorld_Main_Screen.png

12 Big Data, Big Stories

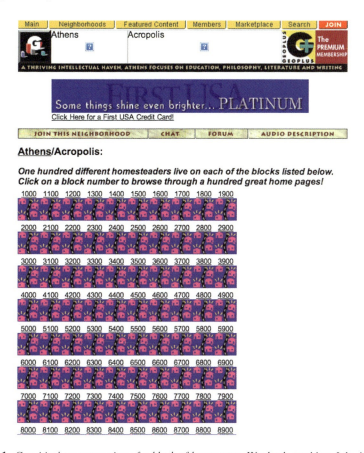

Fig. 12.1 *Geocities' representation of a block of homepages*, Wayback machine, July 4, 1998

New users, referred to by Geocities as *homesteaders* (or more broadly *netizens*, a contraction of "citizen" and "Internet"), would select a neighborhood based on the subject of their homepage. These neighborhoods had descriptive names often corresponding with places that exist in the real world: Hollywood for homepages about movies, and Athens for homepages about historical topics. Once you had selected a neighborhood, you could specify your subject further by choosing a "block" or subneighborhood. Athens/Acropolis, for example, could be a place for homepages about ancient Greek mythology. Within this specific area, you would get a free piece of land (one megabyte initially) to build a homepage.

Although Geocities never enforced the correspondence between neighborhood and content, the model of a city provided a mental spatial arrangement of information. Visually, however, Geocities was represented as a folder structure (Fig. 12.1). By linking your online identity to a homepage located in a specific neighborhood (Fig. 12.2), the Geocities model invited users to add information to the Internet, rather than just extract information from it.

Fig. 12.2 User-made emblem by which residents identified themselves with their neighborhood—found in the *Archive Team* backup

Once you had obtained the necessary land, you could start with the construction of your homepage. Interestingly, the concept of a homepage was not very strictly defined. Geocities provided a page wizard that allowed you to get started quickly using configurable templates, but many decided to build their homepages by hand using html. The lack of a fixed definition of a homepage, combined with the flexibility of a large number of people experimenting (sometimes struggling) with the medium, led to what you might call in retrospect a "homepage culture" (Fig. 12.3). This culture developed in tandem with the medium itself, incorporating

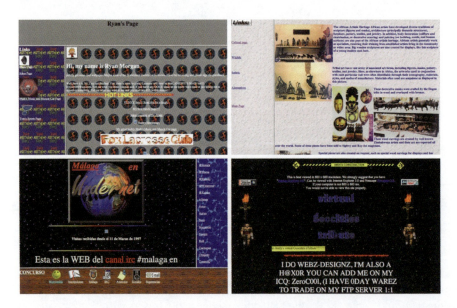

Fig. 12.3 Selection of typical Geocities homepages: Repeating backgrounds, animated gifs, HTML tables, and frames

Fig. 12.4 Collection of "under construction" signs. Found in the *Archive Team* backup

new technical possibilities (audio, moving images) as they became available and new conventions and idioms as these technical possibilities were explored.

A well-known example is the "under construction" message. Originally an apology for a homepage that was not quite finished (perhaps because the author was trying to master HTML code), it became a widely used phrase, found on most homepages. Eventually, it became a concept synonymous for the DIY mentality of homepage culture and was celebrated to an extent that thousands of different visual variations were made (Fig. 12.4). Another expression that has become a part of popular culture as a result of the experimentation by Geocities users is the animated gif. At a time when streaming online video was still mostly a promise, the animated gif allowed authors to create very short animations consisting of just several frames. Animated gifs became a hallmark of the 1990s' homepage, leading to a retro revival in the 2010s. Recently, the animated gif has also become an accepted artistic medium that has found its way into galleries and museums.

The sound of the homepage was defined largely by the use of MIDI files. Originally developed as a digital interface between musical instruments and the computer, the note-based file format was used by homepage builders to synthesize instrumental versions of popular songs. The use of MIDI files as background music for a HTML document reached its high point in the homepages of Geocities, but has rarely been used since.

12.3.1.2 The Rise and Fall

By 1999, at the height of the dot-com bubble, Geocities had become the third most visited Web site on the Internet and had gone public a year earlier. Shortly thereafter, it was bought by Yahoo! for $3.5 billion in stock. In an attempt to monetize on its investment, Yahoo! tried—in what many of the homesteaders saw as a virtual *coup-d'état*—to take ownership of all copyrights that, until then, had always belonged to the authors themselves. This clash of cultures—helped by the emergence of new metaphors of the Internet, such as the social network—gradually led to a mass migration of users away from Geocities to newly founded communities like Myspace, Friendster, and later Facebook. This left Yahoo! with what could be described as a digital ghost town. On October 26, 2009, they decided to pull the plug. With 30 days' notice, all 38 million homepages would be deleted.

12.3.1.3 The Archive

This 30-day time frame was used by archivists around the world to save what could be saved. In a coordinated attempt to download as much of Geocities as time permitted, several groups were able to copy a large (but unknown percentage) of Geocities as it was in October of 2009. One of these groups is the Archive Team. After a month of collaborative downloading, they compiled a 650-GB (compressed) archive, containing most of the tens of millions of homepages that made up Geocities. The resulting file was made available on BitTorrent in order to distribute it to as many computers as possible and save it for posterity.

12.3.1.4 The Installation

The 650-GB BitTorrent file was the starting point for the project *The Deleted City*, A Digital Archeology (Figs. 12.5, 12.6, 12.7 and 12.8). *The Deleted City* is an interactive visualization of what I call a "digital Pompeii": a virtual city frozen in time and preserved in the Archive Team backup. The installation visualizes the millions of homepages as a virtual city map. Monochrome lines, reminiscent of old phosphorescent computer screens, represent the thematic neighborhoods and subneighborhoods within them. After Yahoo!'s acquisition of Geocities in 1999, users were also allowed to build homepages outside the neighborhoods, allowing for

Fig. 12.5 *The Deleted City*, Richard Vijgen, 2012. A view of the neighborhood Athens, surrounded by "vanity URLs"

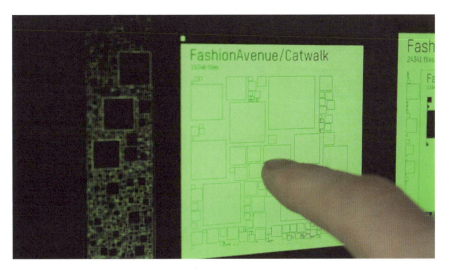

Fig. 12.6 Close-up of the "Catwalk" block located within the "Fashion Avenue" neighborhood

so-called vanity URLs like www.geocities.com/~myHomePage (Geocities was case sensitive).

By allowing this, Yahoo! largely departed from the Internet-as-a-city concept that was epitomized by Geocities. Because in the installation, the layout of the map is generated by a computer program that mimics the growth of an actual city, these small independent homepages are densely clustered near the center of the map

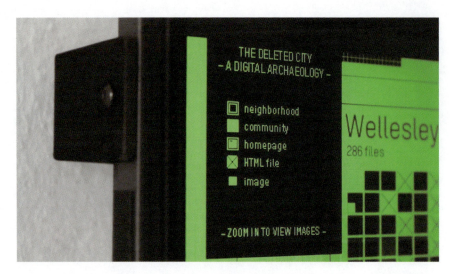

Fig. 12.7 Close-up of the screen showing the legend

Fig. 12.8 Close-up of the screen showing individual images contained within a homepage

(Fig. 12.5) and slowly give way to large "suburbs" as you move away from the center toward the edges of the city.

By using drag and pinch gestures on a multi-touch screen, the user can navigate around the city in a manner similar to Google Maps. Starting with a zoomed-out view, showing almost the entire city, you can zoom in on a neighborhood, thereby slowly revealing the subneighborhoods, blocks, and individual homepages (Fig. 12.6).

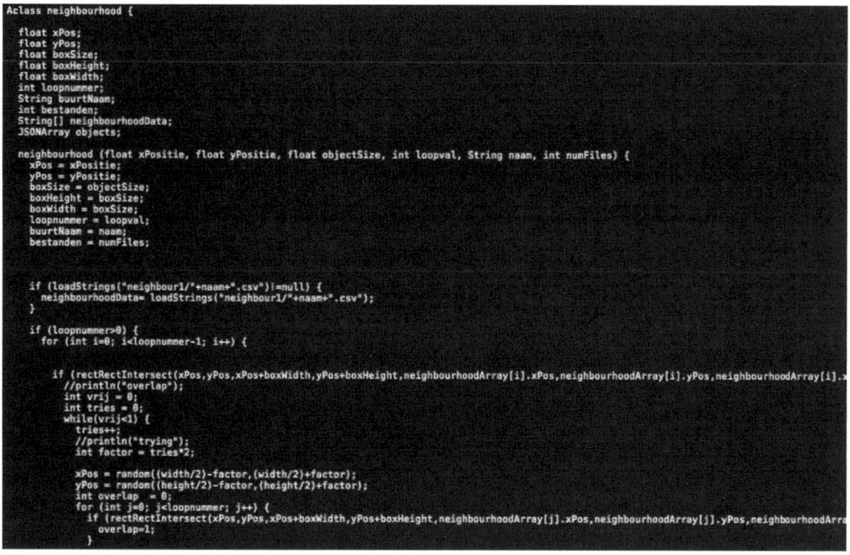

Fig. 12.9 Sample of the file manifest

Zooming in even further reveals the individual files (Figs. 12.7 and 12.8), images, and texts that a homepage consists of. MIDI files become audible when they are in the vicinity, and animated gif files start playing once they are near. Zoom out a little to get an overview of the area, and zoom in again on a neighboring homepage about the same subject.

Exploring a neighborhood like this shows the strength of the concept of the *Internet-as-a-city*, where individual users become neighbors based on the subject of their work. But it also shows the weakness, the lack of enforcement, and of the connection between content and neighborhood, since amidst a group of pages about archeology, you might find a page with a family's vacation photos.

After exploring *The Deleted City* for a while, you might feel as if you have opened a digital time capsule. The gesture-based interaction with the touch screen is inherently comes from twenty-first century, but the contents of the installation look and feel as if they come from a different era (Figs. 12.9, 12.10 and 12.11).

12.3.1.5 Perspective

You might say that the Geocities backup is in fact a relic from a different era. Even if from a historical perspective (let alone an archeological one), it is just from the recent past. However, digital time moves fast, and even though it is often said that the Internet does not forget, once a particular technology has been replaced by a newer version, it seldom survives. Geocities was one of the first successful user-contributed communities on the Internet, and as it turns out, one of the first to be deleted. In a time when the Internet has moved beyond the experimental phase and

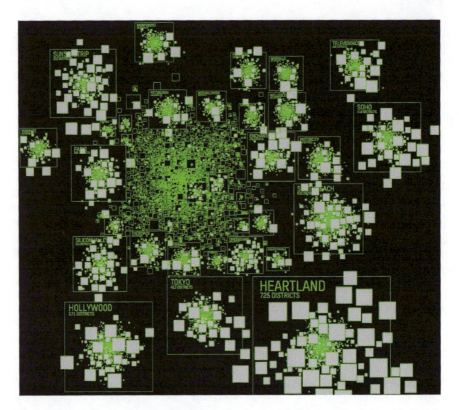

Fig. 12.10 Visualization of the file structure using an algorithm that mimics the growth of a city

has become accessible to a truly global audience, few of its users still have to—or want to—build their own homepages. Even the concept of the homepage itself has been replaced by that of the *profile*: the dominant mode of representation in today's social Internet. Still, many of the concepts and stylistic elements developed during the era of the *Internet-as-a-city* have found their way into the templates and frameworks that define the Internet we use today.

You might argue that the spirit of the global library, where all contribute to a body of knowledge, lives on in Wikipedia, and the personal tone of voice found in many homepages can still be seen in Tumblr blogs, where authors can express themselves beyond the templates and frameworks used by Facebook and others.

Today, however, the somewhat utopian concepts of the Internet as a virtual city or a global library have largely been replaced by more utilitarian notions of the Internet as a social network or cloud. Where Geocities invited you to *contribute*, modern-day networks invite you to *share*. Where hand-coded homepages led to a gradual evolution of conventions, change in modern networks such as Facebook is incremental and centralized. The *netizens* have become *users*.

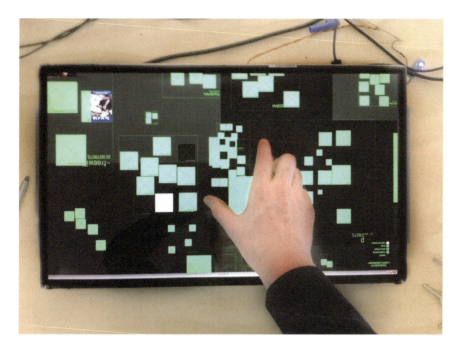

Fig. 12.11 Experimental touch screen setup

The Deleted City aims to revisit a previous incarnation of the Internet and allows you to see how the technology, the esthetics, the metaphors, and the values that underpin it have changed, some for better, others for worse.

12.4 Conclusion

Just as the city was used to describe a radically new environment as the Internet in the 1990s and the metaphor of the horseless carriage was used to describe a radically new mode of transportation in the early 1900s, Big Data is at the verge of becoming a popular reality. We know it is there already, running many aspects of our digital lives, but we have not been able to touch it, we do not know what it looks like, and it has no face. We no longer think of the Internet as a city or the automobile as a carriage without a horse because we got to know and understand these technologies recognizing their distinct qualities and capabilities. Probably, Big Data will eventually be understood not as large amounts of data but as a technology that enables us to do new things and tell new stories. Designing the face we will recognize it by is the task at hand.

Author Biography

Richard Vijgen (Information Designer and Founder of Studio Richard Vijgen) (1982) is a designer from the Netherlands. In 2009 he founded Studio Richard Vijgen, a design studio for contemporary information culture. The studio investigates new strategies to find the big stories in Big Data through research and design. His work is deeply rooted in the digital domain but always connects with physical or social space. Richard designs and produces interactive data visualizations and data installations ranging from microscopic to architectural in scale. He uses code, pixels and 3D printers to convey ideas.

The Studio has worked with a wide range of clients and organizations around the world including The Pulitzer Foundation, VPRO Television and the Dutch Government. His work has been exhibited on the massive video screens of Times Square and the white cube of the Los Angeles Museum of Modern Art. The studio's work has been widely recognized by many professional organizations and publications, including *Wired*, Gestalten, The Smithsonian, Rhizome, The European Design Awards and The Dutch Design Awards.

Richard reflects on and writes about digital culture, he initiates and participates in research projects and teaches Interactive Architecture at the Arnhem School of Art and Design in the Netherlands.

For more information and contact:

http://www.richardvijgen.nl
http://nl.linkedin.com/pub/richard-vijgen/b/b32/844
https://twitter.com/richardvijgen

Chapter 13
Dispositif Mapping

Visualizing Crisis and Complex Problems

Christopher Warnow

Abstract Can information visualization map conflicts, war, or similar complex problems? The method of *Dispositif* mapping approaches this question by collecting actors, institutions, documents, and beliefs involved in a problem and mapping it out as a network. It builds on the idea of the *Dispositif* by Michel Foucault who described it as "a thoroughly heterogeneous ensemble "where" *the apparatus* itself is the system of relations that can be established between these elements." He tried to grasp the a priori or the common ground where a problem arises. As an example, the Greek sovereign debt crisis was examined. Because of the political, financial, and social complexity of the crisis, there is no *Deus Ex Machina* helping with an overview. Furthermore, there are no homogeneous source but scattered beliefs and actors instead. The actors involved were hand-collected, as well as important documents and institutions. They were constructing a graph which soon grew over a one person's memory, which makes it by far not a complete picture but very vast and more detailed than the books about the topic. Instead of a description of the crisis itself, the *Dispositif* visualizes the "*milieu*" in which the actors were taking decisions. Greek shipping magnates and schools teaching market radicalism are prominent, which can explain the tough political course for the average Greek worker.

13.1 Introduction

When thinking about a complex social or political issue, it can feel wicked how contradicted the arguments and involved parties intertwine. For example, it is very hard to grasp the different housing and banking crisis reasons of Spain, Italy, and Ireland which were collected under the umbrella term "Eurozone crisis." But it instead of a limitation, it can become the core of a technique to map such an issue.

C. Warnow (✉)
Wuerzburg, Bavaria, Germany
e-mail: hello@christopherwarnow.com

13.1.1 Wicked Problems

Actually, one could say it is already a vintage idea, because back in the 1970s, the concept of the wicked problem was formulated by Horst Rittel and Melvin M. Webber (Rittel and Webber 1973). They thought, when every actor in a given issue is right and there is no linear causality, one has to go into a loop of describing the overall structure of "*acteurs*" and their arguments, take some action, and then see how the structure changes. Rittel and Webber come up with a ten-point list characterizing wicked problem. The interesting points for this article are as follows:
1. There is no definitive formulation of a wicked problem,
2. Wicked problems have no stopping rule,
3. Solutions to wicked problems are not true or false, but good or bad,
4. The choice of explanation determines the nature of the problem's resolution.

There is a lot of rocket science touch in it. For the quantum physics part, you change the problem by observing it. That is what it says in point 9. It is like Schrödinger's cat in the box who is in a superposition of being dead or alive. A quantum can have both a state of spinning left and state of spinning right. By the act of measuring, or in other words looking at it, it goes from the superposition in a discrete state. It would not have done it if we or the researcher would not have measured it. So most of the time, we do not want to know whether the cat is dead or alive and leave the box closed.

A wicked problem, a financial crisis for example, resides in a superposition as well, a meta-architecture, and an ephemeral and abstract state of relationships between the actors and arguments. By the act of formulating, the complex issue becomes a form and is acted upon (form A), but the form B that derives when formulating it in another way or with another technique is equal to form A. Both are maps and can be acted upon. You see how both would lead to different directions. It is better than no map. The wicked problem is not the ideal idea by the *pagan Greek philosopher* Plato, which has to fit into real-life material. His ideas are fixed, a desk, for example, which no wood can portray as so exact as it exists in the ideal world. A wicked problem has no form per se. This is the quantum physics superstate interpretation of it.

Second rocket science approach is the cybernetic feedback loop. Have courage to read on my dear reader. You will understand rocket science after this chapter. The act of formulating and acting causes the wicked problem to change. Rittel and Webber propose minor changes here and there, taking action at one end of the wicked problem map and seeing how it changes at the other ends. A loop between an action and problem-formulating begins and has no stopping rule. This is the cybernetic touch. This kind of science evolved in the 1940s and becomes second-order cybernetics in the 1970s, thus had an impact on Rittel and Webber. The basic belief is that the world can be seen as a system of parts that are acting with each other in feedback loops. A thermostat measures the warmth of a room and acts upon it, which in turn changes the room's temperature. One of the founders, Norbert

Wiener, found out that the brain and muscles are working this way as well. When you want to grab a pencil in front of you, one could think that the eyes are measuring the distance and angle between your hand and object, causing your muscles contract for a certain time. This is the linear and causal approach, coming from a mechanic view of the world. But Wiener said that the brain constantly measures the hand–pencil relation and tells the muscles to take microactions. A small inch here and there, updating and refining the hands position (Rosenblueth et al. 1943). Feedback loops became hip and meandered into system theory, chaos theory with its nonlinear approaches, think about the butterfly effect. The newest label is complexity science. This is the direction of the wicked problem, the focus on formulating and interacting with it.

13.1.2 Dispositif

Having that, it can be OK for an issue to be wicked. Actually, it can be quite productive. There are numerous ways of formulating; let us try it with data visualization leading to a second ingredient and theory which helps us to map out a complex social issue. We have to cope with the heterogeneity of the parts involved. The financial crisis of 2008, for example, involved banks, institutions, abstract financial products, regulations, deregulations, and a lot of more items you can imagine. How do you collect and store it? Or how do you parse it? There is no application programming interface to the financial crisis. You cannot have a machine learning algorithm to learn which machines were involved. You cannot have a robot jump from Web site to Web site and ask it for insights.

Luckily, there is another vintage idea from the 1970s that can be applied. Michel Foucault, a famous French philosopher, had examined the question of power structures in our society. Foucault asked himself how a group or "*milieu*" does their day-to-day work in disciplining or leading other groups; for example, take the government or schools. He came up with the *Dispositif*, the mechanics that is able to have power in society. He describes it as "a thoroughly heterogeneous "*ensemble*" consisting of discourses, institutions, architectural forms, regulatory decisions, laws, administrative measures, scientific statements, philosophical, moral and philanthropic propositions—in short, the said as much as the unsaid. Such are the elements of the apparatus. The apparatus itself is the system of relations that can be established between these elements." (Foucault 1980) What is so helpful for us is the approach of describing the structure between heterogeneous elements to shed light on their political dimension. The interesting point is how the line is drawn between being inside and outside the *Dispositif*, or how it defines true or false.

One can see a *Dispositif*, the apparatus at work when entering into a huge church. One immediately has to whisper and go slow. One does seldom run fast in a church. It is not the free will that drives our bodies, it is the architecture. The acoustics are demanding silent voices. This is the architectural part of a catholic *Dispositif*, if we want to name an example. Being homosexual is more false in this

Dispositif, while in the *Dispositif* of San Francisco, it is more true. You see how such an apparatus creates a milieu which causes an ideologic framework, or weltbild. It makes certain actions and beliefs more probable than others. Finally, another French philosopher Gilles Deleuze said in the 1990s about the *Dispositif* that "they are machines that make one see and talk." (Deleuze 1992a)

Now, when we want to visualize a complex social issue as a *Dispositif*, it is best to hand-collect the items named by Foucault. But because it is wicked, one can collect other things as well. Ideally, the ideological framework emerges by that work.

13.1.3 Mapping the European Debt Crisis

As a test, I took the European debt crisis with a focus on the Greece debt crisis. As a collecting tool, I used Gephi, which says about itself being Photoshop for networks. It is an awesome attempt of scientists to create an open source tool. I created a node–link structure assigning types to the nodes, for example, manager, company, or book.

In the dynamics of the European debt crisis, two actors stood out between 2010 and 2012. Both demanded cutting of public spending and taxes by the troubled southern European countries. This way, they wanted them to be able to pay back their debt to the creditors. The theoretical framework for that is called austerity measures. Besides being Merriam-Webster's word of the year 2010, austerity means a harsh cut of people's social welfare, wages, and public spending in favor of powerful rich man.

The first was Olli Rehn, the Vice President of the European Commission and European Commissioner for Economic and Monetary Affairs and the Euro. Second to mention is Wolfgang Schäuble, back then Germany's Federal Minister of Finance. The question to build a *Dispositif* map is here: On what is the argumentation based on? Is it a book, a person or common sense? The answer is the important tie to understand their philosophy.

Olli Rehn is crediting his mentors explicitly: "This is very worrying, as the levels of public debt above 90 % tend to have a negative impact on economic dynamism. In their 2011 empirical study, Carmen Reinhart and Kenneth Rogoff found a negative correlation between debt above 90 % and growth."(Rehn 2014)

Wolfgang Schäuble is not talking about the 90 % rate, but cites the same two persons: "Recent studies—most prominently Rogoff's and Reinhardt's book *This Time Is Different*—have shown that once government debt burdens reach thresholds perceived to be unsustainable, more debt will stunt rather than stimulate growth. The recent turmoil within the Eurozone suggests that this finding holds true for growth in Eurozone countries as well." (Schäuble 2011a)—To cope with the 90 % limit, they demanded an austerity course. The problem with this number is that it is plain wrong. The student Thomas Herndon was asked by his professor to recalculate the numbers by famous economic papers as homework. He tried but could not reproduce them. This led him and his Professor Michael Ash to ask Reinhart and Rogoff for the actual Excel sheet they used, and they found mistakes in it, missing countries and unusual averaging for example (Alexander 2013). Europe's leaders are justifying their unsocial reasoning on false data.

Why did they base their assumptions on their studies and not others? The relationship between these actors is the question that leads deeper into the *Dispositif* of the European debt crisis. Olli Rehn knows Rogoff personally, and they met at Harvard for example. Schäuble cited them at a conference at the think tank Bruegel. He met Jacon Funk Kirkegaard there. He is Senior Research Staff at the Peterson Institute for International Economics (Schäuble 2011b).

Rogoff is in the Advisory Committee of this institute. We see a milieu emerge. Going up this chain, we see billionaires like David Rockefeller or Lynn Forester de Rothschild in the board of directors. Rothschild was happy to spend her honeymoon in the White House Usherwood 2008). Maybe Schäuble has spoken to Caio Koch-Weser, the former Secretary of State in the Federal Ministry of Finance 1999–2005. Besides being on the board as well, he is working at the Deutsche Bank AG and is lobbying for Germany. He spoke to the former secretary to Wolfgang Schäuble, Jörg Asmussen at January 15, 2010, and April 26, 2011, for example.[1] Putting these ties together opens a *"milieu"* of *"acteurs"* that are reasoning in favor of rich people. People from labor unions were not asked by Rehn and Schäuble (Fig. 13.1).

13.1.4 Data Propaganda Explained by the Dispositif Map

The president of the European Central Bank Mario Draghi urges the European countries to drive by the austerity course like he did at March 14, 2013. The French President was silenced because Draghi showed statistics to him and other European leaders. The graphs were telling that France, Spain, Italy and Portugal were paying too much wages to its workers (Watt 2013). But his mistake is that he expressed the productivity measure in real terms and the compensation in nominal terms. These are two different scales and show wrong statistics. Interestingly, Germany is a good example within Draghi's wrong measure. But the German workers have to struggle with the low-wage sector Germany has (Marsh and Hansen 2012). Draghi wants to export this idea. Looking at the real measures reveals a fine correlation between productivity and compensation in France (Figs. 13.2 and 13.3).

How could such a leader in finance do such a basic statistical mistake? Maybe the milieu he is acting in reveals a clue. At least he knows Kenneth Rogoff from the Group of 30, "A Consultative Group on International Economic Affairs, Inc." where they both are members. Maybe he knows Carmen Reinhart from the Kennedy School of Government Harvard, where she is a professor and Draghi was a fellow. The other days, he met Olli Rehn at the coalition of European Commission, International Monetary Fund, and European Central Bank, called Troika overseeing

---

[1] URL, March 12, 2014: http://dip21.bundestag.de/dip21/btd/17/123/1712332.pdf.

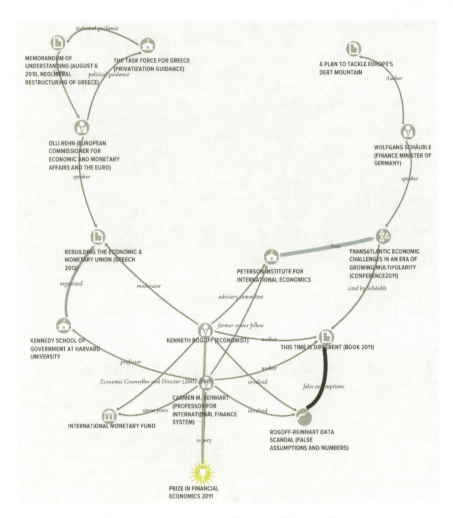

Fig. 13.1 *Dispositif Rogoff—Schäuble—Rehn*, Christopher Warnow, 2011

the austerity measures in Greece. He was working at Goldman Sachs, which is funding the G30, as well as other banks. David Rockefeller, we know this billionaire from the Peterson Institute for International Economics initiated the G30 back in the 1970s with his fund. Again, we see a bias toward billionaires and this time banking institutions. The capitalist-class argumentation toward cutting wages is at work here. If by mistake or not only knows Mario Draghi.

When trying to visualize a crisis as a *Dispositif*, it is not about the arguments but more about the structures of influence. The European debt crisis consists of a conglomerate of events: some more technical in nature and others happened from pure fear. European leaders could have gone in any way. But they chose to hear to the advices of the rich class and cut money from the average people (Fig. 13.4).

13 *Dispositif* Mapping   241

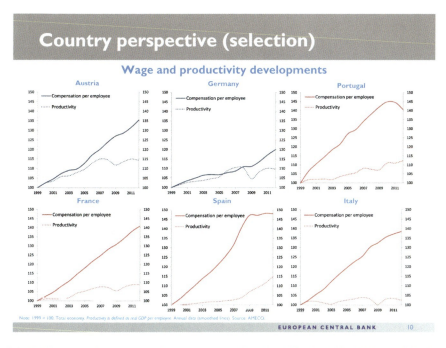

Fig. 13.2 *Draghi Presentation*, ©European Central Bank—AMECO, 2011. URL (available for free): http://www.ecb.europa.eu/press/key/date/2013/html/sp130315.en.pdf

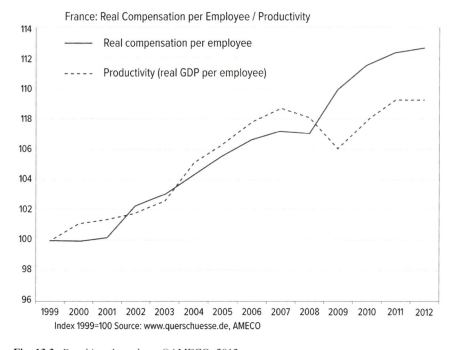

Fig. 13.3 *Draghi real numbers*, ©AMECO, 2012

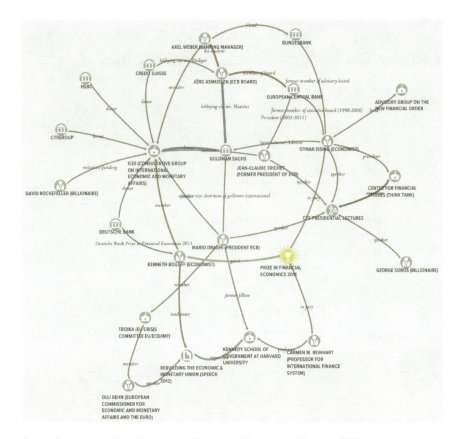

Fig. 13.4 *Dispositif Draghi—Rogoff* billionaires, Christopher Warnow, 2011

13.2 Mapping Influence

Where does the influence take place and who has the power to do so? Two more examples want to show how one can reveal structures of influence.

13.2.1 Germany's Shareholder Structure

The first one takes the "fear of the markets," which was talked about in the mass media between 2010 and 2012 for example. Then, the question is, who are the markets that can be feared? Schäuble, Draghi, and Rehn wanted to gain confidence of the markets in Europe. Therefore, they are demanding austerity measures. In other words, they want the European countries to save money by cutting social welfare to be able to pay back credits. I took the German market as a starting point and had a look at the stocks and its shareholders. The German Stock Index contains 30 companies. I created a network of the company–shareholder relationships. What emerged was quite non-German. Most of the companies are belonging to American and French shareholders. The supernode was a company called Blackrock, the world's largest asset manager. The second one is a scattered network of Capital Group nodes. It is one of the world's largest investment management organizations as well. An analysis of the world's shareholding structure at the ETH Zurich by Stefania Vitali, James B. Glattfelder, and Stefano Battiston (Vitali et al. 2011) suggested that the core of the Western economy is 147 superconnected companies. Capital Group Companies Inc. is happy about the second place. The first place goes to Barclays plc, a more than 300-year-old British bank. The interesting connection here is that this bank bought the rests of the Lehman Brothers' institution and its headquarters, after filing bankrupt in the financial crisis 2008. We remember, the bankruptcy yielded to nervousness in the banking market. Financial institutions did not lend each to each other, and states had to help them out. This in turn raised the nervousness of the financial market because the financial agents thought the states might go bankrupt after helping them out. A weird cycle (Fig. 13.5). "*Voilà*" the European Debt Crisis in a nutshell (Teather and Clark 2008).

Alright, those are the entities in the market that can be feared by the European debt crisis, not German companies. Further, following the money ends at a wall that could be the sky of the Truman show. The only thing one finds about the clients of Capital Group is that they are working with institutions and high net worth investors, billionaires for example.[2] The German low-wage policy not just makes Mario Draghi happy, it generates returns for the high net worth investors around the world as well.

The German sociologist Paul Windolf called this structure "Shareholder Capitalism" (Windolf 2005); Short-termed decisions in favor of the financial market instead of long-term decisions in favor of the company. If the shareholder is not amused, he will opt out. The shareholder network reveals this power structure and points to structures of power and possible influence. Again, there is no labor union in the center of the German economic core.

13.2.2 Mapping Lobbying

The next question leads to last, the example of *Dispositif* mapping. When shareholders are breathing in the neck of German companies, how do they act on and influence the political sphere? How did they convince Angela Merkel to be proud of the low-wage sector? Interestingly, the government gives the answer by itself. Some politicians asked for an official list of meetings between politicians and the banking lobby or automobile lobby between 2009 and 2013. The documents Drucksache 17/14698 and Drucksache 17/12332 are containing a list with these meetings. I wanted to know who is meeting most and was interested in the backbones of German lobbying (Fig. 13.6).

---

[2] URL, March 10, 2014: http://thecapitalgroup.com/our-services.html.

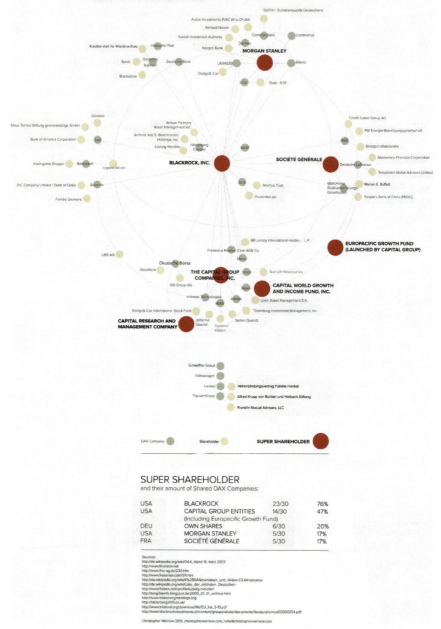

Fig. 13.5 *Shareholder structure*, Christopher Warnow, 2013

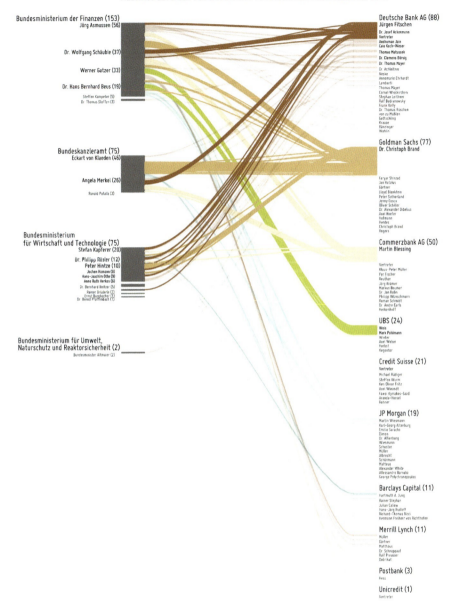

Fig. 13.6 *Banking lobby*, Christopher Warnow, 2014

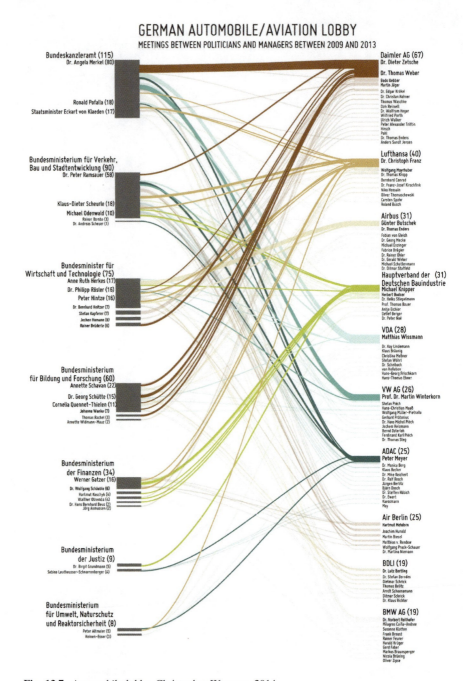

Fig. 13.7 *Automobile lobby,* Christopher Warnow, 2014

And there are. Have a look at the well-connected overview about German Federal Bureaus at the left and banks at the right. Each line represents a meeting. The banks on the right are ordered by their amount of meetings. What makes sense is the permeation of the government by Deutsche Bank AG. The next is Goldman Sachs, and I did not know how well connected it is to the German politicians. Have a look at the route between Dr. Christoph Brand and Eckard von Klaeden. The second route yields to Jörg Asmussen, who later on was sent to the ECB. In 2014, he became Deputy Labor Minister in Germany, where he is able to directly influence and think about the low-wage sector. What we learn is that "*acteurs*" of influence know who to ask. You will not find the names in the mass media. For example, UBS knows that they have to speak to Werner Gatzer and Dr. Hans Bernhard Beus. You do not know it. They do. This is why we have to map those things (Fig. 13.7).

The second table on meetings between automobile/aviation companies and the German government shows a lot of action. What sticks out is a backbone between Angela Merkel and Daimler via its CEO Dieter Zetsche. All in all, the German chancellor likes to speak to the economy. Another study has to show how much she likes to listen to alternative "*acteurs*," for example labor unions.

13.3 Conclusion

What we have seen in the past text is the shift from trying to visualize crisis ingredients to mapping out political, economic, and ideological influence of a ruling class in a neoliberal capitalist regime. To understand crisis in the Western society mostly means understanding the inner workings of the financial system and the use of smart power by the ruling class (Nye 2009).

This way, they set agendas and lobby politicians. They believe in the abstract financial products that caused the financial crisis 2007 and led to the European debt crisis via the following nervousness of banks. Since the ruling class has the power to act upon crisis, one has to know what they believe in. And this is the "strange non-object that is contemporary capitalism" as Nick Srnicek puts it in his text Navigating Neoliberalism: Political Aesthetics in the Age of Crisis (Srnicek 2012). The concepts of wicked problems and visual *Dispositifs* are tools to map this type of economic system. Deleuze calls it a "gas" in his visions of the society of control, where "(…) individuals have become *dividuals*, and masses, samples, data, markets, or *banks.*" (Deleuze 1992b)

> This world view was established in the 1970s with the rise of Milton Friedman's *Capitalism and Freedom* and its implementations in the Reaganomics. It believes in free-floating financial markets that modulate risks and social artifacts into abstract financial products and talks about privatization and free markets but means corporatist states where large corporations and politicians interlock and rotate fast as Naomi Klein puts it (Klein 2008).

How does one map this? Not with photographs or films. They are based on a different material than neoliberal economics. It is data, that flows around the trading desks and builds the basis for risk models of derivatives. Data are the material and oil of our current socioeconomic *Dispositif*. This is why we need images that are based on data too. Designers who are capable of working with machine learning, complexity science algorithms, and statistics are needed. This is the only literacy that makes us understand what Warren Buffet means when he says "There's class warfare, all right (…) but it's my class, the rich class, that's making war, and we're winning." (Stein 2006)

References

Alexander R (2013) Reinhart, Rogoff… and Herndon: the student who caught out the profs. BBC News. URL: http://www.bbc.com/news/magazine-22223190, 12 Mar 2014

Deleuze G (1992a) What is a Dispositif? In: Armstrong TJ (ed) Michel Foucault philosopher. Harvester Wheatsheaf, London/New York, pp 159–168

Deleuze G (1992b) Postscript on the societies of control, vol 59. MIT Press, Cambridge, p 6

Foucault M (1980) The confession of the flesh. In: Gordon C (ed) Power/knowledge selected interviews and other writings 1972–1977. Pantheon, New York, pp 194–228

Klein N (2008) The Shock doctrine: the rise of disaster capitalism. Penguin, London

Marsh S, Hansen H (2012) Insight: the dark side of Germany's jobs miracle. Reuters. URL: http://www.reuters.com/article/2012/02/08/us-germany-jobs-idUSTRE8170P120120208, 12 Mar 2014

Nye JS (2009) Get smart: combining hard and soft power. Foreign affairs. URL: http://www.foreignaffairs.com/articles/65163/joseph-s-nye-jr/get-smart?page=1, 10 Mar 2014

Rehn O (2014) ECMI annual conference—rebalancing Europe, rebuilding the EMU. European Commission. URL: http://ec.europa.eu/commission_2010-2014/rehn/headlines/news/2012/10/20121018_ecmi_conference_en.htm; http://ec.europa.eu/commission_2010-2014/rehn/documents/2012_10_18_ecm_en.pdf, 12 Mar 2014

Rittel HWJ, Webber MM (1973) Dilemmas in a general theory of planning. Policy Sci 4:155–169

Rosenblueth A, Wiener N, Bigelow J (1943) Behavior, purpose and teleology. Phil Sci 1:18–24

Stein B (2006) In class warfare, guess which class is winning. The New York Times. URL: http://www.nytimes.com/2006/11/26/business/yourmoney/26every.html?_r=0, 9 Mar 2014

Srnicek N (2012) Navigating neoliberalism: political aesthetics in an age of crisis. Presented at the matter of contradiction: unground the object, Vassivière. Available online, URL: http://www.academia.edu/1925994/Navigating_Neoliberalism_Political_Aesthetics_in_an_Age_of_Crisis, 10 Mar 2014

Schäuble W (2011a) A comprehensive strategy for the stabilization of the economic and monetary union. Wolfgang Schäuble personal website, URL: http://www.wolfgang-schaeuble.de/index.php?id=30&textid=1456&page=1, 14 Mar 2014

Schäuble W (2011b) The euro area crisis and future global implications (speech). German Federal Minister of Finance "transatlantic economic challenges in an era of growing multipolarity", Berlin. Available online (Bruegel), URL: http://www.bruegel.org/fileadmin/bruegel_files/Events/Event_materials/Wolfgang_Schauble_speech_revised.pdf, 14 Mar 2014

Teather D, Clark A (2008) Barclays agrees $1.75bn deal for core Lehman Brothers business. The Guardian. URL: http://www.theguardian.com/business/2008/sep/17/barclay.lehmanbrothers1, 12 Mar 2014

Usherwood Z (2008) I love my country over my party. Sky News. URL: http://news.sky.com/story/628428/i-love-my-country-over-my-party, 12 Mar 2014

Vitali S, Glattfelder JB, Battiston S (2011) The network of global corporate control. Alejandro Raul Hernandez Montoya, Universidad Veracruzana, Mexico. Available online, URL: http://arxiv.org/PS_cache/arxiv/pdf/1107/1107.5728v2.pdf, 12 Mar 2014

Watt A (2013) Mario Draghi's economy ideology revealed? Soc Eur J. URL: http://www.social-europe.eu/2013/03/mario-draghis-economic-ideology-revealed/, 12 Mar 2014

Windolf P (2005) Finanzmarkt-Kapitalismus. Kölner Zeitschrift für Soziologie und Sozialpsychologie, Sonderheft, 45

Author Biography

Christopher Warnow is a Computational and Interface designer with a strong interest in generative and procedural design processes. While studying Interface Design he experimented in this field via workshops which he organized or as a teacher at Berliner Technische Kunstschule and Hochschule für Technik und Wirtschaft Berlin.

Additional procedural experiments in media art followed which ended up at transmediale and Stedelijk Museum in Amsterdam for example. After having written his thesis "From Augmented to Enchanted Reality" in 2010 he had a mega-inspiring time at onformative—a studio for generative design. Since 2011 he is a research fellow at the Würzburg/Bavaria Research Institute for Design and Systems and data visualization freelancer.

His master's thesis in Information Design ended in 2012 where he mapped power structures of European elites that created to the European debt crisis. Which is a procedural problem since no one knows everything but everyone is acting inside its system.

For more information and contact:

http://christopherwarnow.com.
https://twitter.com/brainSteen.

Part VII
Experiencing Data Through Multiple Modalities

Chapter 14
Sustainability: Visualized

Arlene Birt

Abstract When the goal is to help audiences understand social and environmental sustainability data, communicating the context is of high importance. Data that are designed to tell a story can create a strong connection between high-level, abstract sustainability concepts and the daily, individual actions of consumers. This article outlines the challenges and opportunities that the topic presents and offers key practices in the visualization of sustainability data: starting with the individual, layering information, presenting positively, involving emotions, letting audiences make their own decisions, enabling interaction, and providing access to supporting data. Specific experiences from a variety of the author's own projects illustrate these points.

14.1 Introduction

A cup of coffee contains an entire world within it (Fig. 14.1). So many hands have touched this liquid steeped from roasted beans. Even the water that distilled the essence from the ground beans has had an eventful journey to this cup. As a designer, I am interested in seeing all the ways in which we are connected to the world through the things we interact with on a daily basis—visualized.

Visual stories are an opportunity to educate people in order to motivate long-term behavior change through linking data on sustainability with pictorial stories that provide context. These days, we urgently need improved methods to communicate sustainability to consumers in order to meet global challenges—now and for the future.

A. Birt (✉)
Background Stories, 303 Hickory Drive, 50014 Ames, IA, USA
e-mail: arlene@arlenebirt.com

© Springer-Verlag London 2015
D. Bihanic (ed.), *New Challenges for Data Design*,
DOI 10.1007/978-1-4471-6596-5_14

Fig. 14.1 Author's illustration: Imagining a world where a reminder of a product's entire life story is attached to each experience of it. (All images are attributed to Arlene Birt and background stories, unless otherwise indicated)

14.2 What Is Sustainability? And How Does It Relate to Info Design?

Globalization has brought with it global problems. As a society, we are faced with increasingly erratic weather, tropical storms, water shortage, flooding, drought, bird flu, resource conflicts, inequality, and more. Current estimates put the carrying capacity of earth verging on collapse and we have heard it a million times before: that we—particularly in the western world—have been living beyond our means.

It is hard to know what to do about it: The calamity feels too removed from our daily lives. Even when sustainability-related measures can be described in quantified terms, it is hard for the average citizen to know what such quantities mean. The numbers feel abstract: Only an individual experience can make these numbers real and relevant to our daily life.

Explaining sustainability is a challenge. Driving home what abstract data actually means is an even larger challenge. For example, in an ongoing exploration of carbon dioxide, I am investigating what specific quantities of CO_2 actually means.

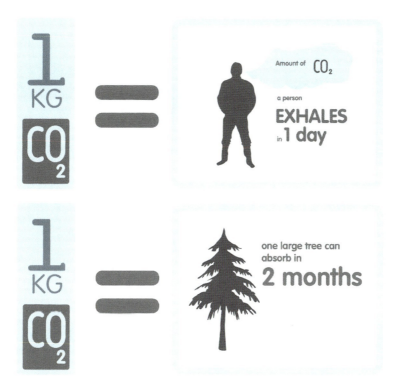

Fig. 14.2 An ongoing exploration by the author which attempts to understand what a kilo of CO_2 means

* **One ton of CO_2** (Figs. 14.2 and 14.3)

Carbon Dioxide (CO_2) is a colorless, odorless gas. It is a greenhouse gas produced by combustion: automobiles and industry. It is also what mammals exhale and plants "inhale."

We hear about tons of CO_2 all the time: but what does a quantity of CO_2 mean? How does it relate to our individual, daily actions?

Numerous calculators exist to help individuals and organizations estimate the CO_2 their activities emit, for example, household use, commuting emissions, or event consumption. So one can arrive at a definite number—albeit estimate—for how much CO_2 a given activity emits. For example, your share of CO_2 emissions on a round-trip economy-class flight from Brussels, Belgium, to Buenos Aires, Argentina, would expel around 2.37 tons of CO_2.

Good to know. But what does 2.37 tons of CO_2 *actually mean*?

This is a difficult concept for even seasoned climate activists to grasp. Given that Carbon Dioxide is a gas, it is hard to weigh. Neither can it be seen or smelled.

To help consumers understand what CO_2 is, means, and does, as well as how our everyday consumption decisions (whether at the store or switching on a light) impacts the world, I have been investigating ways to make visible the invisible.

Using found data, calculations, and statistics, this self-initiated project seeks to visually describe the concept of *one ton of CO_2*—from multiple angles. The meaning of CO_2 is

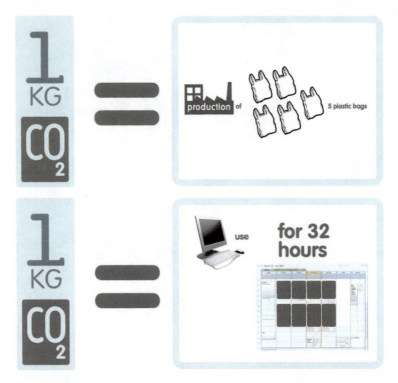

Fig. 14.3 A sample of several perspectives that aim to present a well-rounded idea of what one kilo of CO_2 is. For example, one can look at emissions from production, or electricity generated

explored in terms that people can relate to and put into the context of their own lives. It is an ongoing attempt to address the gap between information (quantities of carbon) and consumer's personal actions, using visual communication in approachable and playful ways to help people develop a well-rounded understanding of CO_2. I believe that the better an idea is understood, the more likely people are to act upon it. Therefore, the result I am aiming for is a creative tool for education.

Estimating the amount of energy use that emits a kilogram of CO_2 throughout the lifecycle of a variety of applications is not a straightforward task; there are a lot of variables at work here. These calculations use sizable generalizations. For example, CO_2 output from energy production is heavily dependent on what kind of energy (coal, nuclear, gas, etc.) is used. The amount of CO_2 absorbed by a tree depends on the size, species, geographic location, and age of the tree—as well as the season.

14.3 [Info] Designing for Sustainability

There are multiple ways to design for sustainability. Ideally, one uses a "whole-systems" or "lifecycle thinking" approach and considers the big-picture social, environmental and fiscal system that a product or idea is part of. Life cycle thinking quantifies the impacts that a product or action has throughout all stages of its life. Whole-systems thinking places an idea into the context of its network.

Throughout the design process, one can plan for energy efficiency or make decisions for the physical product, such as specifying more sustainable or more durable materials, or designing for disassembly (to facilitate recycling/reuse).

Information design can of course come in handy throughout the design process: aiding engineers and scientists to communicate within an internal team in order to select the best design solution. But the phase of info design that I am most interested in occurs after all the usual design decisions have been made. What interests me is how to translate the growing quantity of studies on sustainability to our individual, daily life as a citizen of the earth. In this consumer-facing side of sustainability communication, information designers can play a key role. We can enhance the experience of information for the lay audience in order to connect real life with digital data. Narrative information design is an area that I feel is particularly relevant to the challenges of communicating sustainability, because of its ability to tell a story.

As with all topics on the information landscape, sustainability data are being collected today at unprecedented levels. It is also a topic that is increasingly on the radar of consumers, citizens, and corporations. Among the fast-growing collection of data being assembled on the subject are carbon footprints, water and energy consumption, corporate social responsibility, and a huge range of climate data—not to mention temperature data, weather patterns, and other related data, just to nip the tip of the iceberg.

Despite the growing assortment of data being collected on the subject, the *concept* of sustainability is hard to pin down. This is because of a challenge inherent within sustainability itself.

14.3.1 Challenges: Communicating Sustainability

Part of the difficulty in educating people about sustainability is how complex the topic is. Sustainability is not a prescription. Guidelines for ultimate sustainability change over time, vary based on context, and must endlessly adapt to varying circumstances.

Even the definition is varied. Sustainability is

(…) a balance of social, environmental and economical demands (also known as People, Planet, Profit or Ethical, Green, Fiscal)
(…) "improving the quality of human life while living within the carrying capacity of supporting eco-systems" (World Conservation Union et al. 1991).
(…) "the capacity to endure" (Wikipedia 2014).
(…) "Development that meets the needs of the present without compromising the ability of future generations to meet their own needs" (World Commission on Environment and Development 1987).

Similarly, when looking at a specific product or action, there is no simple answer to the question "How sustainable is this?" The answer would have to be based on a specific use in an exact place and at a precise time. An object or action that may be perfectly suited to one situation as "sustainable" may not be as sustainable when taken out of that context. For example, raspberries are a nutritious and relatively ecofruit when consumed locally, but when they are transported by air from Chile, the emissions associated with transportation make them a much less sustainable option. In California in 2005, imports by air of fruit, vegetables, and nuts by air contributed more than 70,000 tons of CO_2 (National Resources Defense Council 2007).

This complication of location and context-specific criteria for sustainability presents a challenge when designing to communicate sustainability, because designers have to keep in mind the possibility for variables and the importance of communicating context.

14.3.1.1 Lack of "Context Connection"

The best way to understand the reasons behind sustainability is of course to experience the resulting impact firsthand. Without experience, everything becomes the equivalent of a flyover zone: disconnected from our understanding.

Context is the set of conditions that exists around an object or idea, and in terms of which allows the item to be fully understood. Context indicates what factors weigh upon a single idea or object. Because sustainability itself is so heavily dependent on context, I believe we need to help audiences better understand the overall environment that a particular action or object operates in—before we can communicate what sustainability "means" in that particular scenario. Therefore, there is a need for education tools that showcase the "big-picture" in a way that communicates the complexity of the system, while enabling the viewer to envision themselves—and their own role—within it. Systems thinking (discussed further under Opportunities) paired with visual storytelling is well equipped to address this. In this way, understanding of the reasons behind sustainability can be made clear to an individual, enabling one to actively participate in more sustainable choices and to counteract the many challenges that exist when communicating sustainability.

14.3.1.2 Data Deluge

Human rights reports, corporate social responsibility documentation, global warming indicators, air and water pollution, human health studies, toxicity measurements, rates of resource depletion, food miles, water and carbon footprints, etc. The list of sustainability-oriented data being collected goes on and on. Sustainability encompasses more than one issue: It is an enormous and expanding field that will only continue to grow as society increasingly feels the global impacts of issues like climate change.

There is no lack of quantity of information: only negligence in the quality with which it is rendered or framed. Today's audiences are accustomed to massive quantities of information, but they need a bit of guidance to decipher it, in addition to adequate interest to take the time to dig into the info. Information has to be framed to condense complexity into easily understandable—and actionable—ideas.

When communicating on sustainability-related data, there is a need to make information personable to individuals to help people relate to the idea.

14.3.1.3 Being Human

Unfortunately, the capacity of our brain is limited to a degree. The idea that our actions can have greater impacts outside of our immediate circle of influence is difficult to grasp. Our lack of ability to comprehend abstract concepts that associate present events with distant catastrophes is a key challenge in cultivating more sustainable behaviors. We tend to be focused on the more immediate, daily events: We have grocery shopping and kids' activities to worry about—how we will plan for retirement, etc. It is these daily activities and fears that occupy the vast majority of our thoughts. As Deepak Chopra notes, "We have approximately 60,000 thoughts in a day. Unfortunately, 95 % of them are thoughts we had the day before." To engage people with sustainability, there is a need to tie these immediate, everyday experiences to the big-picture.

Another limitation of being human makes it hard for us to grasp our own, distant influence. As infants, we learn about object permanence: the understanding that objects continue to exist even when they cannot be observed. Yet somehow, as adults, we still struggle with the idea that an action we take can have impacts—though they remain invisible to us—that affect the world around us. In seeking to communicate sustainability, the difficulty of this task is compounded by the aforementioned complexity and variation of the set of criteria that defines sustainability in any given situation.

In a society inundated with messages, there are limits to what can be remembered. We group information together in order to better connect it in our memories to previous experiences. For example, the 7-digit telephone number was developed based on research that people can only retain an average of 7 pieces of information at a time.

Fortunately, visual—especially vivid—imagery can help the brain make connections and increase our memory and understanding. According to McKenzie-Mohr in his book *Fostering Sustainable Behavior*, "vivid information increases the likelihood that a message will be attended to initially, a process called encoding, as well as recalled later… furthermore, because it is vivid, we are more likely to remember the information at a later time. This last point is critical, since if the information is only remembered fleetingly, it is not likely to have any long-lasting impact on our attitudes or behavior" (Mckenzie-Mohr and Smith 1999).

Compelling visuals are key to helping us humans consume and remember information; this is compounded by the fact that, according to data visualization expert Stephen Few, "approximately 70 % of the sense receptors in our bodies are dedicated to vision" (Few 2004). These benefits make visuals uniquely situated to communicate context—an understanding of which can aid in establishing a connection between individuals and the "big-picture" of sustainability.

14.4 Opportunities

14.4.1 Visual Storytelling

Buddhist monk, teacher, author, poet, and peace activist Thich Nhat Hanh wrote, "If you are a poet, you will see clearly that there is a cloud floating in this sheet of paper" (Nhat Hanh and Kotler 1987). He weaves a story that extrapolates from nature's own processes: The cloud produces rain that, in combination with sunshine, feeds the farmer who cuts the tree to produce paper.

The idea that everything is connected is not new. John Muir's famous quote "When we try to pick out anything by itself, we find it hitched to everything else in the Universe" (Muir 1979) itself is often repeated.

I believe that we need an interesting story to show us how connected objects and actions within our daily lives really are: A visual path that brings together the invisible threads that tie the world together.

Jo Confino makes the case that visuals add the spice that can evoke a "powerful human response" to the "feelings of disconnection from the planet and its problems [that] preventing people from investing in change" (Confino 2013).

Visual storytelling has been prevalent throughout the development of society. The use of narrative illustrations in stained glass windows rose to prominence during the Gothic period. The images depict scenes from the bible to communicate—and aid the memories of a largely illiterate populace (Fig. 14.4).

Other forms of visual narratives are used today. For example, airline emergency evacuation instructions are illustrated with minimal words in order to be comprehensible to passengers of multiple languages, or those of varied reading ability—such as those too young to read, or the elderly (Fig. 14.5).

In another example, plaques installed on the 1972 and 1973 Pioneer spacecraft map the location of earth within the solar system and include graphics of a man and a woman, the spacecraft, as well as other scientific information about our habitat (NASA 2007). Installed on the first human-built objects to leave the solar system, the visualizations carry a message from mankind to any extraterrestrials that may come across the craft (Fig. 14.6).

Ideas from the above examples can help make sustainability "legible" to a greater population in order to achieve a collective, long-term vision of balanced life on earth.

One of the sturdiest arguments for using visuals is that symbols and images (especially stylized images like icons and pictograms, as opposed to photographs,

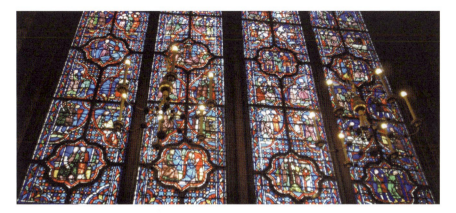

Fig. 14.4 *Sainte-Chapelle* (Paris) window #14: Kings I & II, Solomon I & II

Fig. 14.5 In-flight safety card: an example of highly visual communication

which represent exact conditions) are uniquely situated to show context. In her book, *Thinking in Systems*, Donella Meadows states that to discuss the sustainability of a system properly, "it is necessary somehow to use a language that shares some of the same properties as the phenomena under discussion. Pictures work for this language better than words, because you can see all the parts of a picture at once" (Meadows and Wright 2008).

This suits sustainability just fine: As stated earlier, sustainability is a difficult topic to pinpoint in general. Visuals can quickly ignite interest and provide background context that might otherwise take an entire paragraph to generate. They can also help us remember ideas. Connie Malamed comments, "The imagery used to

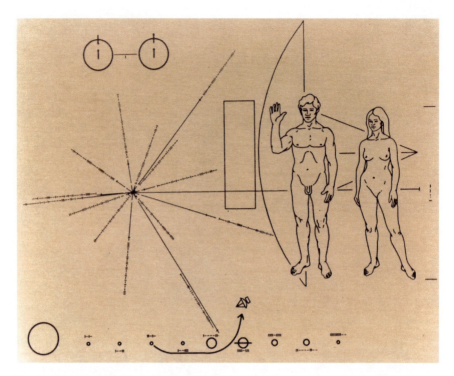

Fig. 14.6 Engraved plaques on Pioneer 10 and 11 spacecraft record human life and the location of earth within our solar system. Photographed by NASA Ames Research Laboratory

represent complex data provides a venue for retaining that information in our memories. While we might not remember exact numbers and percentages, there is a higher chance that the concept or outcome will be remembered" (Malamed 2009).

Because of their precision, words and numbers excel at communicating details, where iconographic images are better adapted to big-picture contexts and narrative storylines. A combination of graphic indication with narrative (visual or text based) and written explanation can create a visual story. This same pattern is evident in language in general, as "all full writing systems are a mixture of phonetic and semantic symbols" (Robinson 1995).

Stories are a part of culture. Jonathan Gottschall notes that stories "help us navigate life's complex social problems" (Gottschall 2012). Stories can also be memorable and motivate action. Made visual, a storyline situates the viewer into the context of a specific landscape, allowing them to understand the situation from their own perspective within the system.

Visual stories also hold tremendous potential for communicating the life stories of products—in particular because a product's life is linear: Its materials are harvested and processed, then the product is distributed and used, and finally, the product retires to the landfill or is recycled.

Fig. 14.7 A section of the repeating narrative of a compostable product

* **Bambu Backstory** (Fig. 14.7)

Developed for bambu, this narrative information graphic (by the author) describes the cycle of a compostable product from harvest through production and consumer use to its return back into the earth. Developed as a promotional piece for the company, the story touts the benefits of the product and shows how ecolabels and certifications fit within the backstory of the product. Designed as a repeating pattern for use on packing tape (among other applications), the structure reinforces the concept of a continuous, closed-loop life cycle.

14.4.2 Systems Thinking and Communicating Life Cycle

Sustainability is a balancing act. It requires an equilibrium that enables a society to endure—to evolve—in a beneficial way. And the balancing scales do not exist in a vacuum: The world we live in is a tangled system in which an infinite number of variables are at work at any point in time. It is impossible to draw boundaries that dictate where a river's influence stops, for example. However, we can describe—and visualize—connections to help us understand the links between this river and the produce available in our local grocery store.

Systems thinking involves looking at the entire situation surrounding a particular idea or object, focusing on understanding how parts within a system influence one another. Knowledge of systems and life cycles is important for sustainability communications because it helps people visualize how their own role fits within the system. Placing ourselves within the galaxy of our own environments let us understand what role we play in the big-picture of our society.

Another, more streamlined, way to look at systems is through life cycles. Life cycles are a linear progression through a system that inherently contains narrative stories within it. In the world of sustainability, life cycle analysis (LCA) is a technical method to quantify the sustainability of a particular product. Working with these ideas alone can create intriguing visual communications to subtly educate audiences on the backstory of a particular product.

Fig. 14.8 The life story of the T-shirt wraps this garment

* **T-Shirt that Tells Its Own Story** (Figs. 14.8 and 14.9)

This T-shirt advertises the story of its own life.

Its individual life cycle is screen printed across the shirt to show specific details related to journey of the garment.[1] The location of the harvested cotton, processing of the fabric, and the journey to the retail store are illustrated. The T-shirt's life after it reaches the consumer is imagined, such as how many times it is washed (and what kind of energy is used to power the washing), and whether or not it is recycled or reused. The T-shirt's own artifacts from the production process (its tags) are cut from the inside and stitched into the relevant part of the T shirt's own life cycle.

14.4.3 In-context Connections

In elementary school, I would often daydream about visible lines connecting my classmates through their invisible associations: Which students might have a crush on others, who might be somehow distantly related, who were neighbors, etc. These lines were codified, red threads that sprung out from the top of our heads and moved with each of us. They could only be seen with special powers—which I, of course, attained in my daydream.

---

[1] The design was developed specifically for T-shirts sold by Droog Design in the Netherlands in 2006.

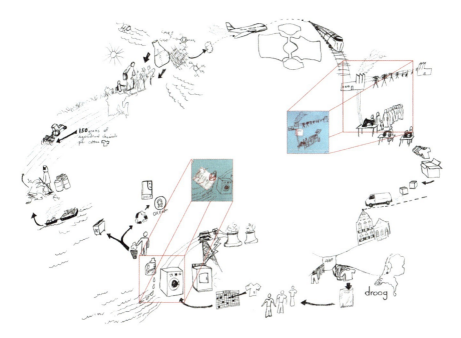

Fig. 14.9 Details of the illustrated, visual story that is printed on the T-shirt

These days, I still want to see the invisible threads that connect everything we interact with to the greater world around us. Author and activist Umberto Eco once said, "There are always connections, you only have to want to find them."

I believe that making these connections visible will contribute to bringing clarity to our complex worlds: Visualization can help us value the system that we live in a wee bit more. My work, therefore, seeks to highlight the threads of connection between our everyday actions and the larger, social, and environmental impacts of these actions on the world around us.

The notable idea of my daydream was that all relevant information was embedded within daily life, augmenting reality and presenting physical visualizations of invisible pieces of collective knowledge in a way that helps clarify the situation.

Likewise, embedding sustainability information into our ordinary lives pulls the research out of the clouds and grounds it in our everyday experiences as consumers and citizens. "When an audience is engaged in the message, a personal responsibility to that message often results" (Ehmann et al. 2012) (Fig. 14.10).

In-context information is that which is experienced at the point it is most relevant. This might be, for example, at point of sale, during a bicycle/auto commute, while turning on a light switch, etc. In context is in contrast to a repository of information that is housed online, in a gallery, or some other sterile space that limits external, "real-world" connections. In-context displays of information force the

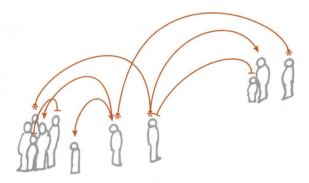

Fig. 14.10 Coded red threads connecting individuals through their shared experiences

Fig. 14.11 A 360° view of the illustration inside the periscope. The viewer only sees a small portion of the illustration at a time, enabling them to discover elements as they navigate the digital space

participant into a direct correlation between his or her own scenario and its larger, global impacts.

* **Behind-The-Scenes Periscope** (Figs. 14.11 and 14.12)

This project enables viewers to, on location, see "into" the hidden systems at work within a neighborhood.

The Västra Hamnen residential development of Malmö integrates impressive systems of ecological sustainability into the infrastructure. I wanted to use design to reveal the energy and environment-saving concepts at work in the neighborhood to help visitors and residents understand more about these concealed technologies.

A 360° illustration is viewed by looking through a rotating periscope.[2] These visuals overlay reality and lead viewers through the landscape to playfully discover a network of behind-the-scenes elements.

---

[2] This project was done as part of an artist residency at MEDEA in Malmö, Sweden with help from Unsworn Industries and the city of Malmö—who produced the periscope.

Fig. 14.12 The periscope on-site in the Västra Hamnen neighborhood in 2010

14.4.4 Bridging the Gap: Science and Emotion

In the words of Plato, "Human behavior flows from three main sources: desire, emotion, and knowledge." So changing behavior needs more than data and information. It is unfortunate then that much of the communication surrounding sustainability has been all about the data. Perhaps this stems from the all too common idea that emotions and science do not mix. Certainly, there is a merit to the idea of presenting facts without choosing sides. But emotions are a key part of the currency that the world runs on. In his book, *Fostering Sustainable Behavior*, McKenzie-Mohr also suggests that information alone is not the solution: "the diversity of barriers which exist for any sustainable activity means that information campaigns alone will rarely bring about behavior change" (Mckenzie-Mohr and Smith 1999). As stated previously, he makes the argument that information that is "vivid, concrete and personalized" (Mckenzie-Mohr and Smith 1999) inspires attention and increases the likelihood of the information being recalled later.

A bit of emotional appeal can go a long way. Visuals can bridge the gap between science and emotion because, when designed accordingly, they can present the scenario without shaming or forcing an idea upon the audience. By educating individuals, people will be able to be more creative in finding solutions that fit their own situations, and better prepared to make future informed decisions independently. When people can start to understand the "big-picture" of their seemingly

Fig. 14.13 An installation of the project on a bike path in Minneapolis, MN (US). The projected visualization updates—in a show of animation—with each passing cyclist

small everyday decisions, they also satisfy the human need to feel part of something larger than oneself as they move up the rungs of Maslow's hierarchy of needs.[3]

Several of the projects I have worked on have significant data behind the scenes, but I try to present the information in a way that makes it relative to the individual.

* **Bicycling Counts** (Figs. 14.13 and 14.14)

Putting data from scientific studies and transportation agencies into context, this project bridges the gap between science and emotional, individual, and real-time experience.

Inspired to celebrate the collective impact of cyclists, *Bicycling Counts* uses bicycle counters to "show" what the number of passing bicyclists means in terms of environmental, financial, and individual social impact. The project has been installed in two locations: each with very different data behind them.

In Malmö, Sweden:
A series of real-time animations was projected on a public graffiti wall as a subtle acknowledgment of the larger contributions of passing cyclists. The visuals animate randomly and change over time. The data for the animations were derived from research done by Malmö's regional transportation organization.

In Minneapolis, Minnesota, USA[4]:
This project counts passing bicycles and, in real time, visualizes each cyclist in terms of individual and collective financial savings: to the individual as well as for society. This

---

[3] Psychologist Abraham Maslow's hierarchy outlines the order in which individual needs must be met before one can move 'up' to other needs. They occur in this order: Physiological, Safety, Belongingness and Love, Esteem and Self-Actualization.

[4] This project was sponsored by Center for Energy and Environment.

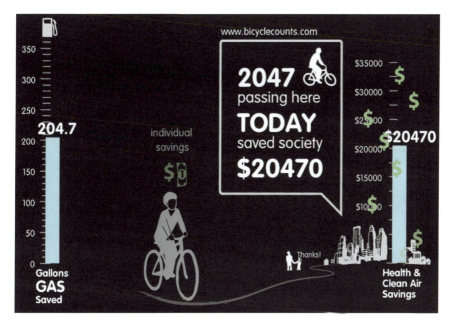

Fig. 14.14 Detail of the Minneapolis projected animation showing the animated elements

mobile version of the project was installed on different bike paths in Minneapolis during Bike-Walk week 2012.

This iteration of the project focuses on the savings to society that result from choosing to bicycle *vs.* driving a car. The project also highlights how individual actions contribute to a healthier environment.

Each iteration of the project involved complex calculations based on existing research and available data. I am no expert in this sort of research, so every decision and calculation made was documented and then reviewed by a team of experts to ensure that all the pieces matched up. The data and decision process were then made available to the public via the project Web site[5] and blog.

14.4.5 Key Practices

There are a handful of practices that I have found to be important when designing data and information to communicate sustainability concepts. Many of these points are not unique to sustainability communication; so I will highlight the ways in which they especially support the communication of social and environmental issues. They are as follows: Start with the individual, layer information, present

---

[5] URL, January 19, 2014: http://www.bicyclingcounts.com.

positively, involve emotions, let audiences make their own decisions, enable interaction, and provide access to supporting data.

14.4.6 Start with the Individual

The vision of a polar bear stranded on a chunk of ice floating out to sea has been used to communicate climate change because it pulls on our heartstrings.

Unfortunately, I live far from the Arctic and its subsequently threatened polar bears. (I assume those residing at the nearby zoo are well cared for.) So although it is a touching image, and I have concern for the welfare of these majestic animals, they are outside my realm of immediate influence. As I go about my daily activities, I do not feel that there is anything I can do about their impending doom. It is debatable if the stranded polar bear approach does much to change behavior because it is so far removed from our individual, perceived realm of influence.

Sustainability communications need to focus on the individual consumer's realm of impact. Borrowing from Stephen Covey's idea, the "Circle of Influence" (Fig. 14.15) contains the things that we feel that we have control over, as opposed to the Circle of Concern, which contains everything that we would like to address (Covey 1989): Helping people understand how they can do something about a particular sustainability topic helps pull an idea into the inner Circle of Influence. Likewise, outlining how a topic can benefit or inflict damage on a personal level can pull such a topic into the Circle of Concern.

In his book *Fostering Sustainable Behavior*, McKenzie-Mohr outlines two approaches to encouraging change: The attitude–behavior approach, which "assumes that changes in behavior are brought about by increasing public knowledge about an issue" (Mckenzie-Mohr and Smith 1999) and the economic self-interest approach, which "assumes that individuals systematically evaluate choices… and then act in accordance with their economic self-interest" (Mckenzie-Mohr and Smith 1999).

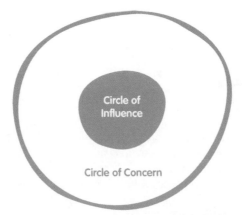

Fig. 14.15 Author's representation of Stephen Covey's Circle of Influence within the Circle of Concern

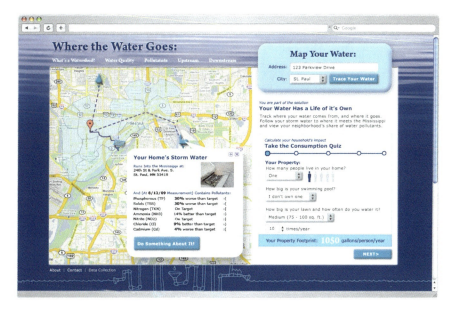

Fig. 14.16 The mock-up outlines how a system could connect household water usage to estimated neighborhood runoff

This second approach highlights the need to connect issues with individual's wallets in order to engage their attention and place a topic within the Circle of Concern. Fortunately, most individuals are concerned with more than their own, singular existence. However, the circle of concern does tend to center first on people, places, and ideas most close (in proximity or psychologically) to the individual: family, community, and region, then nation, world, etc.

This key practice is the need to connect an audience with the impact of a particular outcome based on their personal concerns. People need quick proof that what is presented is relevant to them on an individual level—help them by providing easy entry points into the information. In the following project, I have outlined an approach to connect individual residents with their water runoff in order to educate about water pollutants.

A similar concept—albeit different resource—by the Sacramento Municipal Utility District awards smiley faces on the energy bills of households that score low on energy use (Kaufman 2009). This project taps into the very human trait of competition; households are compared with similar homes.

* **Track Your Water** (Fig. 14.16)

This concept Web site allows Twin Cities[6] residents to trace where their water comes from, and where their wastewater and runoff goes. Real-time data on pollutants within runoff

---

[6] Minneapolis and St. Paul, neighboring cities in Minnesota, USA.

water from the residents' own watershed are included. The system is also tied to a water footprint calculator that customizes the quantities of runoff and pollutants to estimate the resident's individual "share."

The project aims to help people visualize the flows of their water and wastewater, and the processes involved. The mock-up was developed with input from multiple watershed organizations in the Twin Cities. The long-term idea was to develop a campaign to pit watershed "neighborhoods" against each other to compete for the lowest runoff pollutants.

14.4.7 Involve Emotions

Visuals provide context. Text, details, captions, and data can facilitate trust. But information alone does not inspire action.

There is a reason we are moved by music and drawn to beauty. These things tap into our emotions and contribute to our experience of being human. Some research suggests that emotional connections are precursors to behavior change.

In one example, following public outrage over the capture and killing of dolphins during tuna fishing, consumer behavior shifted in support of "dolphin-safe" ecolabels (Teisl et al. 2002). Dolphin-safe tuna worked because it appealed to people's love for dolphins. Who would not want to protect these intelligent creatures that seem to have a permanent smile on their faces? (A campaign to protect endangered spiders may have an entirely different outcome.) The campaign also actively directed consumers toward what they could do to help the dolphins (as opposed to the polar bear example).

Certainly, the smart incorporation of positive emotion into sustainability data can help audiences connect with the information and foster beneficial behavior change. According to game designer and sustainability advocate Jane McGonigal, there are 10 categories of positive emotions: creativity, contentment, awe and wonder, excitement, curiosity, pride, surprise, love, relief, and joy. So there is no shortage of possibility to work with emotions to help make data meaningful.

In my own work, I try to build connections from the subject at hand (whether that be the locality of food purchases or the social benefits of bicycling) to people's own, daily lives. For example, the Malmö, Sweden, installation of the *Bicycling Counts* project visualizes the benefits of bicycling in terms of seasonal Swedish pastries (counting calories burned), money saved, and growing vines in iconic Swedish patterns that gradually flourish as CO_2 disappears from the installation.

The more people can "feel" the data—and understand how that data play into their own lives—the better connected they can become to the topic that the data represent.

14.4.8 Present Positively

No one wants a finger waggled in his or her face. In the 1980s, the ecomovement has tried the shame game: But on an individual scale, it did not work.

It is human nature to take a defensive stance when your own beliefs are challenged. Therefore, presenting information in an approachable, non-accusatory manner is a good way to avoid triggering this unconscious resistance to persuasion. A culture of positivity seems to more effectively nudge people toward more sustainable behaviors. Niki Harré states, "Positive experiences are an important way to inspire and motivate people, as they attract us toward the activity or message being promoted. More than that, however, positive moods bring out important personal qualities that are essential to social progress" (Harré 2011). Positive feelings open us up to new ideas.

Being that my work aims to educate audiences on topics of sustainability, I try to focus on the positive impacts that individuals can have on the world around them, empowering people in their roles as consumers rather than berating them for not yet having done enough. Sometimes, doing this can be as simple as illustrating a subtle reminder of the systems at work that are larger than us as individuals.

* **Water Cycle Dishware** (Figs. 14.17 and 14.18)

An illustration wraps plates[7] and water glasses to tell a positive story of water origins, use, and reuse and remind us that it is a cycle.[8] The visuals become a storytelling tool to educate children (and adults, too) on the principles of the water cycle: flow, evaporation, and precipitation.

Clean water is a necessity for all life on earth. The visual story is a reminder to those of us in the developed world that water comes from, and returns to, nature, and that man, plants, and animals around the world all depend upon clean water. Text on the glass reads, "Clean water… pass it on." Text around the edge of the plate reads "Water. For Life."

14.4.9 Layer Information

Any quantity of information can become too much for an audience when poorly designed. Because of its ever-shifting definition, sustainability is a particularly complex topic to communicate. Overly intricate information can quickly lead to overwhelm. To avoid this, I have found that layering the information helps people—even the most knowledgeable audience—dig into their own level of desired depth.

Presenting information in successive layers of detail provides the opportunity to engage an already knowledgeable audience by reinforcing what they already know—as well as educate a new audience step by step. This is done by first displaying an overview that tells the story and provides context for the message. Then, successive layers of detail can build off the core overview in stepped layers of hierarchy.

---

[7] The plate is produced by Felissimo.

[8] This project was supported by a fiscal year 2009 Artist Initiative Grant from the Minnesota State Arts Board. This activity is made possible in part by a grant from the Minnesota State Arts board, through an appropriation by the Minnesota State Legislature and by a grant from the National Endowment for the Arts.

Fig. 14.17 The story of the life cycle of water wraps a plate

Fig. 14.18 A visual reminder that our water comes from somewhere, goes somewhere, and is a cycle

It is a fine line between clutter and clarity. In the infamous words of Edward Tufte, "To clarify, add detail. […] Clutter and overload are not attributes of information, they are failures of design." Details can help put sustainability stories into context and help audiences understand the nuances of why something that is more sustainable in one area might be less so in another.

14.4.10 Audiences Make Their Own Decisions

Most of my work involves presenting information in an unbiased, fact-based manner. Of course, I do have a not-so-hidden agenda: I aim to help people understand social and environmental issues in order to nudge toward more sustainable behaviors. I believe that by presenting the facts, and by clarifying the connections between this information and the big-picture issues, viewers can better understand the situation. I do not fill in all the lines to connect these dots: I try to leave a bit of space for the audience in every piece that I do. By allowing a viewer space to make their own connections, I hope that they arrive at a decision independently (albeit guided by my work).

This approach allows the individual to make their own choice, thus taking ownership of the decision, rather than feeling that it has come from outside of themselves. The feeling of independently making a choice thereby becomes a more permanent part of the decision-making process. And once we have made a choice, we become subconsciously tied to that commitment. In his book on persuasion, Robert Cialdini writes, "We will encounter personal and interpersonal pressures to behave consistently with that commitment. Those pressures will cause us to respond in ways that justify our earlier decision" (Cialdini 2007).

In my work with *ProductBio*, this principle comes into play with the way we structure information: We provide product details without assigning rankings. Many decision makers within the sphere of sustainability have different opinions and understanding about what is important to communicate when it comes to sustainability—and of course, these vary greatly by product type. By aggregating data from multiple stakeholders, we can begin to understand what is being measured within a particular product category. Consumers are then able to look at what the top impacts are within a particular category and decide which attributes align with their own, individual values.

* <ProductBio.com> (Figs. 14.19 and 14.20)

As a founding partner of *ProductBio*,[9] I am working with a team to visualize the whole story behind how everyday products are made. ProductBio is a system of narrative information graphics that arms consumers with the information they need to gauge the eco-worthiness of products as they align to their own, individual sustainability values.

---

[9] *ProductBio* project, January 19, 2014: www.productbio.com.

Fig. 14.19 An on-package visualization of a product's green, ethical, and local sustainability information

Fig. 14.20 An in-store poster that showcases the story of a company's product and practices and the ways in which those impacts matter

Traditionally, ecolabels or similar "stamps of approval" have helped bridge communications between companies and consumers about environmentally friendly products. But as the movement toward sustainability picks up speed, a more effective tool is needed to clarify how products impact our surroundings. An array of green-marketing claims—organic, local, fair trade, zero emissions, carbon neutral—have surfaced, thanks to growing segmentation within the sustainability market. Unfortunately, this causes confusion among consumers, in part because the sustainability of a product is often dependent upon its context. Big-picture connections between the products in consumers' hands and the outside world are required to communicate the sustainability of a product. Clearly, an evolution of the ecolabel is needed to address the growing needs of the industry.

ProductBio is an integrated set of customizable packaging labels, in-store signage, and a searchable, online database that reveals a product's social and environmental impact. The system visually describes the life cycle of a product—from production to final disposal or reuse—with graphics, facts, and captions in a narrative storyline. Just as nutrition labels provide consistent information on food packaging (calories, total fat, sodium), *ProductBio* notes a product's sustainability practice—and the resulting impact throughout its life cycle. The online experience enables consumers to search for products to match their own values and companies to post data to back up their claims and explain their sustainability initiatives.

The system is built on machine learning technologies: Information is pulled from online sources as varied as ecolabels, corporate social responsibility (CSR) reports, life cycle analysis (LCA), and company marketing information. This automation also enables us to provide vendors additional insight on consumers' values and areas for improvement within their own products' supply chains.

14.4.11 Enable Interaction

Information is much more interesting—and memorable—when it can be explored or personalized to a specific situation. Experiencing data through real-time and in-context modalities helps people relate the information to their own experiences and provides the ideal situation to connect digital data with physical interaction.

Real-time refers to data that is pulled from the moment it happens and is connected it to a specific point of time. The value of this is particularly strong in highlighting an instantaneous link between an individual's action and a "big-picture" impact.

* **<TraceProduct.Info>** (Fig. 14.21)

TraceProduct[10] is a prototype for an in-store, retail system for displaying information on grocery product backgrounds at point of sale. The project helps individual shoppers connect to the backstories of their purchases by putting everyday purchases into global context, focusing on the ways in which these products connect us to the larger world.

During checkout, an animated map interface parallels the usual price list to show the geographic origins of each item. Accompanying graphs visualize local products as a

---

[10] This project was developed in close collaboration with The Wedge Co-op grocery store and was supported by a 2010 [Art]ists On the Verge fellowship from the Northern Lights Foundation.

Fig. 14.21 *TraceProduct* is a prototype for a grocery store point-of-sale interaction that would help people better understand the context from which their products come, and the percentage of overall purchase that is local *vs.* global

percentage of total purchase by number of items as well as by total dollars spent. A ticker of news headlines for the countries from which products originated also scrolls across the bottom of the screen.

Each receipt has a custom code so the map can be explored at home in terms of global statistics, or used to monitor changes in purchasing patterns over time.

I see this project as an iteration toward a "next step" for communicating product-level sustainability: It builds off work I have done to visualize the backstories of individual products, where I recognized that implementing custom labels on all products could take a lifetime. I wanted to explore how product sustainability could be communicated at the next highest level of interaction for the consumer: the retail location.

The challenge with this approach is that point-of-origin information is not yet available in the USA in an easily accessible format. There is legislation that is aiming to encode this information in UPC over the coming years, which would make possibilities for projects like *TraceProduct* more realistic for large-scale implementation.

14.4.12 Provide Access to Supporting Data

Always, cite your sources. Leave a clear and accessible trail to lead curious viewers to your source data.

"Walking the talk" when communicating on topics of sustainability requires being transparent with your own data and processes. In parallel with worldwide calls from the data design community to "free data," consumers are increasingly demanding transparency from companies as part of the sustainability movement. Transparency has the power to engage communities and can lead to brand loyalty. There are also plenty of sustainability skeptics who are already critical of the topic: Do not give them the opportunity to debunk your visualization out of the gate.

In much of my work, it is difficult to place all the background data, sources, and calculations within the experience itself. However, those details can be located on a Web site or other public forum related to the project. In the case of the *Bicycling Counts* project in Minneapolis, the calculations, sources, and assumptions that formed the narrative of the project were first floated by a panel of experts, and then posted on the project Web site.

14.5 Data Design for Sustainability—and Beyond

As a society, we need to evolve new habits that will support a socially and environmentally balanced future. Given the complex and varied meanings of sustainability, designing data is essential to help audiences grasp the information in ways that it can be acted upon.

Data design already does a good job of helping audiences navigate information: Special consideration needs to be given when using data design for communicating sustainability. This includes tactics to give information emotional appeal and make data relevant to an individual's daily life.

Designing information can help consumers understand their role in positive global change, by clarifying the link between the individual and the larger effects of our daily activities. Bringing understanding of a point of action closer to consumers' own perceived realm of influence can help them understand how global solutions relate to seemingly small, everyday choices.

Imagine if your next trip to the grocery store results in a cart full of products that you *know* are good for people and the planet, aided by visual labeling that helps consumers understand at a glance not only the green or ethical claims of a brand, but also the implications of these actions on the big-picture.

Imagine flicking a light switch and knowing exactly how power is being used in your home, where it comes from, and the global implications of the kind of power that is used.

Imagine an immediate sense of pride knowing how your on-progress bicycle trip saves your own, and your community's resources and health.

Imagine if choices didn't have to be about comparing fair trade *vs.* organic coffee but about the role that the coffee plays in neighborhoods, ecosystems and economies and around the world. The data are out there, just waiting for a good design solution to effectively communicate it.

There is a bounty of opportunity to engage and motivate audiences toward more socially and environmentally sustainable practices, whether that be communicating climate change or helping individuals connect on an emotional level to the greater impacts of their seemingly insignificant actions, starting from the first morning cup of coffee and continuing throughout everyday activities.

References

Cialdini R (2007) Influence. Collins, New York
Covey S (1989) The seven habits of highly effective people. Simon and Schuster, New York
Ehmann S, Bohle S, Klanten R (2012) Cause and effect. Die Gestalten Verlag, Berlin
Few S (2004) Show me the numbers. Analytics Press, Oakland
Gottschall J (2012) The storytelling animal. Houghton Mifflin Harcourt, Boston
Harré N (2011) Psychology for a better world. University of Auckland, Auckland
Malamed C (2009) Visual language for designers. Rockport Publishers, Beverly
Mckenzie-Mohr D, Smith W (1999) Fostering sustainable behavior. New Society Publishers, Gabriola Island
Meadows DH, Wright D (2008) Thinking in systems. Chelsea Green Pub, White River Junction
Muir J (1979) My first summer in the Sierra. Houghton Mifflin, Boston
NASA (2007) The Pioneer missions. NASA. URL: http://www.nasa.gov/centers/ames/missions/archive/pioneer.html, 19 Jan 2014
National Resources Defense Council (2007) Food miles: how far your food travels has serious consequences for your health and the climate. National Resources Defense Council. URL: http://food-hub.org/files/resources/Food%20Miles.pdf, 19 Jan 2014
Hanh TN, Kotler A (1987) Being peace. Parallax Press, Berkeley
Kaufman L (2009) Utilities turn their customers green, With Envy. The New York Times. URL: http://www.nytimes.com/2009/01/31/science/earth/31compete.html, 19 Jan 2014
Robinson A (1995) The story of writing. Book Club Associates, London
Teisl MF, Roe B, Hicks RL (2002) Can eco-labels tune a market? evidence from dolphin-safe labeling. J Environ Econ Manag 43(3):339–359. Available online: http://dx.doi.org/10.1006/jeem.2000.1186, 19 Jan 2014
Confino J (2013) The conundrum at the heart of sustainability. The Guardian. URL: http://www.theguardian.com/sustainable-business/conundrum-heart-sustainability, 19 Jan 2014
Wikipedia "sustainability". Available online: http://en.wikipedia.org/wiki/Sustainability, 19 Jan 2014
World Commission On Environment And Development (1987) Our common future, report of the World Commission on Environment and Development. Untitled conference, Annex to General Assembly document A/42/427, Development and International Co-operation: Environment
World Conservation Union, United Nations Environment Programme, World Wide Fund for Nature (1991) Caring for the earth. Union Internationale pour la Conservation de la Nature et de ses Ressources, Switzerland

Author Biography

Arlene Birt is a visual storyteller, information designer, educator and public artist whose work makes background stories visible in order to help individuals connect emotionally to the specific social and ecological-sustainability impacts of our everyday actions. Arlene received a Fulbright grant to research visual communication methods to explain sustainability.

She is invited to speak internationally and collaborates with organizations in the United States, European Union, and United Kingdom. Her work on sustainability—which rides the line between art and communication—has been featured in numerous publications.

For more information and contact:

http://www.backgroundstories.com
https://twitter.com/aBirt

Chapter 15
Encoding Memories

Sha Hwang and Rachel Binx

Abstract Our own personal histories have become balkanized, stored across social networks, Web services, and smartphone applications. With the new affordability of rapid prototyping techniques such as laser cutting and 3D printing, mass customization of physical products is now possible. How then do we design products that tell stories and hold memories as unique as the people that purchase them?

15.1 Meshu

Meshu began as a response to a simple question—what could we build together that encapsulated our shared interests? As we laid our fascinations out, threads began to emerge: working with data, working with memory, and creating physical objects. We were interested in moving away from the digital tools and experiences we had been accustomed to building. The idea for *Meshu*, then, came from our experiences with expressing our own personal data and telling our stories visually.

In designing and building *Meshu*, we became hyperconscious of the fracturing of our experiences online across many social networks. And as we began our tests with manufacturers, the wonder and power of rapid prototyping became clear. But it was when *Meshu* was released out into the world and we started receiving orders and communicating with customers that we truly began to understand the emotional weight of a tool like *Meshu*. *Meshu*, rather than being a store for generative work, became much more about cherishing memories and encoding them into physical form (Figs. 15.1 and 15.2).

S. Hwang (✉) · R. Binx
340 S Lemon Ave #2469, Walnut, CA 91789, USA
e-mail: hi@meshu.io

© Springer-Verlag London 2015
D. Bihanic (ed.), *New Challenges for Data Design*,
DOI 10.1007/978-1-4471-6596-5_15

Fig. 15.1 *Meshu*, Rachel Binx and Sha Hwang, 2012. Laser cut acrylic necklace. Model: Enid Hwang. Photograph: Nadya Lev

15.1.1 The Balkanization of Personal Data

Like many people these days, we both spend a lot of time interacting with people on social networks and building up a large amount of personal data. And there is no data like personal data these days. The rise of companies such as Twitter, Facebook, and Google and the competition between them have turned our own personal data into something more like currency. With our personal data, these social networks can better target advertising, by gender, age, or interests. People talk of data as oil, something to be drilled, stored, processed, and sold. And in this frame, owning data becomes crucial. Less like a collective effort of building infrastructure, competition over online advertising is seen as a zero sum game, and companies behave more like warring states, competing not for land but attention. This balkanization of personal data becomes problematic because the data, in the end, are our own lives.

> Falling in love, going to war and filling out tax forms looks the same; it looks like typing.
> —Quinn Norton

Interacting online is not the cyberspace or virtual reality imagined in the 1980s and 1990s anymore, but more fleeting one, composed of many small glints of interaction. We live on social networks, but we talk with different languages in each

Fig. 15.2 Laser cut bamboo earrings. Model: Bad Charlotte. Photograph: Nadya Lev

one, share slightly different social circles, and engage with different parts of the world. Cyber spaces, a fractured and many layered where.

15.1.2 Visualizing Personal Histories

And it is the desire to piece back these fractured and fleeting memories that data visualization emerges. There are many examples of Web sites that beautifully display activity on different social networks. Alongside.co[1] visualizes Foursquare activity, Intel's Museum of Me[2] visualizes Facebook activity, and Vizify[3] visualizes Twitter activity. All these tools pull down data from these networks, carefully sift and sort them to present back to the viewer. In these cases, visualization acts as a mirror, a way for people to see a reflection of themselves, their activities, and their own personal histories (Fig. 15.3). And oftentimes, this is the first time people have seen their histories on these services.

---

[1] URL, February 5, 2014: <http://alongside.co/>.

[2] URL, February 5, 2014: <http://www.intel.com/museumofme/>.

[3] URL, February 5, 2014: <http://vizify.com>.

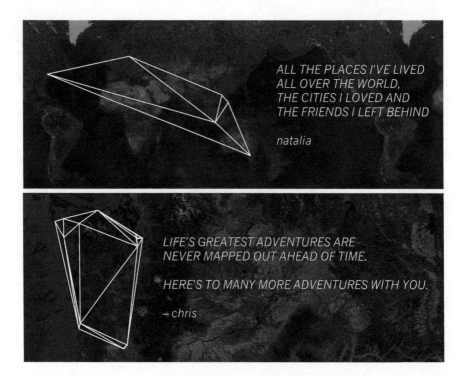

Fig. 15.3 Notes written by customers to accompany each *Meshu*

In 2008, Sha explored this realm by building a tool to visualize a person's music listening habits through their data on Last.fm, an early music Web site that aggregated plays of songs across your iPod, desktop, or the Web. Thousands of Last.fm users ended up creating their own visualizations, which were automatically posted onto a public blog. People with different listening habits ended up with different forms, different color palettes, but the ultimate form still remained abstract. The forms, then, were a sort of generated memory palace, a sculpture whose curves could be learned, contained stories, and held memories.

Later, in 2010, Movity (the company started by Eric Wu, Vaughn Koch, and Sha Hwang), launched a small personal visualization tool of Foursquare checkins, called *Weeplaces*. Unlike the Last.fm project, the data this time were presented literally—a traveling point, moving around on a map, tracing out the path through space of places people had been. The Web site was viewed hundreds of thousands of times, with people sharing their paths. Foursquare at the time did not offer this functionality, and it was exciting to watch people reflect on their own personal travels.

15.1.3 Fractured Memories

At the same time, however, there were sad realizations. Huge gaps in data were common, when Last.fm was uninstalled, when people decided to stop using Foursquare. Most importantly, some of the most exciting stories and paths were lost, as people often did not have access to data plans while traveling. These visualization tools often ended up highlighting the use and disuse of the services, rather than telling holistic stories of a person's life. Just having data did not mean having the moments worth remembering.

In contrast to this sort of service-specific visualization is the work of Nicholas Felton. For his annual reports, his life is the explicit subject, and there is a monumental amount of work done to use an entire suite of self-tracking applications or social networks to record a gapless portrait. In 2011, Rachel produced a project that visualized her entire history of travel, but various APIs could only offer her partial glimpses. She ended up manually entering in her history—to work only with code would have been an incomplete story. As we began to work on *Meshu*, this similar focus on manual entry became necessary to address the scope we wanted. *Meshu* was not built to be peering through a keyhole at someone's personal history, but was built to accommodate telling full stories and not be beholden to any one social network.

Deciding to not wed ourselves to any single data source was a crucial part of *Meshu*. On the other side, the actual algorithm to generate a shape presented its own design challenges. We decided on the Delaunay triangulation, both for its mathematical simplicity and structural efficiency, but mainly because it kept the initial points legible. This legibility was important to us because we did not want to lean on familiar geographies, like *Weeplaces*, nor did we want to have a purely generative work, like the Last.fm visualization. We wanted *Meshu* to be neither random nor literal. To us, there was coldness to a generative approach that made the user feel less like a co-author and more like an accidental participant. On the other end, a far too literal map interpretation felt like exposing too much the personal information entered in. *Meshu*, then, was designed to be abstract but to be easily explained.

15.1.4 Meshu's Matter Battle

Alongside our design and code work for the Web site, we also made many material tests. Receiving our first test prints from Shapeways was a beautiful moment. We had both worked with laser cutters before, and the process there was very direct. We saw the machines working and had both worked with wood and acrylic by hand before, but the silver prints we received felt like magic. The silver pieces, rather than being 3D-printed, were made using a lost wax casting process. In this process, wax is 3D-printed, a plaster mold is formed, and the wax is melted to be replaced by silver. Here then, we had pieces that were removed enough from familiar processes to feel surreal, and the gleaming metal lent real and emotional weight.

Wearing a necklace that we had designed but that had real weight was surprisingly powerful. As our test prints and cuts gathered, the quiet wonders of working with physical materials came back. Rather than digital work, the physical tests we made were casual reminders, things to bump into, and things to catch a glimpse of while walking from room to room. For us, it was exciting to be making work we could touch again, things that could break in a visible way, unlike the quiet and complete nature of code.

In addition to delight, working on physical data visualization produced many new interesting challenges. Designer and architect Bryan Boyer talks about this negotiation with the physical as Matter Battle, the inevitable struggle that occurs when dealing with the real world. Different materials in *Meshu* had different strengths and required different thicknesses. Laser cutting provided more detail, while 3D printing with silver allowed for a cleaner finish. Building a tool that would always export "printable" geometry regardless of input became very important. At the same time, we also needed to be very careful about the esthetics. Slight adjusting of geometry by hand helped preserve legibility through the manufacturing process. It is strange to talk about the hand in a process so otherwise digital and full of technology, but this adjustment acted in a similar way to how exaggerating movement in cartoons made characters feel more realistic. We now refer to this process as the equivalent of kerning in typography, and it is this careful translation between what appears clear on the screen and what will actually return from the manufacturer that remains an important part of maintaining quality on every piece ordered through *Meshu*.

15.1.5 Readymades as Handholds

The tests we made as we were working were of ourselves—restaurants we enjoyed, places visited along a vacation. But we were afraid initially of *Meshu* feeling too obscure—data visualization already a niche, laser cutting, and 3D printing exciting but foreign to most people—so we set about making more generalized pieces like the neighborhoods in San Francisco and New York, or art museum locations in the United States. These premade designs (Figs. 15.4 and 15.5) we hoped to display on the Web site to help people understand the idea of encoding place into geometry, without using personal stories that visitors who did not know us had no frame of reference for.

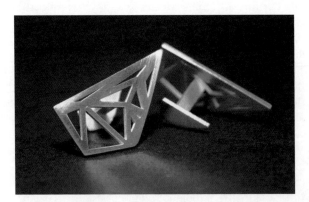

Fig. 15.4 3D-printed silver cuff links

Fig. 15.5 3D-printed white nylon earrings. Laser cut acrylic pendant

We also knew of Nervous System's similar co-creation tools, tools that allowed people to customize parameters on their generative algorithms and order unique jewelry. But we had heard from Jessica Rosenkrantz, one of the partners, that most people played with their customization tools only to buy a premade piece of jewelry. We had hopes of selling personalized pieces on *Meshu* but recognized that generalized readymades would be an appropriate entry point for hesitant visitors and a way to spark ideas on what types of places to enter on the site.

15.1.6 The Shock of Launch

Launching, however, surprised us. The initial wave of orders overwhelmed us, but what became immediately apparent was that people were using *Meshu* in a very different way than we had imagined. We had imagined most of our users to be excited about our work combining data visualization with rapid prototyping, but many of our customers knew about neither. Rather than a shiny new toy, people understood *Meshu* as a way to reflect on the places they had been, and the special moments they had shared with loved ones.

We sold no premade designs. Instead, the pieces that we shipped were ordered by people for their girlfriends, boyfriends, spouses, and friends. When we reached out to our early customers to ask whether they wanted to add a message to the map postcard that came with each order, the outpouring of stories was emotional and overwhelming. Instead of marking favorite restaurants like we had been during our testing, people were reminiscing, tracing routes along first dates, cherishing cities they called home. One boyfriend made a necklace for his girlfriend showing all the places they had met while living on opposite sides of the country. Someone else made cuff links for their husband with all of the stops they made on their retirement cruise. Still others gathered up honeymoons, road trips, or streets they carried with them from their hometown.

15.1.7 A Living Project

And it was here that *Meshu* diverged from being a simple project and turned into a product. With client work, we learned from our mistakes by being more conscious on the next project, but *Meshu* was alive, and ours. We removed the premade designs from our Web site and started asking customers whether we could publish their notes on the Web site. We added the personal note directly into the ordering process because the notes we had received were so powerful. *Meshu* turned from a toy into a memory capsule, and the manual entry allowed people to tell complete stories, not just the stories incidentally recorded on social networks.

And it was this process of retelling a story that helped us understand what *Meshu* allowed people to do. Rather than just commemorating a small facet of their lives, people used *Meshu* by reflecting on their entire lives, drawing out the most important moments, the most influential places, the places called home. And rather than being for themselves, most orders on *Meshu* came as gifts, for birthdays and anniversaries, for weddings, or just surprises. We received stories of how powerful these small pieces of jewelry were and how special and treasured they became. We realized that *Meshu* was less about technology and more about the memories, the act and ritual of reminiscence, and the love expressed in mementos built from shared stories.

15.1.8 Hiding in Plain Sight

The decision we made to hide geography from the final geometry, too, became critical to the stories being told. With street grids or country lines, the jewelry would have no secrets, and the stories would have no privacy. It was the abstract nature of *Meshu* that allowed people to encode quiet and important stories. As one woman told us, she enjoyed the ability for her most precious memories to hide in plain sight and the intimacy possible when finally deciding to share with someone the significance of her necklace. Humorously, another woman requested a set of business cards to hand out on behalf of *Meshu*. Her earrings, she said, attracted a lot of questions and though she appreciated the attention, her earrings told a very personal story that she did not feel comfortable sharing with strangers. And a few times, a friend wearing a *Meshu* necklace being asked, "Where is that?"—a sort of subtle, secret handshake. Something quiet and special being carefully shared.

And in the end, that is what *Meshu* has become, and what it continues to explore. We recently added coasters and rings to *Meshu*, more casual everyday places to hide memories. And rather than a fracturing force, like so many social networks competing with each other, *Meshu* allows people to reflect on their lives, on lives shared with loved ones, and encode memories into objects that both display and protect them. We have realized that while it is a combination of many different emerging technologies, *Meshu* is not about tools or technology at all. No two pieces are alike not because of a generative system but because people lead fascinating and unique lives. And while when we started *Meshu* we jokingly talked about it as the intersection of maps, data, and lasers, it is instead about revisiting memories, treasuring people, and honoring place.

Author Biographies

Sha Hwang (Independent Information Designer) studied and worked in architecture before falling into data visualization and designing for the web. Sha also worked at Stamen, where he worked on projects for CNN, MTV, Adobe, and Flickr.

After Stamen Sha cofounded a startup called Movity, which focused on data visualization and cities. Movity was acquired by Trulia, where Sha worked for the past two years building maps and visualizations. Now independent, Sha continues to explore the intersection of interaction, information, and fabrication.

For more information and contact:

http://postarchitectural.com
https://twitter.com/shashashasha

Rachel Binx (Independent Data Visualizer, Art Historian & Mathematician) studied math and art history before discovering the world of data visualization. Before jumping into Meshu and Gifpop, Rachel worked at Stamen as a design technologist. There she worked on creating realtime data visualizations for clients such as MTV, Facebook, and Oprah.

When not sitting in front of her computer, she likes to be out traveling the world or exploring Tumblr.

For more information and contact:

http://rachelbinx.com
https://twitter.com/rachelbinx

Chapter 16
Changing Minds to Changing the World

Mapping the Spectrum of Intent in Data Visualization and Data Arts

Scott Murray

Abstract The recent explosion in available data sources and data-processing tools has both scientists and artists diving into the world of data visualization. The result is a diverse, interdisciplinary field of practice, in which practitioners cultivate knowledge in other areas: Statisticians are learning about design, while designers are learning about statistics. All of these people are producing visualizations of data—objects of visual communication—but with widely varying intentions and goals for their creations. Several primary goals for visualization—exploratory, informational, and rhetorical—are well-established. But in a field where artists (seeking to produce aesthetic, yet "accurate" work) are learning about science, and scientists (seeking to produce informational, yet "aesthetically appealing" work) are learning about art, how can we delineate between the range of types of intended communications and can those delineations be made in any meaningful way? One of the most exciting aspects of visualization today is the ease with which practitioners from different backgrounds collaborate and engage with each other. By examining the discourse adopted by these practitioners, we can identify what processes they all have in common, and then map where practices overlap and where they diverge.

16.1 Get Started

"How should I get started with data visualization?" This increasingly common question turns out to be quite difficult to answer. Visualization is inherently interdisciplinary; a true mastery of all its forms would require expertise in:

- Visual design,
- Interaction design,
- Data analytics,

S. Murray (✉)
University of San Francisco (USF), 2130 Fulton Street, San Francisco, CA 94117-1080, USA
e-mail: shmurray@usfca.edu

- Statistics,
- Mathematics,
- Psychology,
- Computer science.

That list does not even include the technical skills required for implementing a project with specific tools, such as Excel, Tableau, Illustrator, Processing, R, D3, or —more commonly—some combination of tools, each optimized for a different step in the process.

Yet none of the practitioners I know are experts in all of the subjects and tools mentioned above. Many have formal backgrounds in one subject and then dabble in others. A computer scientist by training may "have a knack" for visual design, or a designer may discover she also excels at statistics. Thus, we pick and choose, and draw from whatever skill sets we are inclined to cultivate within the limits of our available time, interest, and abilities. I find that most people in data visualization are, by nature, very curious; we would prefer to learn everything and be skilled in all areas, but of course life gets in the way.

When beginners ask how they can get started, this interdisciplinary quality of the practice also gets in the way. There is no one best path into visualization; every practitioner has a different point of entry, such as:

- Web design,
- Graphic design,
- Industrial design,
- Architecture,
- Mathematics,
- Cognitive science,
- Computer science,
- Journalism.

With so many possible points of entry, the question is easier to answer on a personalized, individual level. To someone with a highly technical background, I might recommend some design books. To a journalist, I could suggest resources on data analysis and graphical storytelling. But of course even these are generic responses and do not account for the individual's full range of prior experience. An interdisciplinary field can be exciting and stimulating for practitioners already who are already fully engaged. But for those just dipping in their toes, it can be frustrating to ask lots of questions and frequently hear the same answer: "Well, it depends."

16.2 Common Ground

In an effort to provide a more comprehensive answer to such questioners, I want to document the full range of experience and expertise found in the visualization community. As incomplete as this attempt may be, it should be valuable to see our differences mapped out, as well as the practices and language that we have in common.

While searching for this common ground, I also intend to propose an informal *taxonomy of practice*. Much prior work has been done to classify visual properties and common visualization elements (Bertin 2011; Segel and Heer 2010), but here I want to explore the community of practice itself. As the field grows, it becomes increasingly important to understand the range of its participants.

It is my sense that visualization practitioners, despite our diverse backgrounds and the interdisciplinary nature of the field, have quite a bit in common—it is just that we have a hard time describing exactly what that is. As evidence, I observe that many of the same people speak at or otherwise attend the following conferences:

- Eyeo Festival,
- Resonate,
- See Conference,
- Strata,
- Visualized,
- IEEE VIS (formerly VisWeek, includes VAST, InfoVis, and SciVis).

While these conferences appeal to a range of audiences—from primarily academic researchers to decidedly non-academic professionals—there is significant overlap in attendance. How is it possible that a "creative coder" attending Resonate, a small annual gathering for exploring technology role in the arts, could be just as comfortable in that environment as at Strata, a more corporate environment focused on big data and data insights? Why does VIS, the essential event of the year for academics and researchers in visualization, also have an art program for exhibiting "artistic" uses of data alongside "practical" examples? The ACM SIGGRAPH conferences, too, while focused more broadly on computer graphics and not just visualization, have extensive arts programs.

The fact that I have placed "creative coder," "artistic," and "practical" in quotation marks indicates that we have a language problem on our hands. This begins with how we identify ourselves. I have seen practitioners refer to themselves by the following titles:

- Data visualizer,
- Data designer,
- Designer,
- Artist,
- Data artist,
- Code artist,
- Generative artist,
- Creative coder,
- Creative technologist,
- Graphics editor,
- Cartographer,
- Researcher,
- Storyteller.

Each implies a slightly different emphasis—more fine arts, more code, more data—but, at gatherings, these people converse freely, communicate well with each other, and typically avoid using titles altogether. It is a common woe, especially toward the fine arts end of the spectrum, that these titles are essentially meaningless, except as cues to other practitioners already "in the know." The interdisciplinary data artist's elevator pitch is often brief and inaccurate, because the nuances of the process are not easily reducible to summary for outsiders. The result is a more tightly knit (and unintentionally insular, if still friendly) community.

I witnessed this label-aversion play out at a large scale at the first Eyeo Festival in 2011. The conference is held in Minneapolis each June and invites presenters from a range of fields—data visualization, generative art, installation art, design, computer science. Its tagline, "Converge to inspire," is conveniently vague and as such reflects the event's reluctance to pigeonhole its attendees. By the end of the week, I heard many people describing others not as artists, creative coders, or data visualizers, but as "you know, the kind of people who would go to Eyeo." For lack of a better umbrella term, we resorted to self-reference. I think we can explore this phenomenon, look at the principles and practices shared in common, and identify a clearer way of describing ourselves to others.

16.3 Mapping the Field

To frame the discussion, I will propose a series of ranges or spectra upon which practitioners and projects may be situated. For example, in the field of visual communication, there is an ongoing tension between the terms "art" and "design."

Art ←————————→ **Design**

Work deemed to be on the "art" end of the spectrum may, for example, be considered purely aesthetic, have little or no "functional" purpose, and have little commercial or "practical" value (except, of course, as fine art, which, I would argue, is as practical a purpose as any). Work may be on the "design" end of the spectrum if it has obvious commercial value, communicates a specific message, and functions with an explicit purpose. Yet "design" is not without aesthetic value and so shares that element with "art." And "art," such as illustration, may be employed within a "design" context, to communicate a message larger than the art itself.

At what point does an image crossover between art and design, or vice versa? While this distinction is in some sense arbitrary, it nevertheless carries value, at least by forcing us to struggle with the language we use to describe our work and the values we ascribe to it.

To me, the most meaningful way to make this distinction is to identify the goal or *intent* of the creator. For art, the intent may be to elicit a purely aesthetic or emotional experience from the viewer/participant. For design, the intent is typically to communicate a specific message to the viewer/participant. So, regardless of the medium and context, an image made with intent to communicate a particular message or meaning falls near the "design" end of the spectrum. (This assessment is made independent of whether or not the design is successful in achieving its creator's goals.) An image with intent to elicit an emotional experience (without a specific message), can be called art (though, to further muddy the waters, art often has a message). Perhaps this could be simplified even further to say that a work's position on the spectrum indicates only the specificity of its intended message. The more open the message, the more artistic; the more specific, the closer it is to design.

The art/design spectrum, as well as the others I propose below, is presented as two-dimensional, but of course, reality is more complex and not suited to such clean definitions. (This is particularly true given the current rate of change in visualization practice, and the rapid development of new forms.) Please take these proposals as tools for framing discourse about the current state of the field, not attempts to define it in fixed terms.

I will address each of the following in turn:

- Avenues of practice,
- Media,
- Contexts,
- Conceptual structures,
- Goals.

The first spectrum can be used to evaluate either practitioners or projects, while the others are specific to individual projects.

16.3.1 Avenues of Practice

As mentioned above, practitioners self-identify with a range of titles. Broadly speaking, each of these titles could be placed on a spectrum of *data arts* to *data visualization*. Both are visualization, of course, but data arts is more akin to fine arts, and data visualization is more akin to design—that is, it creates visualizations with specific messages, or with the intent to reveal messages intrinsic to the data (e.g., patterns and trends).

Data arts	←――――――→	**Data visualization**
. Artists,	. Designers,	. Data analysts,
. Creative coders,	. Illustrators,	. Scientists,
. Creative technologists.	. Storytellers.	. Journalists,
		. Cartographers,
		. Researchers.

To offer an example, I would file Memo Akten into the data arts end of this spectrum. Akten's project *Forms* (Fig. 16.1), done with the artist Quayola, is

Fig. 16.1 *Forms*, Memo Akten and Quayola, 2012

intensely data-driven or data-derived, yet it is more evocative than explicitly communicative.

That said Akten—through his company Marshmallow Laser Feast—has done projects for corporate clients, such as the *McLaren P1 Light Painting* (Fig. 16.2), which I would classify as a data illustration: It functions primarily as an advertisement for a new automobile, and thus, the communications *intent* is different from that of *Forms*.

Continuing further still to the right edge of this spectrum, we can look at geographic maps, a visualization of practice that has undergone massive changes in the past 10 years. Stamen Design in San Francisco, which refers to itself as "a design

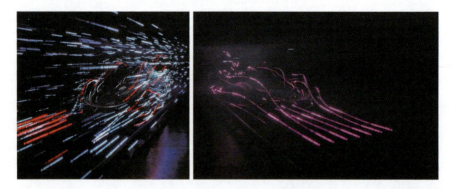

Fig. 16.2 *McLaren P1 Light Painting*, Marshmallow Laser Feast and James Medcraft, 2012

Fig. 16.3 *Toner-style map*, Stamen Design (under CC BY 3.0. Data by OpenStreetMap, under ODbL). URL, December 28, 2013: http://maps.stamen.com/m2i/image/20131117/toner_uwnIcEfPDtw

and technology studio," is known for their wide array of explorations in maps. Their Toner tiles are intended for use when a map will be printed and photocopied (Fig. 16.3). As such, they do not use any color, and gray areas are rendered with a halftone screen, to improve reproducibility by analog means. Yet, even with this constraint, the design functions effectively as a guide for orientation and directions —that is, as a traditional map.

This particular intent and specificity of communication places the *Toner map* squarely on the data visualization end of the spectrum. Contrast that with Stamen's Watercolor tiles, which represent the same underlying data in a completely different form (Fig. 16.4).

The *Watercolor maps* are less precise by design, evoking an abstract sense of place for those already familiar with the place, as opposed to helping orient new visitors to specific locations within a place (e.g., cities, streets, addresses). So I would file the *Watercolor maps* on the data arts end of the spectrum.

But what about Stamen as an entity? Where do its designers and technologists fall, given their influential contributions all along the spectrum?

Fig. 16.4 *Watercolor-style map*, Stamen Design (under CC BY 3.0. Data by OpenStreetMap, under ODbL). URL, December 28, 2013: http://maps.stamen.com/m2i/image/20131117/watercolor__Q2AD5HJgMk

This highlights how difficult it is to classify individual practitioners. I may be acting as an artist today, but I put my designer hat on when it is time to update my Web site, and perhaps I have to think like an analyst when making sense of the data set underlying my next project. I will propose a solution to this classification problem, but first, it will be useful to discuss how to classify individual projects.

16.3.2 Contexts

Visualizations commonly have one primary context of presentation, although essentially every project is now documented and published online in some form.

Gallery Primary, Online Secondary	**Online-Only**	**Print Primary, Online Secondary**
. Prints,	. Personal website,	. Newspaper,
. Video,	. Project website,	. Magazine,
. Interactive,	. Arts blog/site,	. Poster,
. Installation	. Visualization blog/site,	. Flyer,
	. Technical blog/site.	. Academic Paper/journal.

Jonathan Harris and Sep Kamvar's *I Want You To Want Me* (Fig. 16.5) is a computationally intensive installation created for the Museum of Modern Art's *Design and the Elastic Mind* exhibit in 2008, curated by Paola Antonelli. As a

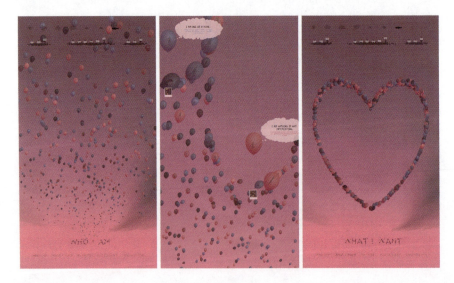

Fig. 16.5 *I want you to want me*, Jonathan Harris and Sep Kamvar, interactive touch-screen installation commissioned for the museum of modern art's design and the elastic mind, February 14, 2008

16 Changing Minds to Changing the World

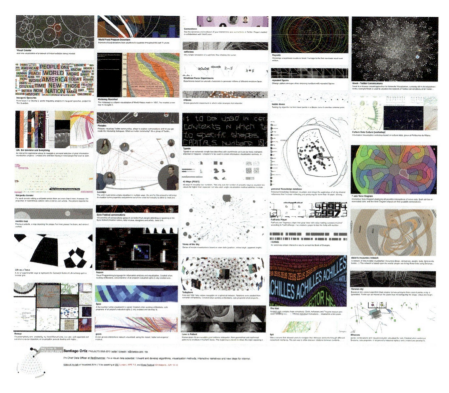

Fig. 16.6 *Portfolio*, Santiago Ortiz, 2013. URL, September 12, 2014: http://www.moebio.com

commissioned piece, it was designed from the beginning for the gallery context. While the artists posted documentation online, it is not feasible to adapt the project for the web, so it remains a gallery-only, in-person experience.

In contrast, Santiago Ortiz's innovative portfolio interface (Fig. 16.6) was designed specifically for the online context and would not make sense in any other medium. It begins as a grid of project image thumbnails, but visitors can drag a round cursor to adjust the visual weight given to projects across three axes: recent, favorites, or all projects. At a glance, we can watch projects resize to reflect, for example, which ones were completed most recently versus which ones Ortiz himself enjoys. This form of representation, unlike Stamen's *Toner maps*, is not intended for print and would cause confusion in paper form, due to clipped images and abbreviated text.

Continuing toward the print end of the spectrum, while many visualizations are designed primarily for print output, the dual approach taken by the *New York Times'* Graphics Desk is slowly becoming more common. At the *Times*, every graphic must work in both print and online. Typically, this means designing a default view that communicates the story. The default view works in the print edition of the paper and also serves as the initial view of the online version. Interactivity can be used to make the piece explorable, enabling readers to dig deeper into specific data values. For example, a recent graphic[1] on drought in the US includes annotations that highlight key trends in the data, but the online version also allows readers to mouse over any section of the graphic to reveal specific drought levels.

16.3.3 Media

Visualizations are often designed with at least one primary target medium in mind. Those media may be considered along a spectrum of *static* to *interactive*.

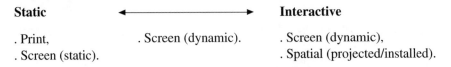

Static		Interactive
. Print,	. Screen (dynamic).	. Screen (dynamic),
. Screen (static).		. Spatial (projected/installed).

Note that the static/interactive spectrum is independent of a work's *context*. For example, not all *New York Times* graphics are interactive, even when published online. *The Drought's Footprint graphic* was only ever intended to be a static image, both for print and on the Web site[2].

Work intended for the screen can be dynamic without being interactive, such as IBM's *THINK Exhibit*[3], which included a large-scale data visualization wall. During its temporary installation in New York, the wall displayed real-time visualizations of data about the city, such as traffic flows, air quality, and water use. Since the display was generated live, it was dynamic, although not directly interactive. This stands in contrast to a pre-recorded video loop, which remains static in the sense that it merely repeats itself and its imagery does not change over time.

Many screen-based visualizations are interactive, of course, but on the far end of the spectrum are works that are so dependent upon interaction with participants that, without it, they essentially cease to exist. *Shadow Monsters*[4], an installation by Philip Worthington, for example, begins as nothing but a silent room with a plain white screen. Only when participants enter, the space does the system spring to life, interpreting their shadows as horned, fanged creatures with creepy hair and nails.

---

[1] See *Drought and Deluge in the Lower 48*, Mike Bostock and Shan Carter, February 22, 2013. URL, November 14, 2014: http://www.nytimes.com/interactive/2012/08/11/sunday-review/drought-history.html.

[2] See *Drought's Footprint*, Haeyoun Park and Kevin Quealy, July 19, 2012. URL, November 14, 2014: http://www.nytimes.com/interactive/2012/07/20/us/drought-footprint.html.

[3] See photo documentation of the IBM THINK Exhibit at Lincoln Center in New York City, URL, November 14, 2014: http://mirada.com/stories/ibm. Project developed by Mirada, opened September 23, 2011.

[4] See *Shadow Monsters*, Philip Worthington, 2004–ongoing. URL, November 14, 2014: http://moma.org/exhibitions/2008/elasticmind/#/229.

16.3.4 Conceptual Structures

Exploratory tools are used to visualize data for the purposes of discovering what is interesting and valuable about that data. Explanatory visualizations take a point of view and communicate to the viewer some pattern, trend, or discovery already observed.

Exploratory ◄─────────────► **Explanatory**

. Flexible views,   . Constrained views,
. Open structure,   . Narrative structure,
. Adaptable encodings   . Optimized encoding.

Exploratory visualizations are often interactive, and many tools are designed primarily for this purpose, such as Tableau and R with ggplot2. Since exploratory designs are geared toward producing insights, they tend to be more literal and specific than purely aesthetic. (A data arts visualization would not be likely to produce valuable insights.) *The Dynamic HomeFinder* was one of the very first such exploratory visualization tools (Williamson and Shneiderman, 1992).

Explanatory visualizations are more focused, with limitations imposed on the viewer and design elements that increase the specificity of the communications value. Discussions of visualization as storytelling are referring to explanatory images and interfaces. Journalistic graphics are typically very strong in this regard, such as in *The Cost of Water* (Fig. 16.7), a piece I worked on for the Texas Tribune, which explores why, despite record droughts, water is relatively inexpensive in Texas.

Some visualizations or tools, of course, try to serve both exploratory and explanatory functions. Often, this employs a structure of a default explanatory view, followed by the use of interactivity to enable independent exploration.

Fig. 16.7 *The Cost of Water*, a collaboration between Geoff McGhee, Kate Galbraith of The Texas Tribune, The Bill Lane Center for the American West, Stanford University, and Scott Murray (published online June 8, 2012). URL, December 28, 2013: http://www.texastribune.org/library/data/cheap-water-in-texas

16.3.5 Goals

Finally, each project is created with different goals, which may be placed somewhere on the spectrum of *inspire* to *inform*.

Inspire ◄─────────────────────►		**Inform**
. Aesthetic experience,	. Motivate,	. Enlighten.
. Emotional experience.	. Upset/disturb/enrage,	
	. Delight/amuse/satisfy.	

An *inspiring* project may, like art, induce a kind of "a-ha moment" absent any concrete information. An *informative* project may communicate specifics of its data, but without any noticeable emotional impact.

On the inspiring end of the spectrum, we may find *Tape Recorders, Subsculpture 12*, a 2011 installation by Rafael Lozano-Hemmer. Sensors track the presence of visitors to the space, and the length of their visits is expressed through the lengths of tape measures. The work is data-driven, but the individual data values are meaningless; the aesthetic and emotional experiences are what matter.

On the informative end of the spectrum, we find wholly uninspiring charts and graphs, like "OMG! Texting turns twenty," from *The Economist*[5]. This is in no way to pick on *The Economist*; when communicating specific data values, it is not necessary to inspire or delight. The chart below efficiently communicates the rise of text messaging and includes several annotations, offering context of historically relevant moments. This chart is intended to inform, and it does so successfully.

Many projects, especially in data journalism, aim for a balance of inspiring and informing—such as when *informing* is essential, but achieving that end requires also engaging the reader on an emotional level.

One such landmark project is *We Feel Fine* (Fig. 16.8), another piece by Jonathan Harris and Sep Kamvar. Made in 2005, *We Feel Fine* is one of the early, online interactive visualizations. Still just as potent almost a decade later, it does not hurt that the data behind the project are themselves all about emotions and the human experience.

More recently, Periscopic's US *Gun Deaths* interactive visualization poetically and powerfully documents lives lost to gun violence (Fig. 16.9)—and projects an alternative future in which victims live out the rest of their lives (as algorithmically projected). The work performs a dual role, both informing us of the scale of tragedy as well as inspiring us to reflect upon and debate the significance of so many lives cut short.

---

[5] Published on December 3, 2012. URL, November 14, 2014: http://www.economist.com/blogs/graphicdetail/2012/12/daily-chart

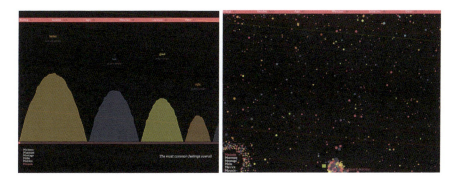

Fig. 16.8 *We feel fine*, Jonathan Harris and Sep Kamvar, 2005. URL, December 28, 2013: http://www.wefeelfine.org

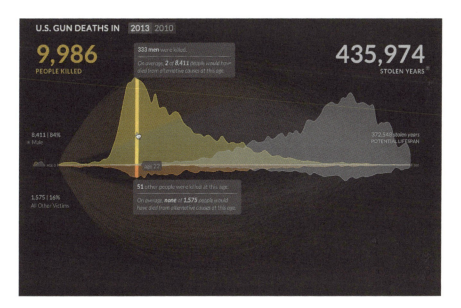

Fig. 16.9 *US Gun Deaths*, Periscopic, published in February 2013. URL, December 28, 2013: http://guns.periscopic.com

16.4 Making Meaning from All This

Given these groupings—avenues of practice, media, contexts, conceptual structures, and goals—it should be possible to (roughly) evaluate and score any given project along each spectrum. If we consider each spectrum as an axis, an arbitrary, normalized value could be assigned to each. For example, a project could be scored anywhere from 0.0 to 1.0 along the axis of *inspire* to *inform*. Scores could be

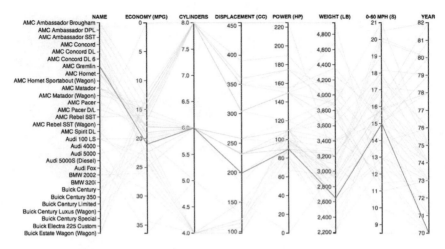

Fig. 16.10 *Ordinal parallel axis example*, Kai Chang, 2012. URL, December 28, 2013: http://bl.ocks.org/syntagmatic/3731388

requested from numerous reviewers and then aggregated to produce mean scores for a project, one value for each axis.

Having converted our collective assessments to data, we could (of course) visualize the results. I would recommend a parallel coordinates plot, with each axis oriented vertically, and horizontal lines connecting the values for each project (Fig. 16.10).

Through interactivity, we could filter the view to show only projects by a particular creator, or by people from a specific subfield (say, only "researchers" or "artists"). This could enable us to discover places in the field where practices converge or diverge, either conforming to or challenging our expectations.

Independent of the visualization, after scoring all projects by a single creator, we could then calculate a "career average" with which to place them along the *data arts/data visualization* spectrum of practice. While acknowledging that we all move between many roles, it could be useful to see how heavily the field skews toward the arts or the other direction. (My sense is that there is so much interest in the field right now, from a diversity of perspectives, that there is a fair balance.)

This approach reminds me of two recent projects. First, a recent visualization by *Pitch Interactive* that visualized artists' careers with colorful star diagrams (Fig. 16.11).

Second, a map by Jeff Clark of visualization practitioners on Twitter (Fig. 16.12).

Whether artists, scientists, journalists, or cartographers, practitioners of data visualization all seem to be in love with data. And what data could be more appealing than data about ourselves? Gazing into mirrors is fun, but beyond that, self-mapping is a great way to understand how we each fit into the field and, perhaps more importantly, it is a tool for explaining to others what the field is all about. With a map, we have a visual interpretation of the phrase heard so often at Eyeo. Now, instead of saying "the kind of people who would go to Eyeo," we could just point to an image and say "data designers are the people on this map."

Fig. 16.11 *McKnight artist fellows: visualizing artists' careers*, pitch interactive with the McKnight Foundation, 2012. URL, December 28, 2013: http://diagrams.stateoftheartist.org/gallery

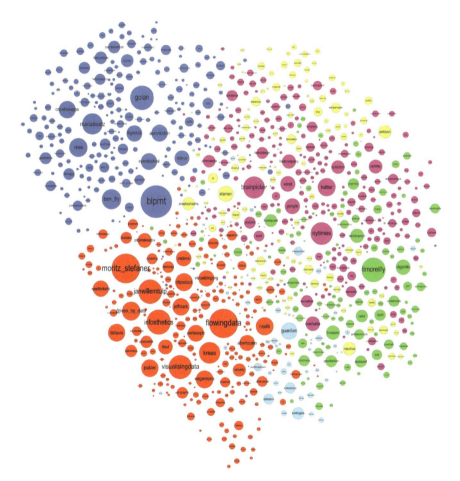

Fig. 16.12 *Data visualization field on Twitter*, Jeff Clark, 2012. URL, December 28, 2013: http://www.neoformix.com/2012/DataVisFieldSubGroups.html

16.5 Common Elements

Other than a love of data and visuals, what do all visualization practitioners have in common?

Tools and Media—We all live and work in the same time and have access to the same tools and media for sharing our work. Fine artists use Processing, but so do data journalists. Statisticians use R, but so do increasing numbers of people from other fields and specialties. Each tool is designed for a different audience or task, but there is a surprising amount of crossover between subfields. Nearly all of these tools involve code, so basic programming ability is a must. It is possible to create great visualizations without code, but it is difficult to articulate new visual forms without it.

Process—We all work with data, defined as structured information. It takes a certain mind-set to appreciate a well-structured, honest data set. Ultimately, we encode that data into visual form, a process that requires another, similar mind-set to appreciate. So we have data and data-visual mapping in common. But governing each of these steps are many rules, usually documented as algorithms in the software we write: scripts to parse data, programs to generate charts and graphs, and applications to share beautiful renderings with our audiences. The algorithm rules every step. Our core value is a love and appreciation for process itself.

Curiosity—I have never met an incurious practitioner. We love learning and we love being inspired by discovering things in the world around us, or perceiving old ideas in new ways. Data visualization is fundamentally about making the invisible visible, a shared goal for all practitioners. Where our work diverges is in the intent of our process, and in what means of visual rhetoric are employed to that end.

16.6 The Value of Interdisciplinary Practice

While the interdisciplinary nature of the practice makes it hard to summarize the field to outsiders, it is also one of our biggest strengths. By drawing on the discoveries and expertise of many fields, we can improve our processes and improve our designs. One concern, of course, is that we may be inclined to learn broadly, but not deeply. Yet, as I described earlier, many practitioners tend to have formal training in one or two different areas, but then more loosely explore others.

Contrast this interdisciplinary approach to a more specialized one. Certainly, there is value in being deeply focused on just a single research area, but such a focus will not by itself produce informative or inspiring visualizations. Domain-specific research—such as in human visual perception, computer graphics, and new visual forms—however, is invaluable for visualization practitioners. The evolution of our own practice depends on the insights developed by such research.

The interdisciplinary mind-set pervades practitioners' selection and use of tools, methods (processes), and domains of operation (uses of tools and methods). Data

16 Changing Minds to Changing the World 309

visualization practitioners are often hired by domain experts (the clients) to interpret and represent the client's data. When Pitch Interactive is hired by Popular Science to visualize historical government projections for energy independence, they are not expected to have prior knowledge on energy independence. When Stamen partners with the nonprofit organization Climate Central to map projected sea-level rise, no one expects them to be climate change experts. When Fathom contracts with Thomson Reuters to map the power dynamics in China's political sphere, it is the client, not the design firm, who is expected to bring the domain-specific knowledge (and data, of course) to the table.

It is an odd role for a consultant, whose area of expertise is not the specific domain at hand, but an expertise in the *process of exploring*—that is, exploring both the new information provided by the client as well as a range of visual forms for representing and communicating that data.

This *exploration*, as fueled by practitioners' curiosity, rigorously structured processes, and media expertise, is what makes data designers so uniquely valuable today.

Does this mean that a data designer, if inserted into any industry or context, could bring value to the organization simply through her *interdisciplinary process*, even without a specific end goal in mind? Possibly. Designing is problem-solving, and the process itself may be just as important and valuable as the resulting product.

Finally, this also explains why practitioners struggle to articulate their daily work to outsiders: A *process* is much more abstract and difficult to explain than a *product*. It is easy to point to images and say "I made these." It is much harder to say "Through years of practice, I have developed a process that guides my decisions and actions, which results in a successful representation of data, more often than not." Good luck offering such an accurate, yet uninteresting explanation in a social context, such as at a cocktail party—you may not be invited back!

16.7 Future Challenges

Looking ahead, what are the future challenges for data design? I see several, each related to the issues addressed above.

Tools—The vast array of tools available will continue to grow and diversify. So the problem is not a dearth of tools; it is cataloging them and making efficient use of the existing tools appropriate to any given task. It seems every week a new framework or library is introduced that provides an improved solution to a very specific problem. Just as we cannot all be experts in every field, we cannot all learn how to use every tool (much as we would like to). We need a better method to identify the best tools for a given task. Whatever that method is, it needs to fit into existing workflows.

Methods—Speaking of methods, we each have our own working process, and our challenge is to develop clearer language around those processes. With better language, we can compare processes and learn what, exactly, certain practitioners do that makes their work more successful (or not) than others. What are data design's best practices? How similar or different are they when making data art, as opposed to data visualizations? I hope that this essay is a small step toward framing that discussion.

Data Design Literacy—It is essential to clarify our best practices so that we can educate new practitioners. I began with the question, "How should I get started with data visualization?" Students and others new to the field deserve better answers to this important question. We need maps and taxonomies of practice (which this essay seeks to introduce), and we need more structure and consistency in our training programs. Although data design has a long history, in this rapidly changing environment, it often feels like we are just figuring things out for the first time.

Data Image Literacy—Practitioners are not the only ones who need to be educated; informed audiences are also essential. The consumers of data design must understand the possibilities and pitfalls of the images we create. Just as media literacy education seeks to ensure critical awareness of film, television, and radio, data image literacy is needed to ensure that the inherent biases of data images are well-understood.

Ethics—While there is a tendency to trust data images as fact, practitioners know that even minor changes to a design can strongly influence how the underlying story or information is perceived. Given the ease with which charts and maps can be made to lie, there may be a need for a professional code of visual ethics, a formalization of already well-known design principles advocating for representations that align with human perceptual abilities.

Given my colleagues' innate curiosity, enthusiasm, and love of process, I am optimistic that data explorers all along the spectrum will engage with each other to tackle these issues.

16.8 Nature of Tools

Among hammers, there are minor variations in form, weight, and size. Yet all hammers share a similar fundamental form. Over time, a builder develops a feel for a particular hammer, sensing how much force is needed to move a nail into position.

Software-based tools are more diverse. They share only fundamental underpinnings, such as the use of computation and some common interface conventions. Despite expressing no obvious physical form, they encourage the development of limited muscle memory, perhaps for common keyboard shortcuts or method patterns. Over months or years of use, a favorite tool or suite of tools will often emerge, and a data designer will gradually develop expertise with that tool, having cultivated a practiced sense for how to strike a particular type of nail.[6]

---

[6] For in-depth discussion of this issue, (see McCullough 1996).

Yet with so many software tools available, it can be overwhelming to know where to begin. New practitioners are not yet attached to any particular tool; they want to choose an approachable tool, the mastery of which will be transferable to other such tools in the future. Unfortunately, software is not as straightforward as hammers. Learning to code in one language may familiarize you with core concepts —variables, arrays, logic, functions—but switching to another language involves different syntax and methods, different best practices and frameworks, often a very different way of approaching the problem entirely. (Worst-case scenario: moving from Python to Java. So many semicolons!) Every time we switch tools, we have to re-learn how to strike the nail.

Even worse, our favorite software-based tools may change themselves right underneath our noses, auto-updating to add new features, remove old ones, modify syntax rules, or change operating requirements. For some people, this would be crazy-making, and certainly, in the physical world, it would be. Imagine a hammer that, after having been used successfully for years on multiple projects, is considered "trusty"—a reliable workhorse that has supported the builder in a variety of scenarios. But this hammer is an open-source hammer, with a core group of five or six dedicated contributors. They actively patch bugs and introduce new features, so every few months or so we get another point release—Hammer 1.1, Hammer 1.2, and so on. With each release, our hammer is still recognizable, but functions a bit differently; we must adjust the angle of our strike. Hammer 2.0 brings new operating requirements; our old, dingy workshop is no longer supported, so the hammer just sits there, inoperable, until we repaint the walls, install better lighting, or move to an entirely different neighborhood. Of course, Hammer 1.9 is still available for download, and we have a hundred copies sitting around on shelves, but it does not drive nails as quickly, precisely, or elegantly. Also, there is market pressure; the hot design firms are not interested in practitioners using old technology.

I present this software-hammer metaphor as further illustration of the *intense curiosity* and *enthusiasm for problem-solving* exhibited by data design practitioners. We enjoy exploring data and learning about the world around us, but we are also excited about new tools, as well as continuous evolution and change in our existing tools. If every project is just another puzzle to be solved, we also secretly enjoy the geekery of solving the *process puzzle*, the ongoing meta-challenge we all share, the operating context inherent to an interdisciplinary practice powered by computation.

References

Bertin J (2011) Semiology of Graphics: diagrams, networks, maps. Esri Press, Redlands
McCullough M (1996) Abstracting craft: the practiced digital hand. MIT Press, Cambridge
Segel E, Heer J (2010) Narrative visualization: telling stories with data. IEEE Trans Visual Comput Graphics 16(6):1139–1148
Williamson C, Shneiderman B (1992) The dynamic HomeFinder: evaluating dynamic queries in a real-estate information exploration system. In Proceeding SIGIR'92, pp 338–346

Author Biography

Scott Murray (Code Artist, Data Visualizer and Teacher) is a code artist who writes software to create data visualizations and other interactive phenomena. His work incorporates elements of interaction design, systems design, and generative art.

Scott is an Assistant Professor of Design at the University of San Francisco, a contributor to Processing, and author of "Interactive Data Visualization for the Web: An Introduction to Designing with D3." Scott earned an A.B. from Vassar College and an M.F.A. from the Dynamic Media Institute at the Massachusetts College of Art and Design.

For more information and contact:

http://www.alignedleft.com
https://twitter.com/alignedleft

Part VIII
Interviews

Chapter 17
Beauty in Data

Jonathan Harris

Andy Cameron's Interview with Jonathan Harris

Andy Cameron (AC): First, aesthetics. How do we think about art which is new, which is doing new things with new techniques? How do we judge it, how do we make sense of it and how do we ascribe value to it?

I think there are two aspects to this—the broader aspect and the narrower one. The broader one is to do with technological art in general—what happens when the technology of art changes and new things happen. For example, the whole thing around interactivity—the opportunity for a representation to speak back, as it were, to the audience. This is a new thing—a new category of beauty as Myron Krueger put it—and it demands a new aesthetics because it profoundly doesn't work in the same way, doesn't do the same thing, as a painting does, or a photograph does?

Jonathan Harris (JH): As a measurement, the simplest rule I can imagine is that the art should be consistent with itself. Meaning, the art should establish a (symbolic, not programming) language of its own, and it should follow that language in every detail—from the largest to the smallest. This is not consistency for its own sake, but consistency as a signal for a complete and well-understood point of view, which you might also call a voice.

Especially with digital work, where the creator is usually more removed from the original act of expression than a writer or painter would be (i.e. code is abstract while words and pictures are direct), having such a consistent voice is doubly important, because that voice—that consistency—is the only thing in the work that

---

Andy Cameron passed away in 2012. He was a digital pioneer, co-founder of the Antirom collective, artist, teacher and, latterly, creative director at Fabrica and Wieden + Kennedy, and also the former of Jonathan Harris. This interview took place in 2010.

J. Harris (✉)
227 Degraw Street Brooklyn, New York 11231, USA
e-mail: jjh@number27.org

belies the presence of the artist. Without this voice, the data-based digital artist is more like an interior decorator or librarian, than like an artist. So this voice is very important. This 'voice' is not the same as 'style'. Aesthetic style is part of voice, but voice is much more—it is a whole conceptual framework for self-expression—a way of understanding the world, posing questions, and depicting answers.

Sometimes people send me links to websites created by others and say: "Hey, this looks like one of your pieces," and I look at the piece and immediately I say: "It certainly does not, and if you had a sensitive eye for details, this would be very clear to you."

There are only a few digital artists whose work I can immediately identify on sight—Yugo Nakamura, Casey Reas, Golan Levin, and maybe a few others. Their work is identifiable because their work is precise—it is consistent with itself. In other work, there is a sloppiness, a laziness, a cutting of corners, a dabbling around—and in my view, this sloppiness is a sign of bad work. If you fully understand what you are trying to say, then you will not be sloppy. If you are groping around without purpose, it is difficult *not* to be sloppy.

Even when mediums evolve very quickly, and when a critical sense of "aesthetic quality" has yet to emerge, a sensitive eye can usually spot consistency, and the presence of consistency is often a clue that there is good work happening therein. So consistency is like an early warning signal.

This idea of consistency also occurs among writers and filmmakers, who design worlds where every detail must feel like it belongs, and if there is a detail that does not belong, the illusion immediately collapses. The same is true with programming digital worlds and experiences—the illusions are fragile and can easily collapse if there are weak elements.

When new worlds emerge, they often require new languages. Likewise, when new languages emerge, they often create new worlds. Artists need to work both sides of this equation.

When culture changes, the old languages are often very bad at describing the new culture. For instance, oil paintings about the Internet seem very silly to me. When there is a new culture that needs to be described by artists, it is often good to use pieces of that culture to do the describing. That is why work about the Internet should draw from the Internet, use the language of the Internet, and be presented on the Internet—the native land will always be most natural.

AC: It's really interesting to hear your thoughts on 'voice'. It makes complete sense to talk of the internal consistency of an articulation—and the sense of a coherent 'character' behind the articulation. It's what an advertising strategist might call 'tone of voice'. At the same time I'm struck by a sort of paradox here—one of the things which defines interactive and generative art is the extent to which the artist isn't really making an articulation—in a sense isn't actually saying anything, or at least, not making 'statements' in the traditional sense. A writer makes up statements like 'the horse stood in the field' and the painter also makes statements visually—summoning things into existence through the illusion of paint on canvas. What you do-in common with other interactive and generative artists, is to stand at one remove from the articulated statement—instead focusing on a way to provide a context for the statements of others, to frame the statements of others.

This it seems to me is the essence of the new languages which you describe—that the artist is once-removed from making a direct statement. So it's not really self-expression in the old sense. Now, I completely agree with you that certain interactive and generative artists—like Levin or Nakamura —make work which is recognizably theirs. But when Levin, for example, makes a sound reactive which responds to other people's voices—and which only exists in a sense when other people speak, it makes the notion of the artist's voice that much more difficult to pin down.

JH: The dilemma you point out is fascinating, and could be the crux of many of the problems facing interactive work (the art world doesn't take it seriously, the interactive artist can't find out how to translate his personal experience and suffering into his work, interactive artists sometimes seem interchangeable, etc.)

When you are not a maker of gestures but a maker of frames, then as a framemaker, you compete on how clever you are in choosing good frames, and on the craftsmanship of the frames you end up creating, but not on the originality and emotional resonance of the actual work, because the actual work is what's inside the frame (in this case, the data), and not the frame that contains it. It's almost like the artists here are not the people like me (who make the frames), but the millions of individuals whose words and pictures show up inside the frames I make.

On the other hand, I love Chekov's idea that the role of the artist is not to answer questions, but rather to pose them fairly. In this sense, the formulation of statements is unimportant—even presumptuous. Instead, it is the ability to pose interesting questions (i.e. create good frames) that defines the artist. In the Chekovian sense, interactive artists are right on the money!

When I was younger, I made a lot of projects that tried to deal with very big themes (*10x10*, *The Yahoo! Time Capsule*, even *We Feel Fine*), but as I get a bit older, I realize how little I really know, and those early projects now seem brash and immature—even a little tacky. At the ripe old age of 31, I'm more interested in posing good questions than in offering answers (Figs. 17.1, 17.2, 17.3 and 17.4).

As for your point about Levin's sound-reactive works only existing when other people speak—that's true, yet still, there is a particular worldview that Levin possesses which dictates the design of his frames, and you can see that worldview (goofy, playful, beautifully crafted) in the situations he designs. Even if he is not making 'statements', you can still read his worldview in his frames (even in his annual Holiday cards), and his worldview is consistent.

AC: Another aspect of the question of aesthetics connects to the emerging field of data visualization. What is the aesthetic of data? How is the aesthetic of data linked to function? What makes data beautiful?

JH: Making data beautiful requires beautiful data. Data cannot be made beautiful by design or by anything else. Data can be made *pretty* by design, but this is a superficial prettiness, like a boring woman wearing too much makeup. Design can only reveal beauty that already exists—hidden beauty—usually by eliminating clutter and rearranging elements. In this way, design is more like makeup remover than makeup.

Fig. 17.1 *We Feel Fine*, Jonathan Harris and Sep Kamvar, 2005

Fig. 17.2 *We Feel Fine*, Jonathan Harris and Sep Kamvar, 2005

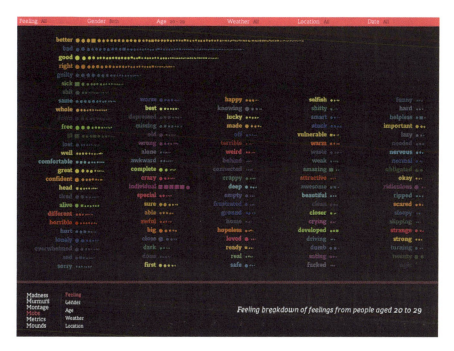

Fig. 17.3 *We Feel Fine*, Jonathan Harris and Sep Kamvar, 2005

I do not consider data visualization to be an artistic genre. It is a tool that has become fashionable, and so it has grown beyond its purpose, claiming an outsize self-importance. Most data visualization work is not interesting because most data is not interesting.

When Sep and I were making *We Feel Fine* and *I Want You to Want Me*, we operated under the premise that the underlying data, presented in plain text format, had to be very beautiful, or else we would not include it. Once we had found data like this, the various visualizations were just playful frameworks for revealing different aspects of that data. But the data had to be beautiful to begin with—that's the part most people forget. It is the same with my photo-based works, like *The Whale Hunt* and *Today*—if the underlying photos are not good, the final interactive projects will not be good (Fig. 17.5, 17.6 and 17.7).

I think of data visualization as a particular technique for expressing particular types of secrets—specifically, superficial secrets that hide on the surface of things (the secrets of charts and graphs and maps and numbers). These are easy secrets, so most data visualization is quite shallow, expressively speaking. There are other types of secrets—I call them "inner secrets"—and these secrets cannot be touched by data visualization. These inner secrets have to do with the heart or soul or subtle essence of things, and they can only be accessed through solitude, contemplation, and personal experience.

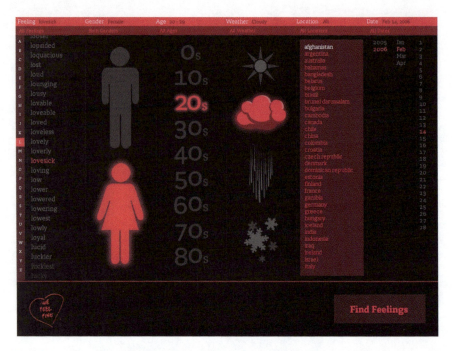

Fig. 17.4 *We Feel Fine*, Jonathan Harris and Sep Kamvar, 2005

After making a number of data-based projects, I became conscious of this limitation, and so recently I have turned more towards real experience (i.e. *The Whale Hunt, Balloons of Bhutan, Today*) (Figs. 17.8 and 17.9), to try to access these other types of secrets. They are much harder to find, but much richer when you actually find them, because they are things that other people can use to deepen their understanding of their own lives. As an artist, if you stumble upon one of these secrets, it is an incredible gift to you and to others, and it can make for very powerful work.

AC: I was really interested in your statement that the beauty of 'beautiful data' comes from the data itself—that some data is intrinsically beautiful, or at least more interesting, than other data. And I was fascinated to hear that you and Sep, when making *We Feel Fine*, set out in the first instance to identify beautiful data. I have to ask—how do you decide if data is beautiful or not? What are the aesthetic criteria you bring to bear on raw data? How do you know beautiful data when you see it?

JH: You have to feel it in your gut. There's no checklist, and even if there were a checklist, it would quickly become obsolete, because it has something to do with originality and strangeness. You have to find data that people have never seen before, but which feels totally familiar when they see it, like you're showing people a part of themselves. This is the kind of data that feels beautiful. It is illuminating, surprising, and personally resonant. I'm always searching for things that are simultaneously familiar and strange—those are the most powerful things (Fig. 17.5).

17 Beauty in Data

Fig. 17.5 *We Feel Fine*, Jonathan Harris and Sep Kamvar, 2005

AC: Much of your work is about providing your audience with a tool—a software application. How can we reconcile use value with aesthetic value? What happens for example when somebody goes to *We Feel Fine* and starts interrogating the system—looking for women feeling wistful or whatever—because we've got quite a complicated thing going on here—we've got you and Sep K as the artists, giving the audience—me—the opportunity to connect with a lot of other people in ways which are really quite interesting. So, where's the art? Where is the art situated within this complex set of interrelationships?

JH: The art is the whole thing—all of it. *We Feel Fine* is a piece of portraiture with many interacting elements. Visual aesthetics are only a very small part—probably the least interesting part. It is more about creating an ever-changing portrait of the emotional landscape of the human world. It is about creating a two-way mirror—where viewers simultaneously experience a God-like voyeurism (spying on the feelings of others) and a bashful vulnerability (realizing their own words and pictures are in there, too). When these two feelings mix together (voyeurism and vulnerability), the hope is that they produce a kind of humbling empathy—demonstrating that individual experiences are actually universal.

Another interesting aspect of *We Feel Fine* is mass authorship. There are now over 14 million feelings in the database, coming from about 4 million individuals, and they deserve to be authors of the piece as much as me and Sep. If the sentences in *We Feel Fine* (written by others) were not so poignant, the piece would be much weaker—it would be less about humanity and more about the impressive acrobatics

Fig. 17.6 *We Feel Fine*, Jonathan Harris and Sep Kamvar, 2005

of data visualization (which would be a selfish, superficial, short-lived goal). *We Feel Fine* is now more than 5 years old, but it still feels quite contemporary, and I believe this timelessness comes from the candor of the sentences, not from the way it is designed. Beautiful self-expression is timeless.

AC: The notion of mass authorship is a fascinating one—and absolutely central to what you do. One can almost think of it as the defining preoccupation of your oeuvre—this balancing of your authorial voice (which is always very clear) and the contributions of thousands and thousands of anonymous collaborators, each with a voice of their own. Now, this is not something which is unique to your work—it's also arguably the defining preoccupation of the age we live in—the shift from the few-to-many broadcast model of communication to a peer-to-peer model where authorship is much more diffuse and widely shared—but also messier, less coherent, less consistent.

Did you set out to do work which has this overarching contemporary resonance? Is this important to you?

JH: Back in 2003, when I was working with you at Fabrica, I remember feeling how non-special I was, and how silly it would be to encapsulate my own particular thoughts and opinions in my work, and how it would be much better to harvest and incorporate the thoughts and opinions of millions of others. Just as I thought I wasn't special, I also thought that no one else was special, so the only sane thing seemed to be to put everyone on equal ground, with equal voice, and that some kind

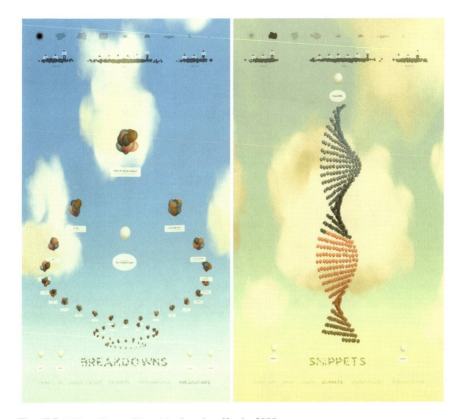

Fig. 17.7 *I Want You to Want Me*, Jonathan Harris, 2008

of 'universal truth' would emerge from that. This is similar to the idea of the Internet as a global brain, where people are interchangeable and individuals don't matter—only the aggregate matters.

Back in 2003, I was enamored with this idea. I think it appealed to the insecure geek in me, who liked the idea that I could learn all there was to know about life from sitting at my desk and designing clever programs—the ultimate revenge of the nerd!

Now I think this approach is deeply flawed, very limited, and dangerous to us as individuals, even as it grows in popularity and acceptance (wisdom of the crowds, etc.)

When people are viewed in the aggregate, individual humans matter less and less, and when systems are designed to deal with the aggregate, those systems become damaging to individuals. As such systems grow in scale and adoption, you start to see the mass homogenization of human identity (everyone filling out the same profiles, choosing from the same dropdown menus, etc.), which is what we're seeing today in the digital world.

The idea that you can learn about life from data is wrong. The only way to learn about life is from life (but this truth is terrifying to programmers, who prefer to sit at desks).

Fig. 17.8 *Today*, Jonathan Harris, 2010

Fig. 17.9 *Today*, Jonathan Harris, 2010

That is why now, I am much more interested in working from real life—incorporating my own personal experience (*The Whale Hunt*, *Today*, etc.) (Figs. 17.10 and 17.11) and designing platforms to activate other people's real personal experience (*Cowbird*)—than in passively harvesting large data sets, as I did in my earlier work (Figs 17.12, 17.13, 17.14, 17.15 and 17.16).

When you interrogate large data sets for universal truths, you end up with a statistical mush that offers vague, blurry, superficial insights (everyone falls in love, everyone gets mad, everyone eats breakfast, etc.). When you're hunting down the universal, the best approach is actually to study the specific and extrapolate—in that way, any insights you find will be grounded in something real. The personal is powerful.

AC: I'm really interested in the notion of sense or meaning in your work, the way in which it appears to be trying to make sense of very large and very complicated sets of data. Linked to this I'm interested in the idea of 'movements' in *We Feel Fine* as different kinds of sense or meaning—from the initial madness of *Madness*, with the mass represented as a proper mass, without meaning, chaotic and messy, through *Murmurs*, *Montage*, *Mobs*, *Metrics* and *Mounds*. Each one is giving a different perspective on the data, a different slice. And each has a very different aesthetic feel about it. What were you trying to do here?

JH: I have always been quite OCD as a person (it runs in my family), and probably the main impulse in my life has been to try to control life's chaos by spotting patterns and organizing the noise all around me. In my personal life I do this with plans, lists, routines, and projects, and I think you can see this impulse carried over into my work.

Fig. 17.10 *The Whale Hunt*, Jonathan Harris, 2007

With *We Feel Fine*, I saw so many different interesting sides of the data—ranging from emotive to analytical—and I could not choose just one at the expense of all the others. So there are six movements that each explores a different aspect of the data:

1. *Montage* lets you see photos of real people—this is the most human and empathetic part of the piece
2. *Madness* mimics the feeling of living in a large city and constantly shifting between total anonymity and extreme intimacy, and what that changing of emotional scales does to an individual—I was living in NYC when I designed that movement, and it really encapsulates how I was feeling living there.
3. *Murmurs* allows you to be passive and witness a scrolling wall of human expression—the Godlike experience begins here
4. *Mobs* is a whimsical way to introduce the idea of statistics into a storytelling context, without being too technical—the Godlike sense is back, here in the form of numbers, appealing to the popular belief that "only if I have enough data, then I will understand" (which is a deeply flawed belief).
5. *Metrics* appeals to the hyper-rational, analytical mind—humans are just numbers now
6. *Mounds* is a playful way of summarizing an entire database—the individual sentences are most abstracted here.

So the movements range from God-like voyeurism / emotional mind, to God-like omniscience / rational mind, but again, together, trying to produce a weird kind of empathy for the human condition, so that viewers end up feeling less like Gods and more like humans.

AC: You've spoken in the past about surveillance and self exposure. Your work seems to be about a kind of poetics of surveillance, finding patterns, creating beauty out of this enormous mass of self-published material.

JH: Yes, I think there is some of that. I'm not so interested in surveillance as such, in any kind of Orwellian way—at least not like some other artists are. For me, surveillance is like data visualization—another contemporary tool we have in our culture, which we can use as artists to say things about our world. Surveillance gets a bad rap (CIA, wiretapping, etc.), but surveillance can also be used to uncover incredible beauty. It can be used to humanize—not just dehumanize—individuals.

AC: It's interesting that you see surveillance as ambiguous—neither good nor bad—but as something which can affirm humanity. In this respect it becomes a kind of anthropology—and a technique for you to uncover humanity and beauty. How do you go about this—I mean what kind of technical decisions do you make to uncover beauty? I guess this connects back to ideas about whether a particular set of data is intrinsically beautiful or not. So, there are ridiculously large amounts of data out there—and you have to make a decision about which subset of it you're interested in. How do you make that decision? I guess I'm interested in how you work with data, in the way other artists work with pigment, or movement, or words or whatever.

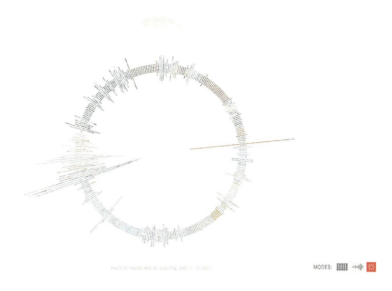

Fig. 17.11 *The Whale Hunt*, Jonathan Harris, 2007

JH: I usually start by deciding what kind of thing I want to make a project about (news, emotion, my own life, etc.), and then I think through all the different aspects of that thing that might leave behind a data trail. Then I start wandering through those data trails, and I see what the data looks like. What I'm looking for is something surprising—some weird pattern, some repetition, something that makes me gasp, something I didn't know, something I haven't seen, some strange subtext, etc.). I often build analytical tools to help with this process, especially to look for patterns. You start to get a feeling for whether something's going to be interesting or not, and if it looks interesting, then you go deeper. For the projects that involve real experience (*The Whale Hunt*, *Balloons of Bhutan*, and *Today*), (Figs 17.17, 17.18, 17.19, 17.20 and 17.21) the process is flipped on its head, because I have to decide beforehand what kind of data I'm going to collect as I go through the experience (temperature, heartbeat, certain questions, etc.). This approach is more about hacking reality and developing hypotheses about which hacks are likely to be interesting. Then I go and put myself in those situations, to see what happens. This is more risky, because you never really know if something will be interesting until you try it.

In both cases, the visual design of the final piece comes much later.

AC: Your work lives on the internet. Why a gallery show? How does the work differ—online and real world?

JH: I love the Internet as an art platform. I love its openness, ubiquity, accessibility, scale, and permanence. I also love the lack of gatekeepers. However, one

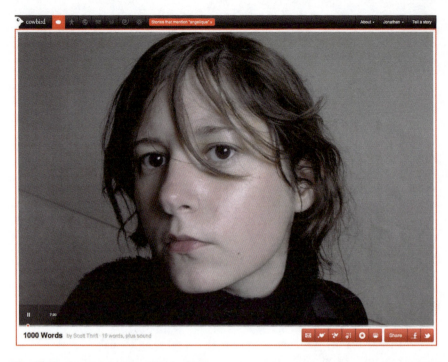

Fig. 17.12 *Cowbird*, Jonathan Harris, 2011

problem with presenting a body of work on the Internet is the fragmented, schizophrenic, piecemeal thing the work ends up becoming. My various projects are scattered across dozens of websites living at different domains, written in a handful of programming languages, some still collecting data, others frozen in time, and others offline entirely.

This makes it very difficult for a viewer (or even for me) to get a sense of the body of work as a whole. I found that seeing the work all together in a gallery has given me a very different sense of it. It feels much more continuous, self-consistent, and slowly evolving than I ever really imagined it to be. I think viewers to the exhibit have the same feeling. Also, we have found that typical visitors to the show are spending 1–2 h there, while other exhibits at that same gallery usually have visitors spending less than 10 min.

So there is clearly a tremendous interest for this kind of work to be seen in an art context. The problem is mainly that the art establishment has not yet found a way to think about it and welcome it (not to mention sell it), so it remains largely fringe—a thing of the Web, but not of the "serious" art world.

One of my goals in doing this show was to offer up my work to the art world, to see if it can even have a presence there, or whether I should forget about the art world and just keep publishing my work online. The show has tremendous appeal among the public, but it's unclear what the impact (if any) will be in the art world.

17 Beauty in Data 329

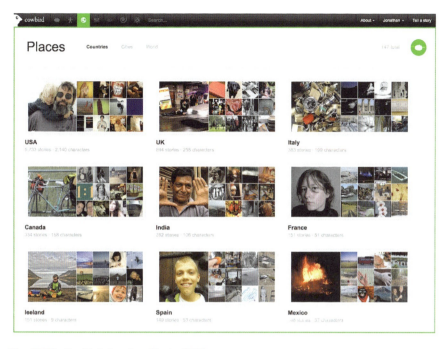

Fig. 17.13 *Cowbird*, Jonathan Harris, 2011

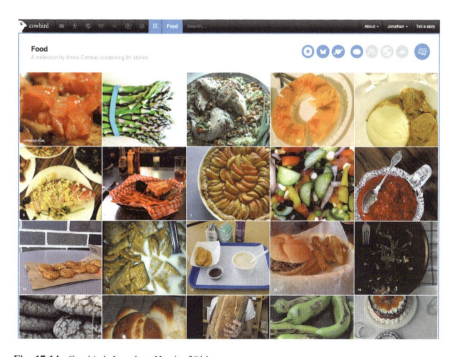

Fig. 17.14 *Cowbird*, Jonathan Harris, 2011

AC: Why do you think electronic art and net art is so disconnected from the broader fine art scene? Do you think this will ever change? Does it matter?

JH: I think people in the art world (especially critics) like to feel elite, like they understand things the rest of us don't. To get anything out of most "good" (i.e. critically acclaimed) contemporary art, you have to have a tremendous amount of domain knowledge or an MFA. This keeps the critics employed, so they can explain the art to the rest of us. With a lot of digital work (including my work), this explanation is not really necessary. Pretty much anyone can understand one of my projects pretty much immediately (which has always been one of my goals). There's a lot of complexity hiding in them, but it's pretty easy to see what's going on right away (Golan's idea of "instantly knowable, infinitely masterable"). I think this approachability scares critics, because there's not much for them to add to the dialogue. This kind of work, when it's done well, doesn't require tour guides. I think critics feel threatened by it, so they try to avoid it, and say it's not art, so they don't have to deal with it.

If digital art were less understandable, more obscure, more abstract, and did more to reference other existing artworks, critics would probably like it more, because they would have more to say about it. But it's unclear whether this would actually be good for the work—probably not.

AC: I'm curious about the links between your work and the movement identified by the French critic and writer Nicholas Bourriaud as "Relational Aesthetics"—where the meaning is in the human relations provoked by the work, as much within and between the audience as between artist and audience. Your work is very different in style from that of Sophie Calle or Rikrit Tiravanija—but seems to me to share similar preoccupations with authorship and participation. Are you a relational aesthetics artist?

JH: I love Sophie Calle's work, and I think there's definitely a relation between hers and mine. I've never considered myself a 'Relational Aesthetics artist' as such, but I suppose my work does end up involving some of those ideas.

AC: Do you think it's possible—or desirable—to be an artist who works solely on the internet?

JH: One hard thing is how to make money. The art world is premised on the fetishization and selling of scarce objects. The Internet (and my work), conversely, is based on abundance. Indeed, websites are often judged by their number of viewers. An artwork's price is unrelated to its number of viewers. A priceless Picasso can hang over the sofa of a hedge fund manager, 100 people will see it a year, and it will still be a priceless Picasso. So for one thing, there needs to be a new economic model for artists working online, or otherwise they will work elsewhere.

Another problem with the Internet is the glazed-over, "I am looking at a screen now" mindset that people go into when they are staring into a monitor and clicking with their mouse. This deadened, distracted, passive mindset (largely brought about

17 Beauty in Data 331

Fig. 17.15 *Cowbird*, Jonathan Harris, 2011

Fig. 17.16 *Cowbird*, Jonathan Harris, 2011

Fig. 17.17 *Balloons of Bhutan*, Jonathan Harris, 2011

by addictive social networking tools) is inconducive to having deep personal experiences, whether with art or anything else. When I see visitors in a gallery looking at one of my pieces, I can see they are having a deeply personal experience—they are very present, in the moment. When they are at home with their laptops, I am not so sure.

As an artist, I am actually moving away from the Internet. I have been doing more work in the physical world, involving strange personal experiences, largely because life is short and I don't want to spend my whole life sitting behind a screen, and there is much to learn from the real world!

Ultimately, I am not interested in the Internet as a subject. I am interested in real people and real experience as subjects. The Internet is just a place where real people gather, and where real experiences are documented, so it can be a good proxy for this kind of portraiture.

Also, it's a great distribution medium.

But no, I'm not married to it.

AC: Tell me about *The Whale Hunt*. It's different from the other pieces. It's a story. It's time based, it's a sequence. So, how can data mining work together with narrative sequence? Is there a contradiction here—between what linguists used to call the paradigm—the set of possibilities, and the syntagm—the sequence of items strung together to form meaning?

JH: *The Whale Hunt* was really about putting myself in the position of the computer, and assigning myself an algorithm to follow as a program would. After

Fig. 17.18 *Balloons of Bhutan*, Jonathan Harris, 2011

creating so many projects that required computers to follow rules incessantly, I thought it would be good to gain some empathy for the computer, kind of like an energy executive spending some time in the mines, digging up coal, to understand what his business is really about. So for *The Whale Hunt*, I took photographs at 5-min intervals for 8 days, and then more frequently when my heartbeat went up, producing 3,214 photographs in all. Once collected, I tagged and classified these photos in a number of different ways, and only then did I create a program to surface the hidden connections between these photographs—connections like color, people, themes, time, adrenaline level, etc.

To me, this is a really interesting and quite unexplored area—using computers to process real human experience and come to a deeper understanding of it. It is like computer-assisted contemplation, or machine-based meditation. I am actually not sure if this can work, but I am interested in trying.

With computers so much is possible, so as artists we really have to ask ourselves, "WHY am I doing what I am doing? Is it just to show off? To show what a good programmer I am? To show how pretty I can make the swirly thing flying around the screen? To show how pretty I can make that generic data set look?" These are the wrong reasons for making things. Instead, we have to ask, "What does this thing give to others? How is this thing improving me as a person? How can I see something no one else can see, and how can I communicate it in a beautiful way? What kind of world do I want to see, and how can I help make it?" These are the kind of questions artists need to ask, but digital artists in particular seem to have trouble asking these questions, because they think the questions are questions for

Fig. 17.19 *Balloons of Bhutan*, Jonathan Harris, 2011

poets and philosophers, not technologists. But if more technologists thought of themselves as poets and philosophers, then very different types of software would begin to emerge, and that software would help to shape the emerging digital world, and keep it from turning into a shopping mall (if it hasn't already).

AC: I'm very curious about *Cowbird*—which as far as I understand, also connects narrative and data-mining in new and innovative ways. Can you say something about this?

JH: I don't want to talk too much about *Cowbird* now, except to say that it is a storytelling platform for others to use to tell stories of any size—from The War in Iraq, to My Day at The Beach. It generalizes many of the principles explored in my earlier works (maps, charts, graphs, timelines, themes, people, simple playful design, etc.), and incorporates them into a storytelling tool that non-programmers can use to tell beautiful interactive narratives. I've been working on it for almost 2 years, and it's nearly ready to share. It is new for me in many ways, but mainly because it directly involves other people. All of my other works are basically portraits, in one way or another, but *Cowbird* is a tool that people use directly. There are all sorts of considerations in tool-making that you don't need to make in portraiture. I've always liked Golan Levin's maxim, "To make tools that are instantly knowable and infinitely masterable—like the pencil and the piano." I've been keeping that rule in mind designing *Cowbird*, but it is very hard!

Fig. 17.20 *Balloons of Bhutan*, Jonathan Harris, 2011

AC: There have been no masterpieces of digital art—or so you famously said at Flash on the Beach 2 years ago. You also said most digital work failed to move you, that much of it is un-emotional. Do you still think this is true? And if so, why should this be?

JH: As for masterpieces—I'm not sure. I guess masterpieces only fully reveal themselves with time, and that the definition of a masterpiece is precisely something that remains relevant over time. But I do still think that in general, digital art occupies an awkward adolescence, still groping around for exactly what it should be, and that the only way to grow out of this awkward adolescence is to make projects that deal with big themes, or that deal with small themes in a big way. Basically, digital artists need to make more serious work. Experimenting and tinkering are great to learn the tools, but once you learn the tools, then you have to use the tools to say something, and the saying something is much harder (but ultimately much more important). It's the only way to break through the digital ghetto and into the mainstream world.

As for digital works failing to touch me—this is something I think about a lot. I think part of the problem (and I mentioned this earlier) is how computer programming is removed from the original act of self-expression, in a way that paint and words are not. My friend Rob, who founded Etsy,[1] used to ask me how I could be a digital artist, and whether I had found a way to channel my real personal

---

[1] URL, January 15, 2014: http://www.etsy.com.

experience / suffering / whatever into writing code. I don't think I have found a way to do that. When I am really upset, or feeling other very strong emotions, it might help me to write or to draw or to paint, but the last thing I want to do is to write code, I think because writing code requires a suppression of my humanity. It's like, in order to write good code, I have to become a bad (unfeeling) person, and to become a good (feeling) person, I have to stop writing code. It is a tradeoff. And I feel this tradeoff very intensely when I go from a few weeks of traveling and writing and photographing, and then I sit down and try to write code again. I can feel the spiritual resistance, because somehow my soul knows that I am a better person when I am not writing code, and it is trying to urge me not to do it again. But as I stubbornly do it (and it usually takes a few days to get back into it) I can feel my living, breathing, human side (really, my body and senses), slowly atrophying and ultimately going away almost entirely.

Writing small computer programs is fun and easy, but writing large programs, with tens or hundreds of thousands of lines of code, is very hard. You have to keep the whole program in your mind at once, and as the program gets large, it takes up more and more of your mind, and you have no space left for anything but the program. It is like a transferal of empathy—from humans to the program. This process always makes me sad, but I do it anyway, because I like what you can make with code.

Anyway, this is something that has been on my mind a lot lately, and I'm not sure which side of it I'll end up choosing.

AC: Programming might not be emotional in the romantic sense of the word emotion, but there are other emotional states linked to the act of programming which are just as important—I'm thinking of the intense sense of losing yourself in the task, the sense that you begin to access a purer realm of abstract thought, a state of mind which becomes a form of meditation in which time and the body begin to lose their reality.

JH: I've heard this idea from time to time—that programming can help you reach some Zen-like state of concentration and bliss. This is probably true of any task or craft, taken intensely, and not unique to coding, though there is something Oracle-like about staring into a glowing screen. However, I'm not sure that forgetting your body is such a good thing—after all, we are human animals having a physical experience here on earth, and coding can make us forget that. Meditation can make you more present—more conscious of your body and senses—but coding takes you out of your body and senses, out of the earthly present, and into some imaginary realm. I think coding makes you *less* present.

AC: Further—getting stuck with a snag in the code can be extremely emotionally debilitating and solving a problem in code can be one of the most emotionally joyful and satisfying experiences a person can have (or is that just me?) (Fig. 17.21)

JH: Yes, I've experienced this "bug/solution/bug" cycle of joy and despair many times, but somehow it's always felt off to me—like it's coming from a place of insecurity and neediness, not wholeness and balance. It's kind of like being in an abusive relationship with a really hot girl who's actually a total bitch and who treats

Fig. 17.21 *Balloons of Bhutan*, Jonathan Harris, 2011

you like crap, but still you can't walk away because the highs feel so good and you like the idea of what you can build together.

AC: So, are you saying that for you digital work can't be emotional and that by extension, a digital masterpiece is (…) unlikely?

JH: No, I'm not saying that. This medium is very young, and we are still learning how to use it to craft statements and situations that could not exist in any other medium. We'll get there.

AC: Do you know how many people in the advertising industry use your work—particularly *We Feel Fine*—as a reference in presentations? Does this translate into offers for commercial work? And if so, how do you respond to this?

JH: I have a vague awareness of this. And yes, I get many, many offers from advertising agencies to do commercial work, but I have never accepted any of them. In general (and please don't take this personally), I really don't like advertising and the role it plays in the world. I think it creates unnecessary feelings of inadequacy and desire, usually translating into excessive consumption, which creates all sorts of other problems. I think a better world would be a world without advertising.

Furthermore, I think most of the digital work that comes from advertising agencies is very bad. I think the reason for this is that too many people are involved. You have one person sealing the deal, another designing the brief, another doing the concept, another doing the design, another writing the code, another doing the testing, and you end up with a big mediocre mess where no one understands anything.

I believe in practicing one's own craft. You learn an astounding amount from the resistance of the medium. My projects always evolve enormously from initial concept to final form. Often you try something and it sucks, so you go in another direction. Often you make a mistake, and the mistake ends up showing you a better way. But unless you have an intimate relationship with the medium and with the piece, you will not notice these cues, and the work will suffer.

To make something really beautiful, you have to treat it like a lover. It has to be personal. It has to obsess you when you're falling asleep. It has to be in your dreams. It has to be with you when you wake. It has to torment you.

If you allow the work to accompany you in those intimate moments, it will reveal itself to you, and the result will evolve like a life form, nuanced and natural.

AC: Have you seen *OkCupid*[2]—the google of online dating?

JH: Yes, I've seen *OkCupid*, and I like some of the findings they publish around the art of successful online dating etiquette. That is an example of beautiful data—data that is useful to real people's real lives.

I think online dating is one of the best examples of technology actually helping people to improve their human lives (by finding love or sex), although most online dating sites are very bad, in the way they squash individuality to classify each person in discrete terms that their algorithms can understand.

When I made *I Want You to Want Me*, I actually went on some online dates, and I ended up meeting someone I dated for seven months, before splitting up. It's weird—I always had a stigma about the way we met, not, I think, because of the stigma of online dating, but because of the stigma of not having a story. I think that the story of how two people meet (especially if the story is improbable and surprising) can be a huge energy source to power the relationship for many years. That may sound silly, but I think it is true. The stories we tell ourselves are the only things that give our lives meaning. It would be too much madness otherwise.

Author Biography

Jonathan Harris (Artist, Storyteller and Internet Anthropologist) (Shelburne, Vermont, August, 27, 1979) is an Internet artist and designer living in Brooklyn, New York. He has won three Webby Awards and was honored as Young Global Leader by the World Economic Forum. His work has received coverage by CNN and BBC and has been exhibited in the Museum of Modern Art in New York and the Centre Pompidou in Paris.

---

[2] URL, January 15, 2014: https://www.okcupid.com.

Combining elements of computer science, anthropology, visual art, and storytelling, Jonathan Harris designs systems to explore and explains the human world. He has made projects about human emotion, human desire, modern mythology, science, news, anonymity, and language. He studied Computer Science at Princeton University. Some of his noted works are the *Yahoo Time Capsule*, which attempted to create a digital finger print of the world in 2006, *I Want You To Want Me*, the *Sputnik Observatory*, *Universe*, and *the Whale Hunt*.

For more information and contact:

http://www.number27.org; https://twitter.com/jjhnumber27

Chapter 18
Tracing My Life

Nicholas Felton

David Bihanic's Interview with Nicholas Felton

David Bihanic (DB): As a way of introduction, could you please introduce yourself and describe your background (e.g. training, work experience, career path, etc.)?

Nicholas Felton (NF): My name is Nicholas Felton. I am an information designer working in New York City under the name *Feltron*. I was raised in California and studied Graphic Design at the Rhode Island School of Design. After a brief stint in advertising after university, I started my own design practice and began focusing on data driven design around 2007. In 2009, I co-founded www.Daytum.com with Ryan Case, and in 2011 we were hired by Facebook to help redesign the profile into what is now known as Timeline. I left Facebook in April 2013 to focus on my own projects.

DB: How did you become interested in the field of data design (data representations/visualizations, infographics, data narratives, etc.)?

NF: One of my favorite books growing up was a book called "Comparisons". It is a visual compendium of the world in a format that would now be considered "infographics". I poured over this book until the pages fell out. In 2004, I began playing with graphic representations of my own data, and this slowly worked its way into my personal projects and ultimately the first personal *Annual Report I* designed. The success of this visual summation of my year leads me to continue the project and to dive deeper (however naively) into data visualization and quantified storytelling.

N. Felton (✉)
20 Grand Ave, Suite 702 Brooklyn, New York 11205, USA
e-mail: nicholas@feltron.com

DB: It's clear that you have a great culture of design, especially in graphic design (as well as a definite mastery of graphic design techniques and processes), and a perfect knowledge of both technical and scientific history of the representation of data; this combined knowledge allows you to realize high quality work—in many ways (…)
How did you achieve such a level of knowledge?

NF: I certainly won't claim to have mastered graphic design, or to have any great technical or historical expertise with data visualization. When I began designing data-driven work for clients, there was not a great degree of scrutiny or sophistication to the graphics being produced, and I was hired simply to make things look nice or interesting. As a result, I was free to develop or discover my own techniques and standards for representing data.

DB: Do you think that data design learning is the ongoing, voluntary, and self-motivated pursuit of multiple and combined knowledge?

NF: Certainly. My work is the end result of a great graphic design education, personal curiosity, an interest in math and code and a fertile moment in which data is both prevalent and popular.

DB: You compose and publish every year (since 2005) an annual report of your personal life. You represent multiple self-tracking data in graphical manner which reflect your own life cycle (life routines and rituals for instance)—in reference to the quantified self, self-tracking or lifestream movement (…)
Could you please tell me more about this project (and your investment in it)?

NF: *The Feltron Annual Report* was first published in 2005. It was intended as a quick project in which to share my interests and habits from the previous year. I had assumed that the Report would be interesting only to friends and family, and was surprised to find it picked up by bloggers and passed around the internet. As a result of the encouragement I received in this first year, I have continued the project and increased its scope each year since (Figs. 18.1, 18.2, 18.3, 18.4, 18.5, 18.6, 18.7 and 18.8).

DB: Why is it important for you to compose a data portrait of your life?

NF: I have found that this project satisfies my personal curiosity about my habits, while also providing me with fertile material with which to design. I had long sought a vein of content with which to design self-motivated projects, and by recording my own life I have found a story that I already own and can record in the most detail.

BD: What are deep meanings (both objective and subjective arguments) you attach to this project?

NF: At the most philosophical level, I believe that this project is a search for TRUTH. There is a thought within me that if I can record just the right data and display it in the right way, I may discover something (…) some crystalline structure, a higher order than I am able to experience. This goal seems tragically impossible to achieve, but the allure of understanding if we are governed by free-will or destiny is compelling.

18 Tracing My Life

Fig. 18.1 *The 2005 Feltron Annual Report*, Nicholas Felton, 2005

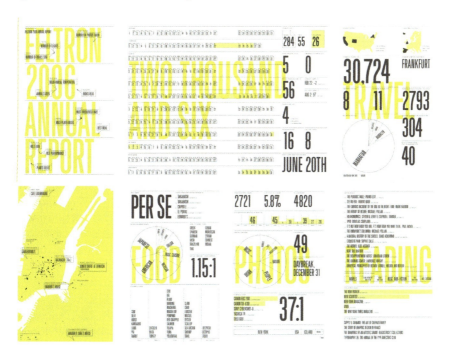

Fig. 18.2 *The 2006 Feltron Annual Report*, Nicholas Felton, 2006

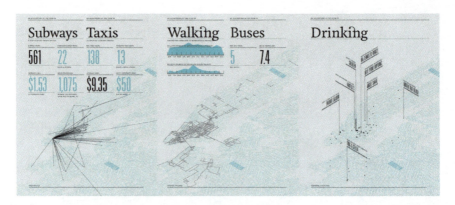

Fig. 18.3 *The 2007 Feltron Annual Report,* Nicholas Felton, 2007

DB: What motivates you to continue this difficult and tedious project, year after year?

NF: The challenge each year is to define an approach that will illuminate new aspects of my behavior. After a few years of straightforward accountings, I started exploring new lenses for recording the year. In 2008, I concentrated on the distances I traveled. In 2009, I asked those around me to report back on my behaviors. In 2010, I lost my father and used this project to describe his life through the fragments of data he left behind. In 2011, I combined two years of data to look at the ways in which my behavior changed depending on who I was with. Most recently I developed an iPhone app for 2012 to ping me at random intervals and record my behavior. I have already planned the methodology for 2014 and will continue the project as long as I think there are new challenges to be tackled.

The project initially served as a self-promotional piece, but has evolved far beyond that point. It is now more of a research & development platform for exploring new dimensions of data collection and representation.

DB: What motivates you to continue this difficult and tedious project, year after year?

NF: The challenge each year is to define an approach that will illuminate new aspects of my behavior. After a few years of straightforward accountings, I started exploring new lenses for recording the year. In 2008, I concentrated on the distances I traveled. In 2009, I asked those around me to report back on my behaviors. In 2010, I lost my father and used this project to describe his life through the fragments of data he left behind. In 2011, I combined two years of data to look at the ways in which my behavior changed depending on who I was with. Most recently I developed an iPhone app for 2012 to ping me at random intervals and record my behavior. I have already planned the methodology for 2014 and will continue the project as long as I think there are new challenges to be tackled.

The project initially served as a self-promotional piece, but has evolved far beyond that point. It is now more of a research & development platform for exploring new dimensions of data collection and representation.

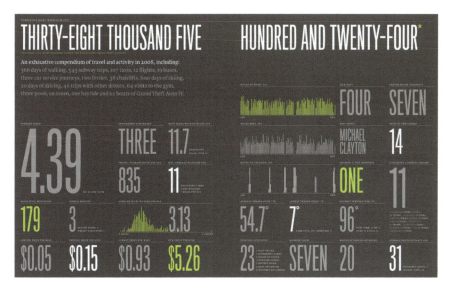

Fig. 18.4 *The 2008 Feltron Annual Report,* Nicholas Felton, 2008

DB: Is it a personal way to defy boredom, inactivity and death like the artist Roman Opalka (towards a new form of struggle of memory versus forgetting)?

NF: To some extent it is an attempt to defy death. This is most clear in the Report I designed about my father's memory. I wanted to collect his life while it was most fresh and preserve his spirit. I wish that I had more information about the lives of my parents at my age and hope that my children and grandchildren will be able to learn about me from the documentation of my life.

Additionally, the act of recording increases my awareness of activities and habits on a daily basis, and I attempt to shape my approaches in ways that improve my behavior.

DB: Of course, these annual reports are primarily 'creative playgrounds' that enable you to renew data visualization experiences… Each year, you manage to create awesome infographics (…)

What process do you adopt to design these annual reports?

NF: Once I have finished my data collection for the year, I spend several weeks cleaning and structuring the information. Generally, this means working in a spreadsheet to make sure that I have the correct metadata for everything (like locations) and that like-objects are named correctly so that they can be accurately aggregated.

Throughout the year and this cleaning process, I am coming up with hunches about what I might discover based on what I have collected. I will start sorting and querying my data to see if these hunches bear fruit and will then start sketching visualizations that would communicate this story in a clear and appealing manner.

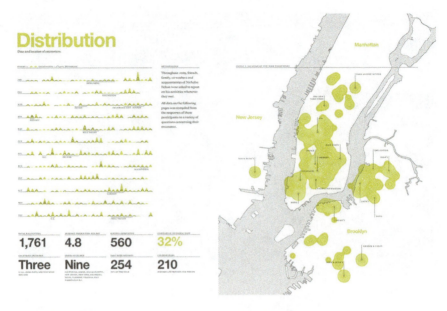

Fig. 18.5 *The 2009 Feltron Annual Report*, Nicholas Felton, 2009

The designs are the result of the visualizations I code using Processing and text or numbers that I extract from the database. I am at once developing the structure of the document, the component stories and a coherent visual approach that can link all of the graphics together.

DB: How do you approach each new report?

NF: Each new Report must be its own object. I scrutinize the old ones to ensure that I am not repeating myself or producing something less interesting than the last one.

DB: What determines how long you should keep personal data?

NF: At this point, I am keeping my personal data indefinitely.

DB: How good is personal data? What is good quality data? (…) All your personal data is held and processed? (…) All your personal data is always useful?

NF: Some of my data is extremely coarse, while other dimensions might be extremely fine-grained (…) but this does not determine how interesting or valuable it is. I might have footstep data for every minute of the day, but I think that 10 daily measurements of who I am with is far more powerful.

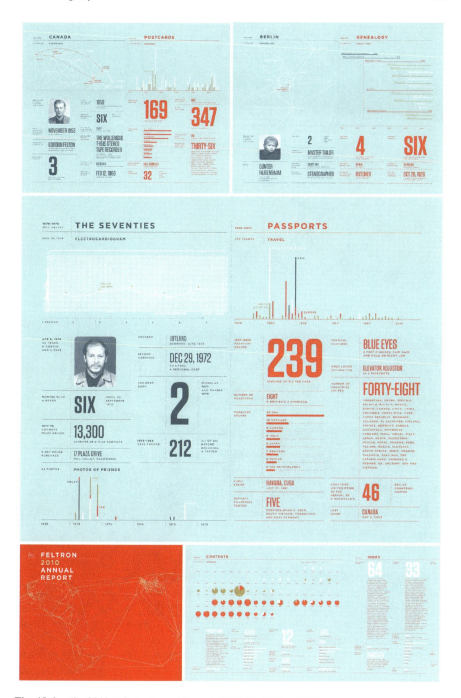

Fig. 18.6 *The 2010 Feltron Annual Report,* Nicholas Felton, 2010

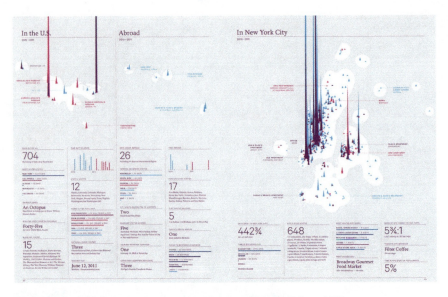

Fig. 18.7 *The 2010–2011 Feltron Biennial Report,* Nicholas Felton, 2011

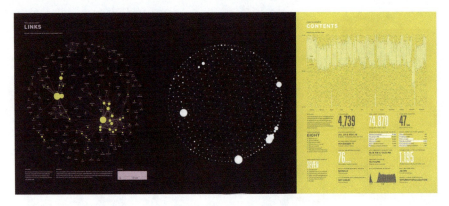

Fig. 18.8 *The 2012 Feltron Annual Report,* Nicholas Felton, 2012

DB: This self-tracking data doesn't seem so personal... I mean these immaterial and evanescent traces of your actions [residual data used here as design material] don't tell us who you are but rather help us to understand how we act, move and feel, collectively (and perhaps where we are heading). Thus, this sort of data should serve to visualize collective phenomena—and not directly individual ones (...)
What do you think?

NF: I do agree somewhat, and the recent challenge of my project has been to delve into more personal phenomena. While it is easy to track how many coffees one drinks during the day, cataloging thoughts and feelings is far vaguer and overwhelming. Developing methodologies for scratching this surface is something that I am pursuing. For instance, in 2009 I asked everyone with whom I had a significant encounter to report back on my behavior, including my mood. Combining mood with topics of conversation and behavior can lead to a much more descriptive picture of who I am.

DB: In other words, don't you believe that your project offers to make visual collective narratives that harness your personal life story (your residual data is used as design material of a collective work)?

NF: I believe that the both the collective AND personal stories are interesting. I am somewhat alone in my self-tracking experiments, but I find that the things I am discovering about myself now are being automated at a rapid pace, and I think that once the data I have access to is available to the masses we will discover much more of the collective narrative and insight.

DB: Graphic design has a powerful leadership role to play in this 'Quantified Self movement' that has profound implications for all areas of life and business (…)
In your opinion, what are the next challenges for graphic design?

NF: Since I attended school for graphic design, data has joined text and image as an elemental material of design. Today our world is being recorded and described with data and I believe that to practice visual communication now demands a fluency in working with data.

DB: You mix (and also combine) multiple existing models for graphical data visualization (e.g. stacked bars, streamgraph, sunburst, node-link tree, treemap, scatterplot, bubble chart, etc.) with your original graphic proposals (…)
Do you intend to invent new data visualization models (especially for large data sets) that can be used by other designers?

NF: I would love to invent a novel method of data visualization, but don't believe that I have achieved that yet.

DB: I'm struck by your ability to design very clean, consistent and minimalist visualizations which also have a very distinctive and recognizable style (…)
What is your creative exploration process (and method) and how has it changed or evolved over the years?

NF: I think that my approach is rooted in minimalism. This is a very 'Tufte-esque' strategy of weighing each mark and its contribution to the story I wish to communicate. Over the years, I have been able to move more quickly and attempt more experimental forms through my use of code. This allows me to iterate quickly and explore new approaches that are impossible with my earlier handmade methods.

DB: Do you always work by a set of guiding principles or creative intentions?

NF: I have my personal standards for clarity and interest that I am always trying to meet (…) but I also have an underlying code of rigor that I try to maintain. I believe that you should be able to measure a graphic and extract the original values that it expresses. I also believe that the graphic should contain some amount of error correction (typically in the form of values) so that you can confirm that the graphic accurately designed.

DB: For you, data visualizations are like narratives, right? They make the story behind data come to life (…)

Could you please tell me how you discover, imagine and tell new visual stories with huge amounts of raw data (self-tracking data and other)?

NF: Each data set requires its own approach and it takes a certain amount of curiosity and sometimes creativity to extract the best stories, but there are some consistent ways to begin. I always like to look at the data in tabular form first, here I can look for the bounds of the data set (…) the number of entries, the dimensions and quality of the data, the largest and smallest and the first and last entries. Then I will typically try to plot the entire set in some basic way, whether as a map or a line graph. This will give me an idea about interesting trends or regions to explore. Additionally, I like to look at any categories that might be included in the data, as this is an extremely valuable way of reducing complexity in a large set. By the time I've done this, I should have a couple of good discoveries from multiple scales and can start trying to connect things I find interesting into a grander and more custom visualization.

DB: How do you find the intrinsic meaning in data and glean from it the stories that beg to be told (stories through data)?

NF: To do this, I like to try and answer a question. *The Annual Reports* all try to answer a question about my behavior, and because the question precedes the data, it ensures that the right data is collected to tell the right story. When the data already exists, having a question to answer will focus the exploration and provide a way of measuring success.

DB: Ultimately, all stories are rooted in data (…) What data do you need to find great stories?

NF: 'Great' is a lofty aspiration (…) but I think that all stories require drama and in data this takes the form of contrast.

DB: Could you create good data-driven stories with poor quality data?

NF: Absolutely. Sometimes you have to work with what you have, but as long as the faults are acknowledged and managed, it should not be an insurmountable obstacle.

DB: Does the interest of the story depend on the quality of data? If so, which qualities are needed?

NF: I don't believe so. The better the data, the more details one can discern. As in all storytelling, details can improve the story, but they won't salvage a dull one.

Fig. 18.9 *Daytum,* Nicholas Felton and Ryan Case, 2009

DB: As you said earlier, you co-founded www.Daytum.com with Ryan Case in 2009. *Daytum* is a ground-breaking self-tracking application (also available for iPhone) allowing people to count, explore and communicate daily data (…)

Could you please tell me more about this project? (Fig. 18.9)

NF: *Daytum* was created to popularize the ideas and approach of *The Annual Reports*. By creating convenient mechanisms for recording and displaying information, users are able to render a quantified portrait of anything they care to track.

DB: Are there many people who are interested in collecting data about themselves?

NF: I don't know that everyone wants or needs to track data about themselves, but I do find that many people already track some aspect of their lives (…) whether it's their gas mileage, their golf scores or the concerts they attend. Among those who are not tracking, I have found that given simple enough tools they can become engaged with their own data.

DB: Each user has access to a personal databoard (widgets) displaying in different ways the everyday data of their life.

Do you see this feature as a kind of mirror to reflect a sort of data-driven self-portrait?

NF: Yes. While the units in the dashboard are a curatorial tool, I think that this quantified approach can represent a more nuanced portrait of a person than a profile that is filled out once from memory. For example, a data-driven list of a person's 10 top musicians would probably differ from their list of 10 favorite musicians (…) and might represent a greater diversity than their own self-perception.

DB: Today, users have access to predefined visualization patterns (…)

In addition to self-tracking tools, do you plan to give users new tools to design their own visualizations?

NF: We may add additional visualizations to *Daytum*, but the data can also be downloaded and brought into other tools for further visualization.

DB: One of the major interests is to cross the data of each user to get an overview of collective trends? Is that right? Is it possible in accordance with the privacy terms?

NF: We do not analyze the entries of our users in aggregate, but it is certainly possible.

Author Biography

Nicholas Felton (Independent Graphic Designer) is the author of several *Personal Annual Reports* that weave numerous measurements into a tapestry of graphs, maps and statistics to reflect the year's activities. He is the co-founder of www.Reporter-App.com, www.Daytum.com, and is a former member of the product design team at Facebook.

His work has been profiled in publications including *The New York Times*, *Wall Street Journal*, *Wired* and *Good Magazine* and has been recognized as one of the 50 most influential designers in America by *Fast Company*. He is credited for influencing the design of Facebook's timeline.

For more information and contact:

http://www.feltron.com
https://twitter.com/feltron

Chapter 19
Multidisciplinary Design in the Age of Data

Stephan Thiel, Steffen Fiedler and Jonas Loh

David Bihanic's Interview with Stephan Thiel, Steffen Fiedler and Jonas Loh

David Bihanic (DB): Could you please introduce yourself and describe your background? How did the Studio NAND come into being?

Stephan Thiel, Steffen Fiedler, and Jonas Loh (Studio NAND: SN): NAND is a design studio in Berlin, which officially launched in Feb 2012. But the three of us have been working together for more than 7 years now. We have met at the Interface Design programme at the University of Applied Sciences in Potsdam at which we have worked on a couple of projects together—most notably *Mæve*[1] in 2008 which was an installation at the Venice Biennale of Architecture. Our path split when Steffen and Jonas did their MA at the Design Interactions programme at the Royal College of Art, London and Stephan at the Bauhaus University Weimar but it was somehow clear that we would come back together to found NAND as finding the right people to work with is a gift.

We founded Studio NAND to provide a basis for our collaboration. We all have many areas of interest which highly influences our work and ideas. We are constantly exploring new domains, building a framework to create holistic design approaches and to have constructive debates about how technology influences us.

---

[1] Project homepage, URL, December 28, 2013: http://mace-project.eu/maeve/.

S. Thiel (✉) · S. Fiedler · J. Loh
Studio NAND, Sophienstaße 18, 10178 Berlin, Germany
e-mail: stephan@nand.io

S. Fiedler
e-mail: steffen@nand.io

J. Loh
e-mail: jonas@nand.io

This is at the core of what NAND does and one might call that a little 'vague'. It is important for us to not have a single, potentially narrow focus, but instead to be able to approach topics as openly as possible. So far, this has led us to create work in large range of media, from web-based visualization to experimental performances, to speculative products. This is our preferred mode of working. And with the setup of our studio, the business side of things now becomes part of this endeavor and we enjoy thinking about how that world can be integrated into that.

DB: How did you come to specialize in the field of data design (data representation/visualization, information/interaction design, creative informatics, etc.)?

SN: It's fair to say this has become one main specialization of NAND because of the increasing relevance of data and algorithms in our societies. Data arrives at the centre of our lives, drives economic, political and personal decisions and yet, most people do not understand the inner workings of the processes and issues that are involved with information technology. We find it highly interesting to explore these issues, the opportunities as well as the conflicts. The most recent adventure in this domain is probably *Emoto*[2] (Figs. 19.1, 19.2, 19.3, 19.4 and 19.5) in which we, together with Moritz Stefaner and Drew Hemment, have looked at a major public event from a data (visualization) perspective. This work raised many questions, such as about the ethical, cultural, political and economic implications of data, access to it, or the cost involved with all of these aspects. Some of these findings are documented on the blog.[3]

A second example which falls within the category of speculative design is Known Unknowns.[4] Being Jonas' and Steffen's MA thesis project at the Royal College of Art's Design Interactions programme, this project raises awareness about randomness as the motor behind many data-driven processes in our societies but also about the nature of randomness itself. With this work Jonas and Steffen continued to follow their fascination about data as a material and its various possible manifestations, an interest which started with *The Digital Identities.*[5]

In addition to that we also like to experiment with 'data' and algorithms, exploring possible new ways of applying visualization to different contexts. An example from this category would be *Understanding Shakespeare*[6] (Figs.19.6 and 19.7) which explores the application of algorithms to literary works and possible visualization forms for drama. This has led to an ongoing initiative in education in which we collaborate with researchers to figure out new ways designing visualizations as a platform for learning. All of these aspects about data are of interest to us, so it was a logical consequence to pursue this further in personal research initiatives and client projects (Fig. 19.1).

---

[2] Project homepage, URL, December 28, 2013: http://www.emoto2012.org/.

[3] Blog site, URL, December 28, 2013: http://blog.emoto2012.org/.

[4] Project homepage, URL, December 28, 2013: http://www.nand.io/research/the-known-unknowns.

[5] Project homepage, URL, December 28, 2013: http://www.nand.io/infosthetics/the-gestalt-of-digital-identity.

[6] Project homepage, URL, January 4, 2014: http://www.nand.io/visualisation/understanding-shakespeare.

19 Multidisciplinary Design in the Age of Data 355

Fig. 19.1 *Emoto*, Moritz Stefaner, Drew Hemment and Studio NAND, 2012

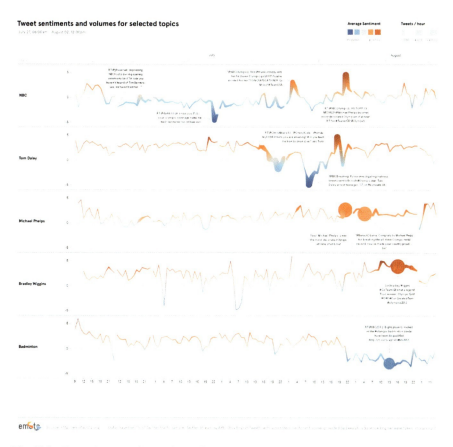

Fig. 19.2 *Emoto* (*sentigraph*), Moritz Stefaner, Drew Hemment and Studio NAND, 2012

Fig. 19.3 *Emoto* (*web*), Moritz Stefaner, Drew Hemment and Studio NAND, 2012

Fig. 19.4 *Emoto*, Moritz Stefaner, Drew Hemment and Studio NAND, 2012

19 Multidisciplinary Design in the Age of Data 357

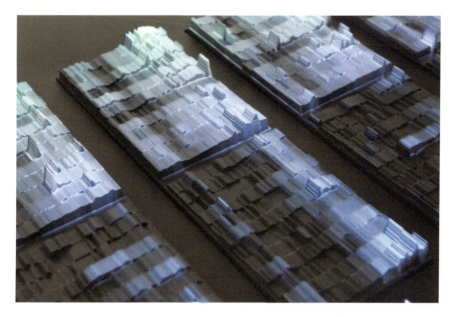

Fig. 19.5 *Emoto*, Moritz Stefaner, Drew Hemment and Studio NAND, 2012

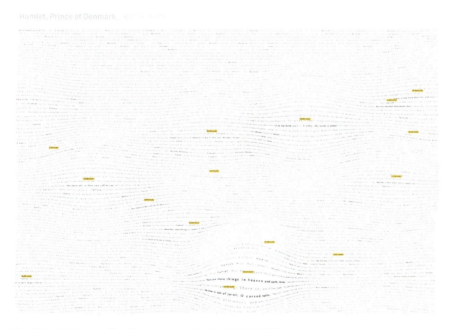

Fig. 19.6 *Understanding Shakespeare*, Studio NAND, 2010

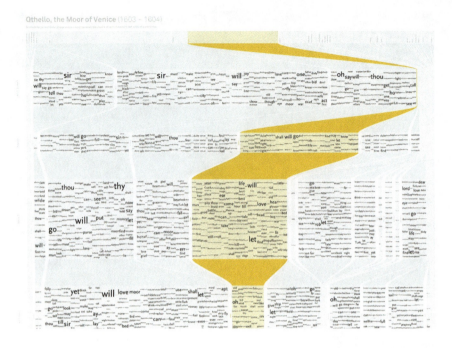

Fig. 19.7 *Understanding Shakespeare*, Studio NAND, 2010

DB: Studio NAND is a completely original data design agency adopting a cross-cutting approach to creation/innovation (Arts, Sciences and Human Sciences); you highlighted very early on the need for a multidisciplinary approach.

Could you please tell me what your design practice process is exactly (a singular creative process that supposes a global view and an interdisciplinary approach to the technical tools, methods, uses and their interrelations)?

SN: When approaching a data visualization project we always work iteratively from the data. This means we are "exploring the problem space"[7] in a prototypical manner with the goal to understand the problem from as many angles as possible. This often happens according to the four phases Spool outlines (plan, implement, measure, and learn). Most prototypes we create in this process will be thrown away since their sole purpose is to contribute to our learning. Only a few, maybe only one, will end up informing the final concept or idea based on which we start to design the visualization principle. At this stage, it is increasingly about exploring the solution space, so classical questions about form, interaction, etc. This is what people probably consider the 'classical' design process. But in the case of data visualization the exploration of the problem space is part of that and consequently phases transition very smoothly or are even intertwined.

From a methodological point-of-view this means we will use off-the-shelf tools or programming languages to get our hands on the data—to "[toil] in the data mines",[8] as Armitage puts it. We will create standard visualizations, such as line or bar charts, for the sake of exploring the data. This helps us to find an interesting perspective on it and the phenomena it may describe. Based on these simple charts we will start to search for interesting formal principles, by altering them or trying out quick ideas which might seems crazy. In the end we often find ourselves working with basic geometric shapes and their possible combinations, different ways to work with positions of such elements or even a completely custom mix of both. During each step, we will evaluate forms, transitions, interactions and decide whether they are an appropriate approach to the problem.

In this phase, the interdisciplinary approach is really in combining the logical world of math and programming with a creative design process. This is firstly reflected in the analysis approach to the data, in which the creativity is probably in the statistical exploration as it would be done by any good data scientist. So, you take over this role while still considering the overall design task. This requires basic statistical knowledge which is a development that is fairly new but becoming increasingly important. A second set of skills is the technical knowledge designers

---

[7] See Spool JM (2012) Exploring the problem space through prototyping. User Interface Engineering. URL: http://www.uie.com/articles/four_phases_prototyping/.

[8] See Armitage T (2009) Toiling in the data-mines: what data exploration feels like. Berg. URL: http://berglondon.com/blog/2009/10/23/toiling-in-the-data-mines-what-data-exploration-feels-like/.

should have if they want to work with technology in a hands-on manner. This includes not only working with a programming language such as Python, Ruby or JavaScript but mainly understanding the algorithms themselves. Databases are another category in this domain. You often need to be familiar not only with one database technology, say MySQL, but to learn about entirely new ones such as GraphDatabases and their rather different logical world. The same is true for the actual formal design work if it involves parametric tools. There it's about algorithms and form. But this is fairly common for designers working in data visualization and related design fields which use parametric tools to create visual output. But such a diverse set of tools is nothing new to design as a discipline. Design was always about the combination of different fields to create meaningful and compelling solutions or perspectives. Whether we deal with crafts, communication or technology and data.

What is becoming the greater interdisciplinary challenge nowadays is really a multidisciplinary one. We start to enter a level of complexity in the problem space, where we—as multidisciplinary designers—have to decide more often, when we need to involve experts into a project. Since in most projects this usually applies to several experts (statisticians, social scientists, biologists etc.) it is increasingly a designer's responsibility to create a basis for exchange.

DB: Do you think that the future belongs to multidisciplinary designers?

SN: Absolutely, but not only to designers. We think every discipline can only benefit from a multidisciplinary approach. One reason is that our world is already becoming incredibly complex due to the acceleration of technological progress. Therefore the demand for experts specialized in a specific domain will grow as well, as the demand for people who are able to work as a kind of "umbrella" spanning the disciplines. It is just an intrinsic value of art and design to communicate between people in a very direct and visual manner.

Furthermore, technology is becoming more accessible while it progresses increasing the connectedness between the disciplines. From synthetic biology to manufacturing, we see people empowering themselves with these technologies in a DIY manner, which is fantastic. We support this new direction with workshops and our platform Creative Coding as much as we can.

If you look at academia, there are a lot of new postgraduate programmes focusing on the intersection of design, art, science and technology already. Jonas helps to develop one of them at the moment as a collaborative project between the Academy of Visual Arts, the Technical University and the University of Applied Science, Dresden.

DB: You said that you use "design as a method to craft engaging experiences, stories and visualizations." Does this mean that design offers practical, methodological (and theoretical) tools/ways to create connections (and also a synergistic effect) between different fields/disciplines?

SN: Yes, most importantly the prototyping process described earlier, and in the context of different fields/disciplines it is about opening up this process and using it as a foundation to talk about ideas and aspects of various topics. This includes expressing concepts visually, physically and interactively. Designers learn to control that process as part of their profession. The methods are as simple as sketching on paper, or low-fi prototyping[9] to bring ideas to life quickly and make them concrete for discussion. We haven't yet found software tools which are perfectly suitable for such a scenario in the context of data visualization, at least none which do not require any familiarity with procedural processes; a "hot topic" and interesting problem right now.

There probably exists no golden rule or method for the creation of connections between different contexts—'the creative process' if we follow Csikszentmihalyi's argument.[10] But possible aids here are methods shared with other disciplines as well: playful settings that involve association, combinations of the seemingly uncombinable etc. The attempt has been made by various design practitioners to put that process into a framework, most notably design thinking. But interestingly enough this has been regarded a 'having failed'[11] the purpose of true cultural innovation as it is only a normative method originated from a very certain cultural area.

DB: It's very surprising to see the diversity of your work (data visualization, data narratives/stories, physical data visualization, data design tools, etc.). More than a new interdisciplinary research method, you seem to invent a crossover design approach. Is that correct?

SN: We are constantly exploring methods and tools to handle this design approach just because we observe that the level of complexity involved in design briefs rises dramatically, and will continue to do so. But we are surely not the only designers thinking about this. It is true that we are not so much interested in specializing in one design discipline only.

We think it is very much appropriate to support the crossing of the design fields. Both design cultures will continue to be important. A *crossover* design approach will become more and more important in high-level design work such as product and service design and development, design research and academia. But designers specialized in a certain discipline will always be more efficient, economically and conceptually, when it comes to working out the nitty-gritty in a concept. So, yes, various different (specialized) design fields will and should continue to exist, but a new 'layer' of conceptual design responsibility will emerge on top of these specialists

---

[9] Which refers to the process of sketching physically using really cheap but versatile materials.

[10] See Csikszentmihalyi M (1997) Creativity: Flow and the Psychology of Discovery and Invention. Harper Perennial, New York, p.79.

[11] See Nussbaum B (2011) Design Thinking Is A Failed Experiment. So What's Next? Fast Company. URL: http://www.fastcodesign.com/1663558/design-thinking-is-a-failed-experiment-so-whats-next.

which will bridge all disciplines. That's an exciting domain to work in if specialization is not your main interest but rather innovation on a much broader level.

DB: Is it a way for you to bridge the gap between Arts and Sciences (not just to reveal the relationships between the two)?

SN: The intersection of the Arts, Design and Sciences is a very exciting field to work in. Particularly because there are so many new opportunities and interesting things to discover within it. Really good artists, designers or scientist have always shared methods with the 'other' sides as well. Consequently, sustainable innovation has always happened at this intersection. We need to further strengthen these connections and figure out a way of doing so in a bit more structured way.

This will not be about creating a system of rules but rather about the identification of 'landmarks' in this creative process. This is part of the reason why we constantly also observe our own process in creating projects.[12] And we are always trying to figure out new ways of describing that process systematically. This way we hope to learn more about it and to identify additional methods that help us in pushing work at the intersection of the Arts, Design and Sciences further.

DB: As you know, the fictitious separation of Arts and Technology, Engineering, Sciences is a recent phenomenon. Arts and Sciences are so distant in the twenty-first century because we live in the age of specialization. But you propose to consider new transversal paths for creation in Arts and Sciences (…)

Today we are not only keeping Arts and Sciences separated: we are maintaining countless specializations within the Arts and within the Sciences. Do you try in a sense to change attitudes, and thus to reverse this trend? Ultimately, are there significant differences between Arts and Sciences?

SN: There are differences and we think that they are important and have to be kept. One significant difference is the term 'research', although both disciplines rely on it as an essential foundation. Julian Klein wrote a brilliant essay about artistic research, describing art and science not as separate domains, but two dimensions in the common cultural space. In the introduction he reframes his main question "What is Artistic Research? 'to' When is a research artistic?".[13]

As mentioned above, every major cultural innovation happened between disciplines in a way. Whether you advance simulation techniques in the field of computational fluid dynamics or propose a new form of speculative philosophy, until you re-apply that knowledge (creatively) to new contexts, the knowledge will remain within the isolated canon of its field. Most researchers do this as part of their

---

[12] See the process videos for *Understanding Shakespeare* for instance or how we try to document the creation process for projects in general: project homepage, URL, December 28, 2013: http://www.understanding-shakespeare.com/process.html.

[13] Klein J (2010) What is Artistic Research? Gegenworte 23, Berlin-Brandenburgische Akademie der Wissenschaften. Available online, URL, December 28, 2013: http://www.julianklein.de/texte/2010-artistic_research-JKlein.pdf.

research already, e.g. with the goal to use simulation techniques to develop new forms of bionic motors for transportation systems. But to which extend you look at consequences and implications of your research beyond your field is what is becoming a new important factor. Through specialization, disciplines have developed into small isolated islands. Bridging to their immediate neighbors might be easier than changing the continent.

The progress of specialization with each field cannot be reversed, and it shouldn't be, if we want to advance the scientific disciplines. But with upcoming technologies such as synthetic biology, the impact of any innovation cannot be foreseen, whether it will be positive or negative.

DB: In the *Othello* description project, you write that you aim to find traces and patterns that reveal cultural, historical and social fluctuations (…)

Do you think that one of the biggest challenges for data design is to help us better understand the world we live in?

SN: In our opinion, describing the world through data is one large, iterative, collective experiment mainly motivated by the urge to understand more about it. We apply quantitative measures orders, categories and order to observed phenomena and processes, evaluate these through statistics, visualization or re-contextualization in general to then refine our measures. Quantification is a cyclic process, probably never 'finished'.

The design of tools[14] to be used in this process, the kind of insight they could possibly reveal, the questions they could possibly provoke, is fundamentally important in this context. That is also why the design process starts with the data, how it is created, processed etc. This sets the stage for all steps to follow afterwards and consequently influences what can be possibly understood.

DB: More broadly, do you think that data may have a real influence on the real world (and also affect our lives)? How do you work with data? How is it appropriate to use data?

SN: Data itself is hardly going to influence our lives. If it is not put to use, it is just 'there'. It's how we use the data which influences our lives. If you look at the stock market for instance, it becomes very clear how data, being an abstract representation and aggregation of market activity, has an influence on the real world. For instance, think about the influence algorithmic trading has on the lives of real people through fluctuating food prices. We have to keep in mind that data and how we use it, already has an influence on our real world today, and that this can get dangerously out of control.[15]

---

[14] And hereby we refer also to finished design products or services, they are just tools in our constant approach to data and understanding.

[15] See Slavin K (2011) How Algorithms Shape Our World. TED talk. URL, December 28, 2013: http://www.ted.com/talks/kevin_slavin_how_algorithms_shape_our_world.html.

In other areas, such as the 'Quantified Self movement', the situation is less clear, mainly because data is a rather new component in such domains. Just because we track our activity, weight, blood sugar levels, etc. it does not mean we are catalyzing this information into action. So it is about the design of services, products and platforms that aid us in this process. More interestingly, we have not yet defined ethical standards of how to use data, especially data related to individuals. Questions such as: "Can I use the data just because it is public?" are not yet answered, and probably can never be.Boyd and Crawfords paper *Six Provocations for Big Data*[16] has some interesting points in that direction.

As much as our work is concerned, we generally like to get our hands on all kinds of data and experiment freely with it. But we also try to respect issues with privacy, e.g. if we publish work.

DB: You regularly collaborate with research laboratories/teams and you participate in various experimental research projects (practice-based research projects)—at the intersection of different disciplines (e.g. *Othello*, *Botanicus Interacticus*, and *Understanding Shakespeare* projects, etc.) (Figs. 19.8 and 19.9).

Does the scientific research approach (academic research) make it part of your creative design approach? I mean, do you defend a new kind of 'translational research' for creative exploration that encourages and promotes interactions between research and practice (between academic research and design practice for instance)?

SN: We think the creative practice in some areas has arrived at a point where it would definitely benefit from a more structured form for the sake of exchange between the practitioners. That doesn't mean that the creative process itself should be structured, but to advance the field of design as a whole, it is worthwhile to look at certain phenomena from a scientific research perspective. It will become increasingly important to reflect on questions such as: when has a certain design decision been made in the process and why? Another interesting aspect is the credits. Who has done what in a project? In this case we like to be very specific by crediting everybody for their involvement. Both things could be improved in the commercial design world. A better documentation would benefit the exchange about the process involved in the creation of a project and consequently help not to reinvent the wheel in some cases. Crediting would improve the situation of how openly we deal with innovation. The creative genius is a myth, it is all about the longer, hard work of a multitude of people.

DB: I am very impressed by your ability to deal with various topics that seem very different—the topics you cover in your projects seem as varied as the technologies you use (…)

---

[16] Boyd D, Crawford K (2011) Six Provocations for Big Data. Oxford Internet Institute Decade in Internet Time Symposium. Available online, URL, December 28, 2013: https://papers.ssrn.com/sol3/papers.cfm?abstract_id=1926431.

Fig. 19.8 *Botanicus Interacticus*, Studio NAND with Ivan Poupyrev (Disney Research Pittsburgh), TheGreenEyl, Philipp Schösslera and Christian Riekoff, 2012

One of your strengths is to brilliantly succeed to show that these different topics have in fact many relationships—the data that is processed, represented, visualized, materialized (…) becomes the 'glue' that holds all these topics. In some ways, if the topic of your projects is not directly the data, it is the raw material. What do you think?

SN: This is probably true for many projects. But some are also discussing the impact of a certain technology onto our societies not looking at data in particular (is there technology without data?). The projects that would come to mind are *Krebs Faraday Collaboration*,[17] (Fig. 19.10) *Cruiser Charisma*[18] and *Amæ Apparatus*[19] (Figs. 19.11 and 19.12). These projects deal with new habitats for synthetic life-forms, techno-utopian scenarios and ways of emotional expression. But in most other cases, one could say that data is a raw material underlying a work.

DB: This is also because you seem to consider each project as a place for discovery and experience. Thus, the unity of all these topics is in short the consistency of your creative exploration approach—which reveals your adventurous nature? Is that right?

---

[17] Project homepage, URL, December 28, 2013: http://www.nand.io/research/krebs-faraday-collaboration.

[18] Project homepage, URL, December 28, 2013: http://www.nand.io/research/cruiser-charisma.

[19] Project homepage, URL, March 11, 2014: http://www.nand.io/awareness/amae-apparatus.

Fig. 19.9 *Botanicus Interacticus*, Studio NAND with Ivan Poupyrev (Disney Research Pittsburgh), TheGreenEyl, Philipp Schösslera and Christian Riekoff, 2012

Fig. 19.10 *Krebs Faraday Collaboration*, Studio NAND, 2011

SN: Yes, probably. We are as curious as many other people. But this is not always easy to be 'adventurous' in a commercial context, but it has always been beneficial for us and the clients. We are a commercial studio at the end of the day.

DB: What is really interesting to notice through your various projects is the ability of data to reveal and alter our emotions (e.g. *Emoto* project), sensations (e.g. *Botanicus Interacticus* project), feelings (e.g. *Amæ apparatus* project), behaviours (e.g. *Unreliable Machinery* project) but also relationships, meanings (e.g. *Understanding Shakespeare* project). The ability of data to also highlight phenomena on different scales (…)

Consequently, it seems legitimate to ask: what is actually data? What is data quality? What can we do with data—are the possibilities endless?

SN: As mentioned earlier, data is the manifestation or temporary result of our approach to describe the world. The question of legitimacy of data in a certain context consequently always depends on the intended purpose of usage. Data gathering and data analysis have this empirical appeal. A major reason why data visualization is often being understood as "pertain(ing) to reality"[20] And this is true to a great extent, but it is equally important to always question the data, how was it gathered, what is its bias?

---

[20] See Kuang C (2009) The Irresistible Appeal of Info Porn (words of Aaron Koblin). Print Magazine. URL: http://www.printmag.com/article/the-irresistible-appeal-of-info-porn/.

Fig. 19.11 *Amæ Apparatus*, Studio NAND, 2011

Fig. 19.12 *Amœ Apparatus*, Studio NAND, 2011

If we do not ask such questions we can do great harm. One example is the "fascinating map of the world's most and least racially tolerant countries" published by Fisher at the *Washington Post* in May 2013.[21] The strong criticism it has caused[22] shows very clearly how one map, and a lot of problematic assumptions involved in its creation can cause great conflicts.

DB: You are very involved in the Open Source Culture. As creative coders and data designers/artists, your commitment to creative practice of appropriation and free sharing of found and created content is very important. You have built a dedicated platform (e.g. *Creative Coding* project) that enables (especially) students to learn programming and also to develop their own digital creative approaches (…)

What is your vision for the deployment of Open Source? Do you consider it as one of the key levers for the renewal of digital creation?

SN: All three of us learnt most things we do on a daily basis through Open Source knowledge and tools. *Creative Coding* is a way of giving something back to the community, but it also should serve as a platform for our teaching and learning initiatives in the future. In the future we will extend *Creative Coding* to also include topics beyond programming such as fabrication, crafts, and workshop documentations including the sharing of briefs and assignments and so on. The motivation for all of this is simple in principle: access to and exchange about knowledge and

---

[21] Fisher M (2013) A fascinating map of the world's most and least racially tolerant countries. Washington Post. URL: http://www.washingtonpost.com/blogs/worldviews/wp/2013/05/15/a-fascinating-map-of-the-worlds-most-and-least-racially-tolerant-countries/.

[22] See the most prominent example: Fisher M (2013) The cartography of bullshit. Africa Is A Country. URL: http://africasacountry.com/the-cartography-of-bullshit.

tools should be open and free; and they should remain that way. Luckily, we are standing on the shoulder of giants in that regard.

Looking at how projects such as Wikipedia, Processing and many others changed our use of digital technologies it is very likely that many other Open Source projects will do the same.

DB: Can we consider the deployment of Open Source as a way to implement technological interoperability: moving towards open standards that contribute to interoperability of systems and products?

SN: Interoperability is sometimes the biggest problem in the multitude of Open Source solutions. For instance, web standards and their support across browsers are a mess. But in a way this multiplicity is also a motor for innovation. And eventually, technologies which originated from this ocean of initiatives will become the kind of standards which advance an entire generation of systems and products.

The important things for Open Source remains that every individual or group of individuals is able to contribute to technological developments or provide alternatives to them.

DB: For Jan Ljungberg, "Open Source movement"[23] is one key to the understanding of future forms of organizations, knowledge work and business"[24]; the shift from the industrial era (proprietary software industry) to the era of sharing and use (free and open source software).

Do you think that this new age based on Open Source culture will allow us to create a sustainable future: towards an "Open Sustainability" where the free sharing of knowledge can provide a fertile ground for creation, and also for new collaboration, creation and innovation?

SN: Definitely. we liked Wolfgang Wopperer's argument that Open Source, the maker movement in particular, is a form of education in the twenty-first century,[25] not an industrial revolution or form of autonomy for all of us. This also includes that not everybody is going to take part in this movement. For some purposes or in some areas people will continue to prefer commercial systems if they just require them 'to work'. The important part is still that any person can take part in the further development of certain aspects of our civilization if they want to.

DB: The new era of sharing and use that I mentioned earlier announces a new age of industry where Open Data (an essential component of a sustainable future) will play an important role (…)

---

[23] Where communities of programmers collectively develop software of a quality that outperforms commercial.

[24] Wopperer W (2000) Open Source Movements as a Model for Organizing. In Proceedings of ECIS'2000 (Paper 99).

[25] See Wopperer W (2013) Making, Sharing, Caring—Schöne neue Autonomie oder Dritte Industrielle Revolution? Slides available online, URL, December 28, 2013: http://www.slideshare.net/wowo101/making-sharing-caring-tu-dresden.

Do you think that Open Data—which is a liberating tool for Government transparence—can also empower designers, developers (and other creative and innovative people) to create a better experience for the public? What are the Open Data challenges and opportunities?

SN: The main challenge to Open Data is data literacy, or a lack thereof among the general population. People need to be educated about data to truly leverage its potential but also to demand the data to be open in the first place. This is a huge gap in the society which has mainly been created by the rapid advancement of technology itself. This does not mean that people need to learn programming at a semi-professional level, but rather that they need to learn about the procedural processes underlying data.

Another challenge is the lack of standards and formats and the general lack of knowledge about Open Source technologies within the government institutions themselves. In addition to that we need standards and formats which allow documentation of how data was gathered. The usage of the Open Data will directly re-inform the way it is being gathered and processed in the first place and we need to create a platform to enable that process.

If we master these challenges we may create the opportunity for Open Data to significantly contribute to discourse about all aspects of our public organizations. How governments operate and make decisions, how the healthcare system is operated and so on.

DB: Are we actually on the brink of a new digital revolution fuelled by Open Data that could enable technology innovation to meet social innovation?

In other words, can Open Data be an opportunity for stimulating social innovation (Open Data as a key to citizen empowerment)?

SN: The excellent example of Local Project's *Change By Us*[26] shows that a smart platform design can actually have a big impact on social innovation, by redefining how we identify and approach problems. Even if this does not (yet) include Open Data per se, it shows the potential of design and technology.

DB: What about crowdsourced data (Free Data)? As you know, the data-sharing capabilities of the current generation of computers are huge. The amount of data generated and shared by people doubles roughly every 1.5 years. The plummeting costs of creating, gathering, storing, distributing, and indexing data have a tremendous impact on the expectations of individuals as they create and share information about themselves and their relationships with others (…)

Do you think that one of the big challenges for data designers is to work-create-innovate with crowdsourced data (towards new discoveries via social/crowdsourced data)?

---

[26] URL, December 28, 2013: http://www.changeby.us.

SN: *OpenPaths*[27] shows nicely how crowdsourced data can be put to good use for researchers and data activists. The even more surprising thing is that this originated from hacking activity, a good example for how such valuable initiatives can be created 'out of the blue' if people have a certain understanding of technology.

DB: What I find particularly interesting in *Emoto* is the fact that you combined different modalities of data representation—which become complementary—and created a new kind of data experience that mixes visual, physical and sensory perception (…)

Do you intend to renew the way people experience data through new modalities?

SN: As mentioned earlier, we like to look at projects from all possible angles. *Emoto* is an example of how that could look like, but we even consider it an intermediate step in this overall endeavor. The physical, data-sculpture part of *Emoto* was definitely the most interesting one in terms of new modalities. Making data physical, embedding it in the real world, making it responsible there is definitely a huge interest of ours right now. This is again due to the combination of various projects, such as *Emoto* and *Botanicus Interacticus* and its underlying technology *Touché*.[28] We think there is great potential in this haptic and physical approach to data (…)

The interesting part for us is the 3-dimensional, object-based approach which adds so many dimensions as compared to the screen. Not just the third dimension of course, but materials and their aesthetic qualities, haptics and so on. While the screen has many other advantages and is generally the de-facto medium over which we engage with visualizations today, it's interesting to look at older media and their qualities which have gotten lost to some extend. I think the combination of the physical with the digital will create new exciting experiences, but making them meaningful takes a lot of time.

For *Emoto*, this means that the screen-based, online visualization had its own quality because we were able to reach many more people via a rather conventional platform, at least as people's familiarity with computers are concerned. Nicely complementary to that was the installation, because it was created for a different audience that is not primarily interested in online media. For this audience the different, seemingly non-digital nature of the installation enabled different points of engagement with the overall virtual and digital nature of the project. So the potential to bridge the digital world with the real is within the physical, haptical, materialized media.

DB: Is it correct to say that you try to organize data and to structure users' experience of this data accordingly?

---

[27] URL, December 28, 2013: https://openpaths.cc.
[28] URLs, December 28, 2013: http://www.disneyresearch.com/project/botanicus-interacticus-interactive-plant-technology (Disney Research); http://www.ivanpoupyrev.com (Ivan Poupyrev).

In other words, could we say that the biggest challenge of your work is to find the right balance between these two intents-goals: define the data structure and improve (at the same time) users' experience of this data?

SN: As outlined earlier, the special situation with data visualization is that you often explore the problem space and the solution space at the same time. In other words, you are trying to understand the data and the potential insight you can derive from it, while thinking about possible experiences to this data. This can only be tackled in small iterations which prioritize the discussion among the team and with the client or project partner to move forward.

DB: I feel that you attempt to explore data in relation to embodied human experience and data aesthetics. Is it true? If so, for what purposes?

SN: For some projects, yes. This is just an interesting area for us to work in. Screens might be the most common medium at the moment to present digital work, but data will move away from the screen into our environment and so will the representation of data.

DB: Do you intend to explore the role of data aesthetics in constructing human knowledge?

SN: While the visual form is an important part in data visualization, it is not our primary goal to only explore this. Of course we like to experiment with form, but other aspects, e.g. communication and readability, are more important when it is about constructing knowledge.

DB: In your opinion, what are the biggest challenges for data design?

SN: Data will enter our daily lives even more. So data visualization will face the problem of needing to address problems and requirements of a larger audience. This process already takes place, especially in the area of (data) journalism where visualizations need to develop more techniques to deliver certain topics. One of these techniques is storytelling—very hyped at the moment—but some useful design principles are being developed there. The challenge here will be how to push that even further.

Another challenge will be in the entire physical domain, how to find interesting ways of embedding data in the physical.

DB: What is your vision for the future of data design?

SN: Tools will improve, visualizations will be ubiquitous. And so people's ability to talk visualization—referring to data visualization literacy—will increase making this a language of sorts. The improvement of tools will contribute to how people can intervene with certain data related issues, may it be through products, services or public domain issues (governmental etc.). So, bottom line: data visualization will hopefully become part of how we communicate at a broader scale.

DB: How do you intend to tackle these challenges?

SN: Educate people, create compelling and/or provoking design proposals for discussion; contribute to the field through open knowledge and tools.

Author Biographies

Stephan Thiel (Designer and Co-founder of Studio NAND) is a Berlin-based interaction and data visualization designer with a focus on computational analysis and visualization of text and literature as well as the design of interfaces to public archives of cultural data.

With Jonas Loh and Steffen Fiedler as Studio NAND, he is currently directing various design projects spanning the fields of information visualization, interaction and exhibition design. They advise corporate clients and research institutes on how technology can help understand and improve businesses, social relationships, knowledge management and communication.

An additional part of Studio NAND is the discussion of the impact of technology on our modern societies by creating design fictions and speculative products or by lecturing at universities and conferences.

For more information and contact:

http://www.nand.io
http://de.linkedin.com/in/stephanthiel
https://twitter.com/stphnthiel

Steffen Fiedler (Designer and Co-founder of Studio NAND) is one of the co-founders and Interaction Designer at Studio NAND—a multi-disciplinary design practice that explores interactions between society, science and technology. They use design as a method to craft engaging experiences, stories and visualizations for private and public sector organizations.

He has taught/lectured at University of Arts Berlin (UdK), University of Applied Arts Vienna, Copenhagen Institute of Interaction Design (CIID), Dresden Academy of Fine Arts (HfBK), University of Applied Sciences and Arts Northwestern Switzerland (FHNW), University of Applied Sciences Potsdam (FHP), Berlin Technical Arts School (BTK), Academy of Visual Arts Frankfurt a. Main (AVA).

For more information and contact:

http://www.nand.io
http://de.linkedin.com/in/steffenfiedler
https://twitter.com/steffen_fiedler

Jonas Loh (Designer and Co-founder of Studio NAND) is co-founder and designer at Studio NAND, a studio for research based design exploring interactions between society, science and technology.

As part of the staff at the Dresden Academy of Fine Arts he is responsible for establishing a new postgraduate programme at the intersection of art, science and technology. He graduated of the Design Interactions MA course at the Royal College of Art, his work has been recognized internationally through various exhibitions and publications including the MoMA New York, the Venice Biennale, *Wired*, V2, and the Siggraph Emerging Technologies exhibition.

For more information and contact:

http://www.nand.io
http://de.linkedin.com/in/jonasloh
https://twitter.com/jonas_loh

Chapter 20
Designing for Small and Large Datasets

Jan Willem Tulp

David Bihanic's Interview with Jan Willem Tulp

David Bihanic (DB): As a way of introduction, could you please introduce yourself and describe your background (e.g. training, work experience, career path, etc.)?

Jan Willem Tulp (JWT): My name is Jan Willem Tulp, I live in The Hague in The Netherlands and I work as a freelance information visualizer. I have a one-man company called TULP interactive. I have a BSc in interaction design. I was very much interested in the combination of software and design, which was the main reason I studied interaction design. It turned out I was a fairly good programmer, and after my graduation I found it hard to find a job where I could both work on software and on design. Looking back at my career I have switched quite a bit between various companies. I enjoyed all of my jobs, but reasons to get a job at another company were that at more creative companies (web-design, web-marketing, etc.) I missed the technical challenge, but at more technical companies I thought that the things we created were primarily exciting at the back-end, not necessarily at the front-end. Eventually I was a Software Architect leading a group of software developers at a client, and giving workshops for colleagues about the latest technologies and agile and lean ways of writing software.

In my spare time I was often creating graphics with software like Photoshop and 3D software. And eventually I learned about the field of data visualization, there was not a single epiphany for me, more a gradual awareness. At some point I realized that this was what I wanted to do, and so I made the decision to prepare for my freelancing business. I decided to work 4 days a week instead of 5, and use this one day a week to immerse myself into data visualization, and use my savings to compensate for the loss of income.

J.W. Tulp (✉)
TULP Interactive, The Hague, The Netherlands
e-mail: janwillem@tulpinteractive.com

© Springer-Verlag London 2015
D. Bihanic (ed.), *New Challenges for Data Design*,
DOI 10.1007/978-1-4471-6596-5_20

So I started reading the books from Stephen Few and Tufte, and became more and more excited about data visualization. For me it felt like all the pieces of the puzzle came together, the field I had been looking for all these years. With data visualization there are so many aspects, graphical, technical, statistical, interaction, etc. that to me it is challenging in many ways. I started contacting freelancers like Moritz Stefaner and Ben Hosken from Flinklabs to learn about working as a data visualization freelancer. I also participated in challenges, especially the ones organized by http://www.visualizing.org (and winning one as well!), and I read more books, like the ones by Colin Ware, but also about Processing, graphic design and interaction design.

When I had 3 clients lined-up who were interested in working with me I made the decision to start officially as a freelancer. That was September 2011. And ever since I started freelancing I have been overwhelmed with projects

DB: As many great data designers (who have strong artistic and technical skills), you explore the intersections of Art-Design and Science. You aim to discover relationships that may exist between different materials and technology in order to create data visualizations that inspire and engage people (…)

Today, it appears clearly that it's a mistake to separate these two disciplines (design and technology): technology and design perfectly complement each other (…) Technology has now entered the realm of design enabling a better understanding of its emotional aspects and implications (…)

Do you think that a closer alliance formed by design and technology could help to create better experiences through data?

JWT: Creating data visualization indeed involves both artistic and technical skills. But I think that there is more.

First there is the dataset itself. The quality and the potential of the data is of crucial importance. You can be the greatest visualization designer in the world, but with low-quality data with little potential you can design what you want but there the result will never be great.

Secondly, it's also what you do with the data. The dataset itself can already be interesting, but you can also do some statistical analysis, maybe enrich or combine the dataset with other datasets to make it more interesting or to create a richer context.

And finally the visual representation and interaction design play an important role as well. Even with a great dataset and some insightful results after a statistical analysis, if you as a designer make some choices so that the visual appearance is not appealing or effective, or both, it also does not work.

DB: You explore, transform and visualize data as images (or interactive visualizations) to gain insight into phenomena—of course, the purpose of data design is insight not images (…)

How do you find insights in data?

JWT: It is true that usually it's about images, but for me there are some other factors that play an important role as well. I make different choices if I create a visual analysis tool that is of crucial importance to a small group of specialists that use a visualization on a regular basis inside a company for instance. Effectiveness of a visualization is very important here. I make other choices if the visualization is to inform and engage with a wide audience. Also other factors might play a role as well, like the available size of the canvas, the medium (print, web, poster, iPad, etc.). I have done one project for Nielsen where I have created visualizations that were going to be part of their new corporate identity where visualizations were, even though based on real data, only used for aesthetic purposes in brochures for instance. So, for me it is a bit more nuanced when you say that it's insight, not images.

With regards to finding insights, for some client projects the insights are already known. The work I do for *Scientific American* is usually based on research that has already been completed and a paper has been written. So the conclusions and insights are already there. My job here is to create a visualization that communicates these insights in an appealing way.

For some other project I look for insights myself. This can partially be some statistical analysis I do, sometimes with tools like Tableau, or Gephi if it is a network based dataset. But quite often I write custom Python code to do some data analysis. I really like this because it also allows me to wrangle a dataset, restructure it, combine, and use any other package that is available in Python if necessary.

But another important part is also the fact that creating visualizations is an iterative process: I try to make a dataset as soon as possible so that I visually get a sense of a dataset. And then based on what I see, I start making choices, improvements, other approaches, refinements, etc. During this process I create lots of interim visualizations that I critically look at if it reveals some insight or not, and how well it reveals this insight. And based on my conclusions I make adjustments.

With interactive visualizations I create the means for the end-user to find insights by himself. In this case I have provided the required interaction possibilities that allows for filtering, clustering, changing encodings, changing configurations, etc. in order to reveal insights. The amount of interaction and analysis possibilities depend on the type of project. For general audience type projects you usually want some basic interactions, but visual analysis tools require more ways of interacting to reveal insights.

DB: Do you think that data visualization allows to gain insights that one could not get in any other way?

JWT: I do think so yes. I also do think that some insights can be communicated in different ways, but the famous example is of course the Anscombe's Quartet, where 4 small datasets have the exact same summary statistics. So, just based on this you would conclude that these 4 datasets are the same, but one you start visualizing the datasets, you see that each of them has a different shape. So, you would definitely have missed this information if you would not have visualized this dataset.

I also believe that especially complex patterns can be communicated visually very well. Take your own social network on LinkedIn or Twitter. You could describe all of your connections and how all of your connections might be linked together, but showing a network diagram, perhaps with some clusters gives you an overall sense of the network that is hard to get in another way.

I also believe it works in a very strong way if you have multiple visualizations that are linked, and an end-user can brush in one visualization and sees how that affects the other linked visualizations. That reveals some relationships that are also otherwise hard to communicate in an effective way.

DB: What is your personal method for finding new ways to visualize and make sense of data?

JWT: I am not sure if I have really found new ways to visualize or make sense of data. What I do tend to do is to constantly critically look at my own results and improve on that. And these improvements can just be combinations of other visualizations, or parts of other visualizations. You might look at The *Sociopatterns* project (Fig. 20.1)[1] I did for *Scientific American* as a radial network diagram, enclosed by a donut diagram with custom endings for the links (circles inside circles). I still don't know if this is a new way or not, but it is a custom visualization that combines different components or ways to visualize information into one representation.

I also believe very strongly that it is very important to pay attention to details, in general but also with regards to the visual appearance. I quite often spend quite some time making minor adjustments just to make sure that labels are optimally readable, elements are well positioned and have good proportions, transitions don't go too fast or too slow, picking the right colors, curves are curvy enough but not too much or too little etc. It involves a lot of tweaking of very small things, but this absolutely contributes to the overall appearance and makes the final result look much better.

DB: As a result of your data design projects, the audience gains the power of understanding data easily and quickly (…)

Do you also intend to improve public awareness on critical issues (e.g. *Urban Water Explorer and The Economic Value of Nature* projects, etc.)?

JWT: Sometimes yes, but that is mostly only a possibility when I initiate personal projects. *The Urban Water Explorer*[2] (Fig. 20.2) and *The Economic Value of Nature*[3] projects were both part of a challenge organized by http://www.visualizing.org and they provided the dataset of the partner they were working with at that time. And for most client projects there is also an intention or a goal a client has, and client projects are in many cases based on the dataset from that client.

I would like to do more projects to improve public awareness or any other way a visualization could make a contribution, but I must admit that so far my self-initiated projects were driven by other intentions, like an interesting question I had myself I wanted to answer with a visualization, or an interesting dataset I found and wanted to explore, an idea for a specific type of interaction, or just simply to try out a new technology or concept. And sometimes it is also driven by a dataset itself.

---

[1] Project homepage, URL, March 12, 2014: http://tulpinteractive.com/projects/sociopatterns/.

[2] Project homepage, URL, January 12, 2014: http://tulpinteractive.com/projects/urban-water-explorer/.

[3] Project homepage, URL, January 12, 2014: http://tulpinteractive.com/projects/the-economic-value-of-nature/.

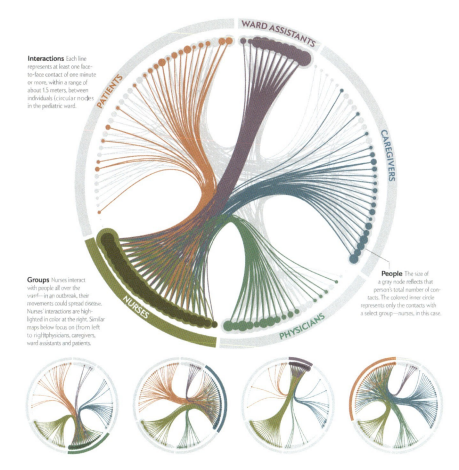

Fig. 20.1 *Sociopatterns*, Jan Willem Tulp, 2011

DB: Do you wish to communicate the value of behavior change and help inspire your audience to action? Do you consider that designers must have active citizen participation and encourage more people to also participate?

JWT: I think that would be great, yet at the same time I realize that my personal passion for data visualization itself is stronger than an ambition to encourage active citizen participation. I think it would be fantastic if that is the effect of a visualization for some people, but usually I am more driven by other motives, as mentioned in the previous question.

In The Netherlands when there are elections there are always a few websites where you can fill out a few question and get some advice on what party you should vote on (the last elections there were over 20 political parties in The Netherlands to choose from!). A few years ago before I started working as a freelancer I created a different visual interface to answer those questions and to get some other types of

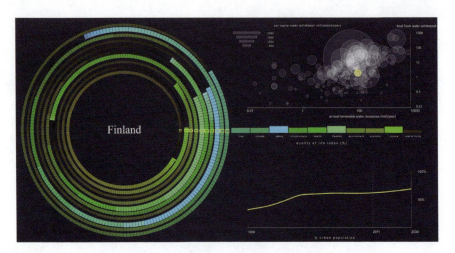

Fig. 20.2 *Urban water explorer*, Jan Willem Tulp, 2011

answers, like overlap between political parties. Also, my version gave direct feedback of your answer to a question instead of showing the final result at the end which was done on the original website. For this project I was driven by the fact that I thought that the user interface was really lacking context, and not giving you direct feedback so you were not really aware of the effect of your choice. So, I just wanted to make something that would give you better information, initially just for myself, but then publish it and see if other people could benefit from it too. And I did get some fantastic results, people were making screenshots and posting that on forums to show that the found insights they were not aware of. A friend of mine even decided to vote on a different political party than he has done so far, so he really made a drastic change in his political preference based on this tool. I think it is fantastic that this can be the effect, but I was primarily driven by the fact that I could make a better user interface, not necessarily to change as many people's mind on what to vote.

DB: Does data visualization provide the opportunity to pass knowledge to the masses?

JWT: I think it does, but there are some crucial factors here if you want to do something for the masses. It really has to be about a subject that is relevant to people at the time they look at it. And I think you can improve the effect if it an original and/or an engaging design that people enjoy playing with (and learn about the insights on the side!).

In general I think the general public might just have a quick look and then go on. I recently created a network visualization of ingredients that go well together for *Scientific American*. This visualization received a lot of attention, and was written about on quite a few websites with large number of visitors, like *FastCo Design* and *LifeHacker*. Of course, attention from these websites helps, but one differentiating factor compared to other visualization I created for *Scientific American* is that food and ingredients speaks to almost everybody, makes a meal every now and then. It also helped that the dataset really revealed some surprising combinations of ingredients that appear to go well together. Some people said that they were really amazed by the insights, or that they were going to use the visualization to try out some new recipes. I think this is an example of a visualization that was using a dataset and showed insights that are relevant, surprising and fun to almost anybody, and that's really crucial if you want to pass knowledge to the masses.

DB: You manage to create simple visualizations that communicate complex data (…) Every designer knows that keeping visualization simple lets the data tell its own story (…)

Is it difficult to keep it simple?

JWT: I had to think about this question, what do I do to keep it simple?

One of the things that I can think of is keep asking yourself if the visualization still serves its purpose or communicates the intended insights. Quite often you can communicate many more types of insights based on the same dataset, but actually keeping the communicated message simple helps.

I also tend to use a limited number of colors, or gradients of colors. Using too many colors not only becomes harder to understand for people, it can also become overwhelming, especially if the colors don't really go very well together.

What's also really important here is that you pay attention to details to make it look simple. For instance, if elements or components are not neatly aligned, the overall appearance becomes noisy.

DB: Let's keep it simple to increase cognition (as said, in substance, Ben Shneiderman). May data visualization have a significant role in the amplification of human capabilities (perception, sensation, intelligence, memory, etc.)?

JWT: It at least amplifies cognition in the sense that you externalize information that you otherwise had to keep in your head. So in that sense a visualization truly contributes to amplification of human capabilities.

Also, representing information in a visual way, for instance properly using pre-attentive variables, allows for really quick identification of exceptional elements. This is also a much more effective and faster way than for instance reading textual information.

DB: In a recent conference, Ben Fry was referring to a collaborator's comment who said about a new data visualization in progress [I quote him]: "this data visualization proposal is too pretty, it doesn't make sense (…)" [it's a very telling comment, isn't it?]

Do you think on the contrary that the aesthetic consistency of a visualization (which determines its beauty) can improve its understanding?

JWT: Based on my own experience I think it does. Research should show if this is actually the case or not. I do know that Nick Cawthon and Andrew Vande Moere published a paper called "The Effect of Aesthetic on the Usability of Data

Visualization"[4] in which they conclude: "More specifically, the results illustrate that the most aesthetic data visualization technique also performs relatively high in metrics of effectiveness, rate of task abandonment, and latency of erroneous response. We argue that these results show that aesthetic should no longer be seen as a cost to utility."

DB: Do you think that data visualization opens new windows to see the world in new ways?

JWT: Yes, data visualization is a very good way to make the abstract and complex understandable, as it communicates information in a way that is very natural to us as human beings. We are very good at processing visual information, and are much better at pattern recognition than computers currently are.

Personally I find it one of the most rewarding moments the first time you visualize a dataset. I am always really curious what a dataset 'looks like'. This is something you cannot comprehend just by looking at the raw numbers. It is even more rewarding if it is a dataset that has never been looked at, for instance because you have collected the data from many sources and merged it into one, or you have enriched the dataset.

DB: Can it help people to better understand/interpret the world that they live in?

JWT: I think it does, for the same reasons mentioned in the previous question. However, the fact that a visualization can open a new window to see the world in new ways does not necessarily mean that it is better understood. The role of the visualization designer becomes important to guide the user though the visualization in order to understand it, depending on the prior knowledge of the user.

DB: What impact can data visualization have on our lives? Can it affect our daily lives? If so, how?

JWT: It can. I mentioned the alternative interface that helps you pick a political party before. That has really helped some people make a better informed decision. But key here is that the visualization should be relevant at the time the user looks or interacts with it. And of course, scientific visualization, for instance MRi scan visualizations have an impact for sure. And the more traditional diagrams that are often used in Business Intelligence tools also help make analysts better decisions. But both the MRi scan visualizations and Business Intelligence dashboards provide information that can be acted on directly, and is relevant at that moment, which are very important factors in order to have an impact.

DB: You constantly invent new visual metaphors, new visual figures and patterns that expose the data in creative and effective ways (…)

---

[4] Cawthon N, Vande Moere A (2007) The Effect of Aesthetic on the Usability of Data Visualization. In Proceedings of IV'07. Available online, URL, January 12, 2014: http://infoscape.org/publications/iv07b.pdf.

Is it essential for you to create new adapted visualization ways to new types of incoming data?

JWT: It is of course not essential to create new visual metaphors, figures and patterns. But it is part of the creativity that I enjoy very much about the way that I approach my work. In some cases traditional statistical diagram could also suffice, but the fun is to create creative approaches to represent a dataset. And that's also one of the reasons why clients approach me, because I give a unique twist to a visualization with my own style.

DB: Is it fair to say that a visualization pattern applies only to a specific set of data?

JWT: I think you can call it affordances of the data if that's what you mean. So, timestamps in a dataset would suggest a line graph, geographical coordinates would suggest a map, etc. I do believe however that it's not only about the affordances of the data that matters. In the *Close Votes* project (Fig. 20.3), I have visualized cities in The Netherlands. Size and color of each circle that represent a city changes based on the similarity of voting result with the selected city. Now, I have 2 different configurations of the visualization, one is a map of The Netherlands. This naturally allows you to reveal geographical or location based patterns. But I have also included a radial configuration. The visualization still shows cities, but not positioned on a map but radially. The reason is that for this layout I wanted to focus on similarity but on a slightly different way, and also allow the user to more easily compare cities based on number of people, which is easier to spot in this view than the geographical view.

So even if the data itself has some affordances and suggests some way of visual representation, it is also important to keep in mind what you want to show, and if that affordance is the primary type of insight you want to reveal or if it should be context only, or perhaps not even relevant to the types of patterns you want to show.

DB: You use various tools and programs to design your data design projects. Among them is the famous JavaScript library for manipulating documents based on data using HTML, SVG and CSS. You count today among one of the best experts in this solution.

Could you please tell us more about it? Could you please indicate what its main advantages are? Why choose D3.js for presenting complex data into perceivable information?

JWT: Yes, D3 is currently the most popular framework for creating data visualizations. Even though the learning curve might be a bit steep for some, it has some very strong features. One of them is that it is a framework that has very important visualization concepts, functions and utilities built into the framework, while still offering flexibility to make custom visualizations. The fact that it is web-based is also great, because it allows to run the visualization easily in the browser, and it allows you to make combinations with other Javascript frameworks (for instance underscore.js which I use a lot too) very easily. Moreover at this moment there is a

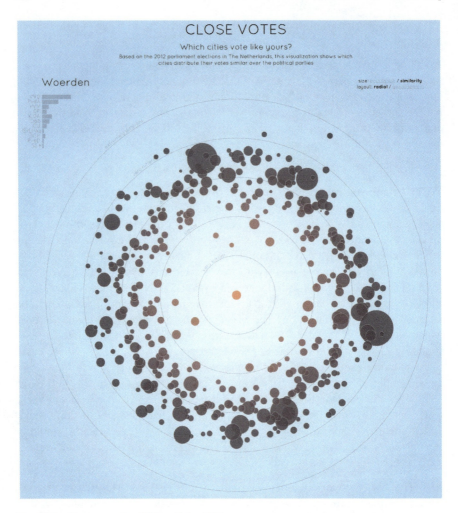

Fig. 20.3 *Close votes*, Jan Willem Tulp, 2012

very large user base for D3, there are many inspirational examples freely available, some people have written books, there are a lot of questions asked and answered on StackOverflow, etc. Finally, since D3 primarily works with SVG, you are essentially creating vector graphics. These can easily be extracted from the HTML page, and saved, so that you can open them in vector graphic tools like Illustrator to make the final tweaks.

DB: Today, you are one of the leading experts in Big Data visualization (visualization of very large amounts of data)?

What is the specific design approach and process for extremely large datasets?

JWT: I don't necessarily consider myself a Big Data visualization expert. Especially when creating a web-based visualization, you have to make sure that your data files are eventually small enough so that performance is good enough in the browser (in case of an interactive visualization). I do work with large datasets, and even though storing them in a database could sometimes be an option, my approach is usually writing custom Python code to analyse the data, but also optimize and simplify the dataset in various ways (aggregating, filtering, geometry simplification, pre-calculation, removing unneeded data, etc.).

I could say that in my case many times when working with large datasets it is all pre-processing and optimization of the data. The final dataset is usually rather small, about 5 Mb is the maximum I recently had for a web-based project.

DB: How to use visualization in order to support the understanding of very large data sets?

JWT: I think visualization for big or small data is not fundamentally different. Human perception still works the same way, regardless of the size of the data you are visualizing. I do think that there can be technical challenges, like how to display large amounts of data in a limited space, or how to keep a good performance with large data sets. But these are more technical challenges than visualization challenges. The human visual system is able to see patterns really well, and that works for both smaller datasets and larger datasets.

DB: Do you sometimes see yourself as pioneering into previously unknown data territories, facing vast areas of undiscovered data landscapes to explore (e.g. *The Data Centric Universe* project)?

JWT: Yes I do, both with regards to data has not been visualized before, and with regards to domains that I am unfamiliar with. Today almost every organization works with data, and scientists have been working with data for much longer. This means that there are a lot of potential clients for me, and my clients also belong to very diverse market segments. So, for one project I am working with data from Amsterdam Airport, the next project I am working with dataset of everything that has been discovered in the Universe so far (the data centric universe).

DB: Do you intend to uncover the secrets of Big Data?

JWT: Not necessarily. I do think that visualization can really help to get a deeper understanding of a dataset. At the same time, as mentioned before, I do think that visualizing data is not necessarily fundamentally different for large datasets or small datasets, except for the technical challenges perhaps. Visualization in general can help to communicate in find insights in complex data, and this can work very well for Big Data as well.

DB: Does visualisation uncover the big picture of Big Data?

JWT: In a way yes, as it can provide an overview of a dataset. But this is also true for smaller datasets. But as mentioned before, visualization is a good companion for Big Data (Fig. 20.4), as it helps to find and communicate insights and patterns.

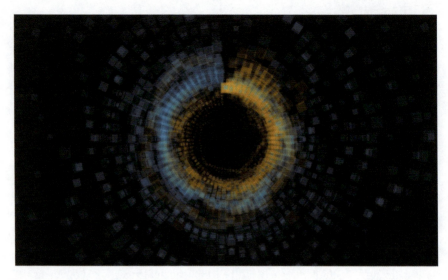

Fig. 20.4 *Nielsen identity design*, Jan Willem Tulp, 2012

Perhaps with statistical analysis you have to know a bit more what you're looking for, and visualization can have a more serendipitous effect, where you can be surprised by things you did not look for in the first place. Visualization can therefore be a nice companion in the visual analytics process where the visualization can generate some additional questions you didn't think of before.

DB: How to extract precious insights from Big Data?

JWT: In situations where you have to deal with many dimensions, you will probably have to do some statistical analysis or dimensionality reduction before visualizing. With hundreds or thousands of dimensions, you are simply overwhelmed, and you cannot just pick a few dimensions to visualize and hope for the best. After all, there are only a few visual mappings you could apply, like size, color, length, position, etc. And if you have many more dimensions than you can visually encode, then you have to prepare your dataset first, and also perhaps create some designs that allow for combinations of various dimensions. Scatter plot matrices or parallel coordinates are traditional examples of visualizations that allow for visualizing multiple dimensions at the same time, but you can also think of linking several visualizations to interactively explore a dataset. But still, statistical analysis might be very useful here.

DB: Are you specifically interested in 'data deluge' visualization (the Big Dataflow)—some, like the datajournalist Simon Rogers, called the 'Tsunami of data'?

JWT: Not necessarily. I think I am more interested in other factors that might make a dataset interesting, not necessarily the fact that it is Big Data. Personally I find it a challenge to come up with interesting and new ways to display a dataset and its insights. Especially if there is some interactivity where a user can explore a dataset a little bit, those are fun.

DB: You create aesthetically appealing visualizations that do not distort data and distract the viewer from real information (…)

How can aesthetics qualities help to enhance understanding of data? Can beauty promote understanding in one way without undermining it in another? In a few words: is beauty (sometimes) useful?

JWT: I think it is. I think that when a visualization is aesthetically pleasing, looks original, it might speak to the emotional side as well, not just to the analytical or logical side. So, it becomes more enjoyable to experience a visualization. At the same time a visualization can become more memorable if it has a unique, and aesthetic appearance, which might be a great benefit if you want users or readers to remember the insights or stories you want to communicate based on the story. I also am convinced that if you pay attention to details and to aesthetics, that it will result in a less noisy visualization. Paying attention to alignment, balance, color, proportions, etc. will make it look nice and clean. And a clean visualization is easier and more pleasurable to comprehend than a noisy one.

DB: In your opinion, what are the big challenges for data design?

JWT: The process of creating a visualization requires you to switch between data, software and design continuously. You constantly have to verify visually if your visualization shows what it needs to show, and if it does so in a clear (and aesthetic) way. But also if there is enough context to understand the message, and if the interaction makes sense. So you're constantly checking various aspects of the visualization. And for me, working on my own, or at least specifically on the visualization part of a project, this is doable, since I do everything. But I am really curious if this is a scalable process where multiple people can work on the same visualization at the same time. Two people, both sitting behind the same screen, can possible, but when it becomes more than that, or you are not behind the same screen, it becomes difficult to continue this process of continuously evaluating your design, changing code, improving data, etc. And in many situations you cannot delegate the work of a part of the visualization to someone else (except for user interface controls or if there are multiple linked visualizations), because it is the entire visualization that needs to work as whole. So, how to scale up this process?

DB: What is your vision for the future of data design?

JWT: I think visualization will become more common. More people will be creating visualizations, and people will be more familiar with visualization. I also think that the need for visualization will increase, as Big Data, open data, quantified self, data driven journalism are all data related trends that could benefit from visualizations. You can see this already today, but it's still in its infancy right now.

DB: How do you intend to trackle these challenges?

JWT: To be honest, scaling up the process is something I haven't figured out. I do think it is great that more and more people are becoming interested in becoming a data visualization designer that has knowledge of coding, data and design. But that would mean that they could work in individually or in small teams. But working on a visualization with a large team remains a challenge I think.

Author Biography

Jan Willem Tulp is a data visualizer from The Netherlands. Since 2011 he works as a freelancer (TULP interactive) for clients all over the world. His visualizations do not just provide insight, but are well crafted and beautifully designed. His work ranges from custom visual exploration tools that help a client explore a dataset to visualizations that are more explanatory in nature. Some of his projects can be considered data-art.

Every now and then Jan Willem initiates a project by himself, just to explore new ideas, technologies and to have some fun. His work has appeared in magazines, books and exhibitions. Jan Willem works for clients such as *Scientific American*, *Popular Science*, Nielsen, Amsterdam Airport, Philips and *World Economic Forum*.

For more information and contact:

http://www.janwillemtulp.com
https://twitter.com/JanWillemTulp.

Chapter 21
Process and Progress: A Practitioner's Perspective on the How, What and Why of Data Visualization

Moritz Stefaner

David Bihanic's Interview with Moritz Stefaner

David Bihanic (DB): As a way of introduction, could you please introduce yourself and describe your background (e.g. training, work experience, career path, etc.)?

Moritz Stefaner (MS): I have quite a diverse background. Right after school, I applied for art schools, but they wouldn't accept me. I then proceeded to do a one-year 'multimedia producer' crash course, and worked in web agencies for a few years, mostly doing Flash websites. In my mid-twenties, I decided to go back to University to study Cognitive Science—understanding how the mind works, and how we represent information has always fascinated me, and this study program was the perfect stimulation for a nerd like me. After a few really inspiring years in Osnabrück, I went to Potsdam to study Interface Design, where I also worked as a part-time research assistant for a few years. The last few years, I have been working as an independent consultant and designer of information visualizations, and quite successfully so.

DB: How did you become interested in the field of data design (data representation/visualization, information/interaction design, creative informatics, etc.)?

M. Stefaner (✉)
Eickedorfer Damm 35, 28865 Lilienthal, Germany
e-mail: moritz@stefaner.eu

MS: I always had a knack for design, and structures, and numbers. I read Hofstadter's Gödel, Escher, Bach[1] when I was 18 and it pretty much blew my mind, so you could say that laid the foundation for my later interests in the beauty of data and code. The first real data visualization I produced was probably the Organic Link Network, which I coded in 2002 as a sort of gimmicky addition to my website at that time. But then it took until 2005, when I did my B.Sc. Thesis in Cognitive Science about mapping document spaces,[2] and when we built a haptic compass in form of a belt,[3] that I ultimately got fascinated with the field and its potential and realized this was the one thing I wanted to pursue further. For instance, the experience I gained in linguistics made me think about design in terms of a visual vocabulary. How can you shape the single elements of a visual design such that it is easily comprehensible? What is the vocabulary, what is the syntax of interactive visuals, of informative aesthetics?

DB: You have a broad knowledge of artistic and scientific issues of data design due to your transversal path (…) In a sense, in choosing to specialize in the field of data design, you have found the right balance between Arts and Sciences, between [I quote you] "Beauty and Truth." Was it a natural choice for you?

MS: To me, truth and beauty are equally important. In a visualization project, if you have only one without the other, you are not done yet.

Buckminster Fuller, the famous designer and systems theorist, said once, in essence, that he didn't think about beauty when he started a design, engineering, or architectural project. He was just concerned with its function—he wanted to find the right way to devise the product. But then, in the end, if the solution he came up with was not beautiful, he knew something was wrong. For Buckminster Fuller, in some sense, beauty was an indicator of functionality and of truth.

Design is much more than mere decoration. Often, people think of design as decorating a pre-existing structure, the sugar coating you can add as a last step. That is the wrong approach. Good design is tightly intertwined with the content it presents. It consists of thinking about what to show; what to leave out; what to highlight; how to structure information; what rhythm, visual flow, and pace, you want your story to have. That's what design is all about.

DB: With your MA thesis focused on *Visual Tools for The Socio-Semantic Web*[4] (which fits in the field of Web Science developed especially by the Tim Berners-Lee), you were aiming to engage an interdisciplinary dialogue between science,

---

[1] Hofstadter RD (1979) Gödel, Escher, Bach: an Eternal Golden Braid. Basic Books, New York.

[2] Stefaner M (2005) Projection Techniques for Document Maps. B.Sc. Thesis (supervisors: Dr. Petra Ludewig, Dr. habil. Helmar Gust), University of Osnabrück. Available online, URL, January 11, 2014: http://moritz.stefaner.eu/write-talk/b-sc-thesis/.

[3] Project Homepage, URL, January 11, 2014: http://feelspace.cogsci.uni-osnabrueck.de/.

[4] Stefaner M (2007) Visual Tools for The Socio-Semantic Web. M.A. Thesis (Supervisors: Prof. Boris Müller, Prof. Danijela Djokic), University of Applied Sciences Potsdam. Available online, URL, January 11, 2014: http://moritz.stefaner.eu/write-talk/ma-thesis-visual-tools/.

engineering and design. Are you actively pursuing the same objective today? If so, how are you getting there?

MS: Yes, absolutely. Great data visualization is always about the interplay of analysis and synthesis, and the interplay of our minds and our senses. I think many of the great data visualization artists and researchers today are actually great role models for scientists and artists also from other fields.

Looking back to my Master's Thesis, many of the themes I treated in it are still relevant for me today. For instance, the general idea of mapping the 'real worlds', i.e. the world of actions, relationships, possibilities, social constructs, instead of merely physical infrastructure has become quite important to me again. In many ways, San Francisco and New York are closer together than San Francisco and Kentucky. What counts today, is who and which information you can reach and with which efforts, not where you are located physically. This general theme has re-occurred in a few recent projects—from mapping the digital shape of cities[5] (Figs. 21.1, 21.2 and 21.3) to a new geography of Germany in my *Electionland*[6] visualization (Fig. 21.4) based on voting behavior. In my view, untangling the complexity that arises from an interconnected world, is one of the most exciting challenges in information design and cartography today.

DB: Could you tell us a bit about your work process? How do you come up with your design solutions?

MS: Sure. First of all—it depends a bit if I work on a client commission or a self-initiated work. In the latter case, I might just be inspired by a certain phenomenon, data set or technique that I want to try out. When working for a client, there is usually a bit more context, goals and constraints to consider.

Usually, at the beginning, I ask the client for two things: a data sample and a some answer to a few questions clarifying the context and basic motivation of the work.

The basic set of questions I usually ask are:

- Why are we doing this?
- What are you hoping to achieve?
- Who are we targeting?
- How is the end product going to be used?
- How are we publishing?
- What data do we have available?
- Which other existing materials should we take into account?
- Which constraints do we have?
- Who is responsible for what?
- Who else is doing something similar?

---

[5] See *Stadtbilder* project, URL, January 11, 2014: http://stadt-bilder.com.

[6] *Electionland* project, URL, January 11, 2014: http://www.zeit.de/politik/deutschland/electionland.

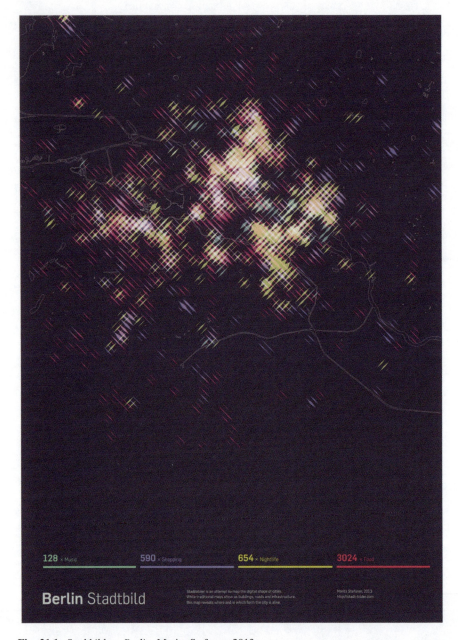

Fig. 21.1 *Stadtbilder—Berlin*, Moritz Stefaner, 2013

To me, answers to these questions are really important to understand why the client thinks a data visualization is important, and also to understand when the project is done, and successful. Often, both the client and I realize that half of these

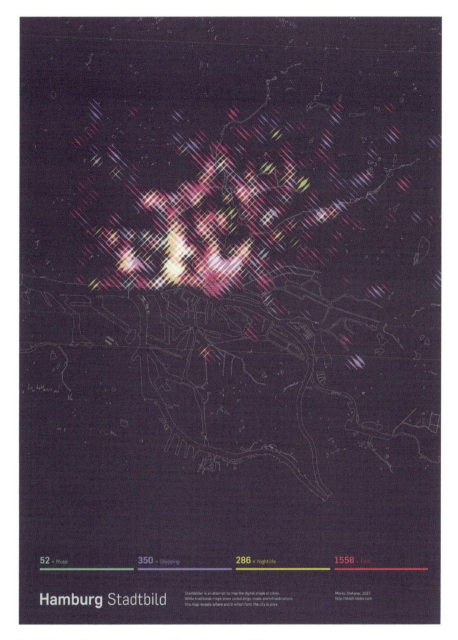

Fig. 21.2 *Stadtbilder—Hamburg*, Moritz Stefaner, 2013

questions cannot be answered yet, but that's fine, as long we make sure to answer them along the way.

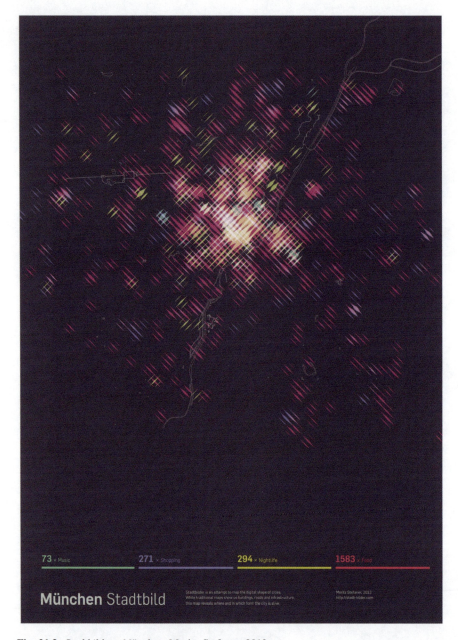

Fig. 21.3 *Stadtbilder—München*, Moritz Stefaner, 2013

As mentioned, the other important component in this first conceptual phase is to have a data sample. On the one hand, we want to know very early if the data is interesting enough to create a great visualization—of course, rather than trying to 'blow up' dull data with spectacular visuals (which I despise), I try to achieve a

21 Process and Progress …

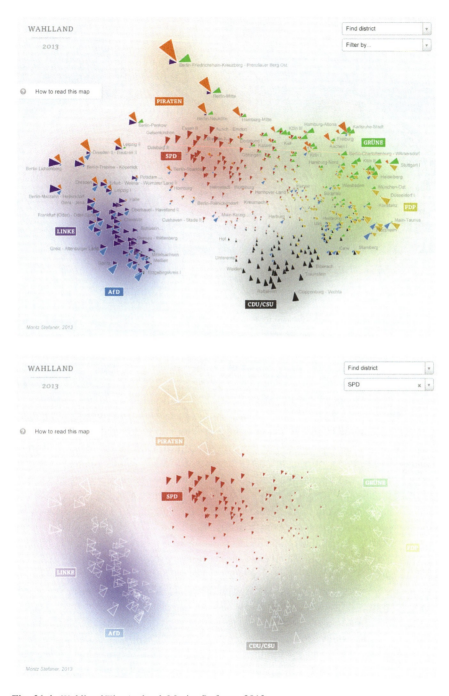

Fig. 21.4 *Wahlland/Electionland*, Moritz Stefaner, 2013

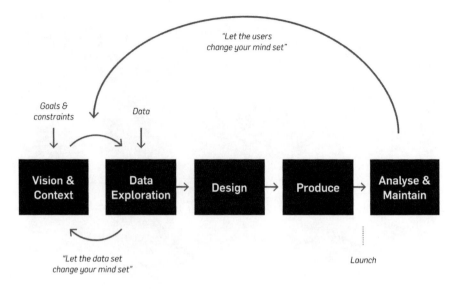

Fig. 21.5 Workflow, Moritz Stefaner, 2013

position where we have much more data than we want to use, in order to be able to edit down, put into perspective and distill. The other important information to gather is if the data seems sufficient to reach the project goal at all. Very often, my clients overestimate the depth, and completeness of the data they have available, and it is good to determine that right away.

The third reason why I need data early in the process is that my design approach requires that I immerse myself deeply in the problem domain and available data very early in the project, to get a feel for the unique characteristics of the data, its 'texture' and the affordances it brings. It is very important that the results from these explorations, which I also discuss in detail with my clients, can influence the basic concept and main direction of the project Fig. 21.5. To put it in Hans Rosling's words, "let the data set change your mind set."[7] Why? Well—some ideas sound great on paper, but are dull, when we look at them using real data. Othertimes, totally new ideas can come into play from the close dialogue with the data, based on things we discover and learn along the way. So, it is really a process of continuous exploration—creating a view on the data answers a few questions, but raises new ones, so I create new views on the data to answer these questions and find new ones again. In this phase, it is really important to move fast and don't get too married to specific solutions yet, so usually, I make really simple, generic charts using Tableau, Gephi or quick custom scripts in D3.

After a while, when the data has been explored sufficiently, it is time to sit down and reflect—what were the most interesting insights? What surprised me? What were recurring themes and facts throughout all views on the data? In the end, what do we

---

[7] See Rosling H (2009) Let my dataset change your mindset. TED talk. URL: http://www.ted.com/talks/hans_rosling_at_state.html.

find most important and most interesting? These are the things that will govern, which angles and perspectives we want to emphasize in the subsequent project phases. Often, I will also estimate and price the concept and data exploration phase separately from the second, more clear-cut second design & production phase. Sometimes, we will also let the project end after the data exploration phase, because the data is less interesting than we thought or does not match the client's expectations—but, in fact, I consider these projects 'successful failures' and a real service to the client, as I prevented them from spending money on something they don't want or need.

When it comes to coding and producing data visualizations, it is important to keep in mind that not all design decisions can be made in advance. So, also during the production phases, a vital ongoing discussion between code, design and data analysis is really important. The other important thing to note is that in the end, the details can make a great difference if people enjoy and use your data visualization, or are confused by it. As they say, the last 20 % are the second 80 %, and a lot of work can and should be put into getting a help section, legends, annotations, an introduction etc. into proper shape and test the whole product with a few users.

And, once the product is out, a lot can be learned from observing people interacting with it. How much time do users spent with your visualization? Which options do they discover and which are they missing? When they comment on or link to it—which parts of the project did they find most interesting to mention and refer to? Quite often, I would love to do a second iteration after the first launch, because only once the product is out "in the wild", one sees more clearly its strengths and weaknesses.

To sum up, my main advice is to use data as early as possible in the process, take it seriously and not just treat it as a vehicle for your ideas. Accordingly, it is a good idea plan with a long data and concept exploration phase and to accept that data visualization projects are just a bit more non-linear than the production of a brochure or a simple website.

DB: Most of your projects, including the static graphics, appear like an exploratory "Gesamtkunstwerk," with no real and/or accurate narrative that pops out immediately. That said, I know that you are keenly aware of the importance of storytelling in order to enrich the user experience and make data more accessible for all. As there is an endless possibility of data combinations in your works, is there also an endless possibility of hidden stories that users can gradually unveil?

MS: Absolutely. In my work, I never try to tell a single story. I try to tell thousands of them.

The trick is to not present them all simultaneously or with the same priority, but deliberately establish a hierarchy and sequence of perception events. The design of a good visualization is all about knowing the data set well enough to make sound decision on what to prioritize and which macro-patterns are interesting enough that they should pop out immediately, and then, having the skill set to apply the right visualization technique and establish the proper visual hierarchy, on order to make that happen (Figs. 21.6 and 21.7).

You could say, I try to explore every little corner of the data and flip every single stone, in order to then be able to be the perfect "guide" for the user through the new territory, and provide them with a short, but also scenic route to the most important

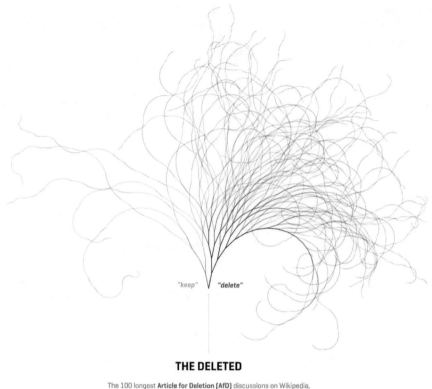

Fig. 21.6 *Notabilia*, Moritz Stefaner, Dario Taraborelli and Giovanni Luca Ciampaglia, 2011

structures and patterns in the data, while letting them also the freedom of self-exploration.

If successful, then, thousands of individual stories can be presented, waiting to be discovered through the interface of your presentation. I think this richness and openness is the key difference of the genre of interactive data visualization compared to, let's say, more traditional information graphics, and which should celebrate and exploit that capability of the medium.

But there's also another reason why I like these open information experiences as opposed to single linear "here's what you need to know about X" narratives: No knowledge sticks as well as the knowledge we elaborated ourselves, after a first hunch, further looking into evidence, maybe pondering counter-evidence, and finally formulating a certain grounded belief. The other really powerful mechanism is learning by playful exploration—much like kids learn, through probing a certain possibility space, trial and error, action and reaction. This type of active information discovery is something I really like to promote, so we become and stay critical

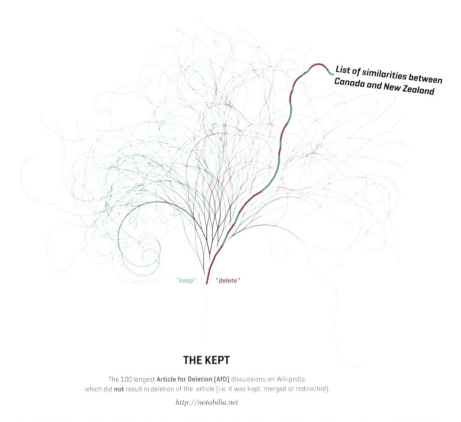

Fig. 21.7 *Notabilia*, Moritz Stefaner, Dario Taraborelli and Giovanni Luca Ciampaglia, 2011

consumers of the information that surrounds us, and interactive visualizations can help us train that muscle (Figs. 21.8 and 21.9)

If we take that thought further, we move away from data visualization as a communication device towards a new kind of glasses, that we can use to explore the world around us in novel ways. Just as the microscope allows us to see that very small, and the telescope enables us the very far, data visualization can act as a macroscope (term and concept going back to Joël de Rosnay, "The Macroscope"[8]) which allows us to bring the 'infinitely complex' to human scale and allows us to investigate nature, humanity and society at large.

DB: This reminds me of Lev Manovich: "[…] data visualization art is concerned with the anti-sublime. If Romantic artists thought of certain phenomena and effects as un-representable, as something which goes beyond the limits of human senses and reason, data visualization artists aim at precisely the opposite: to map such

---

[8] De Rosnay J (1979) The macroscope: a new world scientific system. Harper & Row, New York.

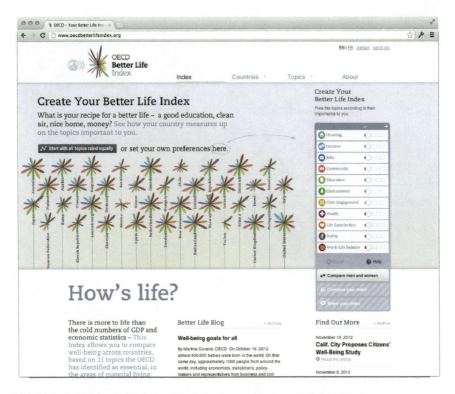

Fig. 21.8 *OECD better life index*, Moritz Stefaner with Raureif GmbH, 2011

phenomena into a representation whose scale is comparable to the scales of human perception and cognition".[9]

This idea seems fundamental to appreciate the contribution of data art and more broadly to understand the key issues of data visualization (...) What do you think?

MS: Indeed, I think this observation is spot on, and very connected to the macroscope concept. Another reason, why this line of thinking is important, is that many of the most important issues and developments today lack "photographability": algorithmic trading, market speculation on natural resources, climate change, the credit crisis, tax fraud and evasion—all these important issues are very hard to put in pictures. Data visualization can help us both to understand these complex issues a bit better, but also to provide images to debate about, and refer back to.

I sometimes compare our work as data visualizers today as the "new photojournalists": we travel to foreign "data countries," and, having an open mind, first try to collect as many different impressions and images of the experiences we make on the way—the exploratory, highly iterative data exploration and visualization

---

[9] Manovich L (2002) The Anti-Sublime Ideal in Data Art. Available online, URL, January 11, 2014: http://users.fba.up.pt/~ldcag01015/anti_sublime/beauty.html.

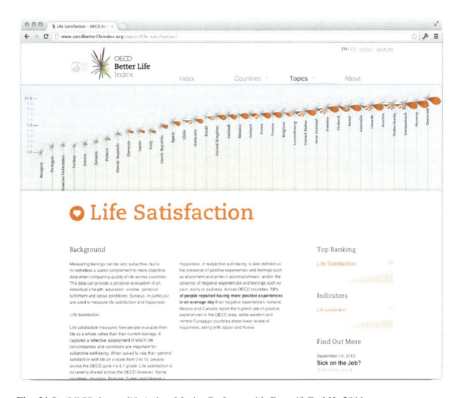

Fig. 21.9 *OECD better life index*, Moritz Stefaner with Raureif GmbH, 2011

sketching phase which should stand at the beginning of each data-heavy project. In this phase, sometimes you hit a dead end or get bitten by a snake (or Unicode errors in Python). However, back home, when the editing, and story-formulation part of the project begins, however, we need to select the best, succinct representation of the phenomenon as a whole—what is the single image or diagram that represents the essentials of what we found to be interesting and true in the best way? And exactly in this editing process lies the main editorial contribution of the designer as author. Let's make no mistake—even a very data-heavy, sober representation of data has an author who made clear decisions on what to include or not, what to combine, or not and what to prioritize. And the same holds for the underlying dataset, of course. So, acknowledging the role of authorship, with all the journalistic responsibility it brings, is an important result of this line of thinking. After all, we are creating views of the world that shape people's world views, and continuous, critical investigations of both the 'how' as well as the 'what' of data visualization are important here.

Author Biography

Moritz Stefaner (Independent Designer and Consultant) works as a "truth and beauty operator" on the crossroads of data visualization, information aesthetics, and user interface design. With a background in Cognitive Science (B.Sc. with distinction, University of Osnabrueck) and Interface Design (M.A., University of Applied Sciences Potsdam), his work beautifully balances analytical and aesthetic aspects in mapping abstract and complex phenomena. He is especially interested in the visualization of large-scale human activity.

In the past, he has helped clients such as the OECD, the World Economic Forum, Skype, dpa, FIFA, and Max Planck Research Society to find insights and beauty in large data sets. He was nominated for the Design Award of the Federal Republic of Germany and is a multiple winner of the Kantar Information is Beautiful awards. His work has been exhibited, among others, at Future Everything, SIGGRAPH, and Ars Electronica.

He is part of the advisory board for the Places and Spaces Exhibit and has been serving as a juror for the Malofiej award for information graphics, as reviewing expert for the Future and Emerging Technologies programme of the European Commission as well as advisor for the Digital Communities category of prix Ars Electronica. He has co-authored books for publishers such as O'Reilly and Springer and has spoken and lectured on numerous occasions on the topic of information visualization.

For more information and contact:

http://moritz.stefaner.eu
http://well-formed-data.net
https://twitter.com/moritz_stefaner

Chapter 22
The Art & Craft of Portraying Data

Stefanie Posavec

David Bihanic's Interview with Stefanie Posavec

David Bihanic (DB): As a way of introduction, could you please introduce yourself and describe your background (e.g. training, work experience, career path, etc.)?

Stefanie Posavec (SP): My background is in graphic design for print: I completed an MA in Communication Design at Central Saint Martins College of Art & Design, in London, England, in 2006. This MA was where I first became interested in visualization, as this is where I completed my *Writing Without Words* project. After I graduated, I didn't publicly promote this project, but instead took a job as a book cover designer at Penguin UK in their non-fiction and classics imprint, which is where I expected to spend the rest of my career, as it was a "dream job" for me. While working there, some parts of my *Writing Without Words* project (Figs. 22.1, 22.2, 22.3 and 22.4) were exhibited in a gallery in Sheffield, England, where someone who wrote for a large US design blog saw my work and posted it online. After a few days my work had been linked and re-linked to by loads of different websites and I had a hundred emails in my inbox! It was at this point that I decided that a career in data might be something to consider, and I am now self-employed so I could explore working with data. While I occasionally work on book and book cover design projects, I tend to focus on information design, data visualization, and data art projects for the bulk of my design work.

DB: How did you become interested in the field of data design (data representation/visualization, information design, data illustration, etc.)?

SP: When I was 19, I came to study to London for a semester (I had barely taken any design classes at university at this point). My parents came to visit in the middle

---

S. Posavec (✉)
59, Delawyk Crescent, London SE24 9JD, UK
e-mail: stef@itsbeenreal.co.uk

© Springer-Verlag London 2015
D. Bihanic (ed.), *New Challenges for Data Design*,
DOI 10.1007/978-1-4471-6596-5_22

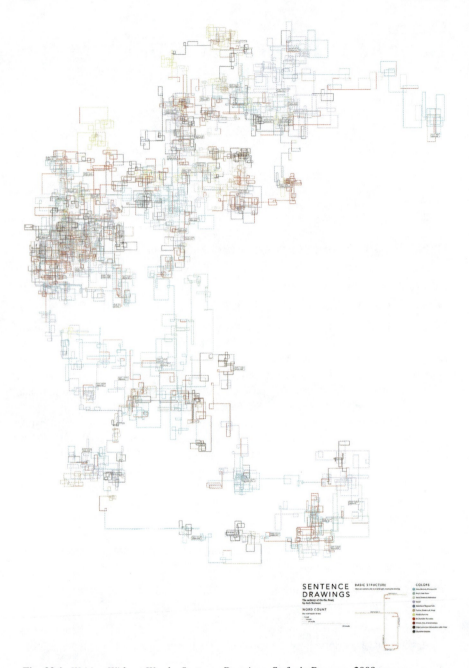

Fig. 22.1 *Writing Without Words*: *Sentence Drawings*, Stefanie Posavec, 2009

Fig. 22.2 *Writing Without Words*: *Sentence Drawings*, Stefanie Posavec, 2009

of my studies and my mother offered to buy me an art book as a gift, and I chose *Maeda@Media*.[1] I don't think I realized it at the time, as I was still in the early stages of my design learning, but I think my interest in logical design systems was spurred by looking through this book.

DB: You deeply explore the complex relationship between textual and visual forms of expression (…)

What exactly are you looking for, and how? Have you made new creative discoveries?

SP: For many of my text-based projects I am trying to find methods of exploring something that is subjective (literature) using quantifiable, methodical means (the gathering of data).

I'm not sure if I have made any ground-breaking creative discoveries; rather, my discoveries tend to help me access subject matter that I have a deep emotional connection with and understand it more thoroughly.

DB: In your projects, you show, in particular, that a linguistic sign is first a visual sign, an inscription included in a formal and symbolic representation system of meaning (…)

What are main issues of this creative exploration? What is the ultimate goal of your creative approach?

SP: The goal of my creative approach across most of my personal and more illustrative projects is to create a visual that, from afar, is aesthetically appealing and communicates a message or illustrates the emotional "tone" of that subject matter. However, when one discovers that the visual was created using data, I make sure that the data is presented clearly enough that they have the option to come closer to the work and gain some subtle insight from it.

I think I am always striving to balance these two viewpoints within a project, and I enjoy the challenge of trying to balance these two sides of the project while always representing the data truthfully and faithfully.

DB: What is your creative approach/process/method? Can you describe the various stages of a work's creation, from when the idea appears in your mind to its final realization?

SP: I'll describe the process I use for creating what I would call 'data illustration': for data illustration, I am not creating an information graphic or data visualization, but instead am using data to communicate a secondary message where data insight isn't the main goal of the project. For a project like this, the goal of the project is similar to traditional illustration: communicating an emotion, message, or feeling through imagery. In short, I'm communicating a more subjective message. This illustrative method of working with data is something that I do for artistic commissions, commercial illustration, or personal projects.

---

[1] See Maeda J (2000) Maeda@Media. Rizzoli International Publications, Milan.

Usually, if I am given specific subject matter to respond to through the use of data, I immerse myself within the subject, researching the subject and searching for interesting, meaningful patterns of data that I can use and visualize. Through this process of research, I will often find datasets that I find intriguing, appealing, or beautiful, and I will use the project as a way to shed light on these hidden patterns locked within that subject.

Alternatively, another approach I take is to research the subject matter, draw my own conclusions about that subject, and then I look for data within the subject that I think will best communicate my intended message. While this seems similar to the creation of an editorial piece of information design, this is slightly different: here, my conclusions are often emotional and subjective, and I am using the data to craft a specific visual and conceptual 'tone' to the illustration.

Next, I try to find the data that I need to enhance my main concept, and spend time searching for beautiful, meaningful datasets that will be able to create intriguing forms. When visualizing the data, I try to represent the data in a shape that alludes to the subject matter, connecting the data and the subject and creating a visual metaphor in its final visualized form.

DB: In your projects (e.g. *Writing Without Words*, *Sentences Drawing*, *Rhythm Textures* projects, etc.), you question the place of the visual within the activity of reading and writing…

Would you say that you aim to explore the complexity in language of forms—which is sensitive to the balance, the rhythm, the harmony and aesthetic consistency (meaning in form)?

SP: Besides my interest in language and literature, I think the main reason that I am interested in visualizing these subjects is twofold: firstly, I work in this way because of my personal interest in language and literature. Obsessively gathering data from a piece of literature that I love provides me with the opportunity to more directly engage and immerse myself within this literary work. It provides me with a space where I am able to methodically explore the boundaries of a literary work I am curious about.

Secondly, I enjoy visualizing literary and linguistic data because I am interested in highlighting the hidden patterns and 'inner workings' of a modular system (language) that can be reconfigured and shuffled around and, depending on the combination of words, will become a literary masterpiece or a literary failure. My goal is to highlight the hidden complexities found within something that target the more subjective, emotional part of ourselves, and approach it from a different viewpoint.

DB: Representing words or sentences visually demands an act of translation where signs are translated into other signs, thus constructing, displacing and altering meaning (…) Do you intend to reveal the systemic basis of language (its logic and structure)? In others words, are you looking for ways to discover the very essence of language (understood here as a specifically human means of making sense) behind the forms?

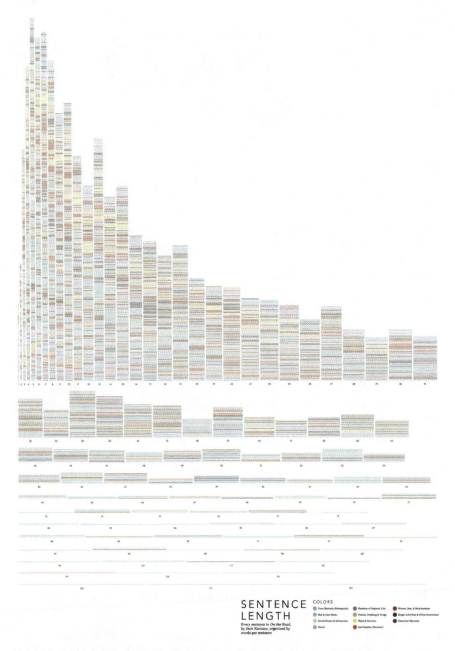

Fig. 22.3 *Writing Without Words*: *Sentence Length*, Stefanie Posavec, 2009

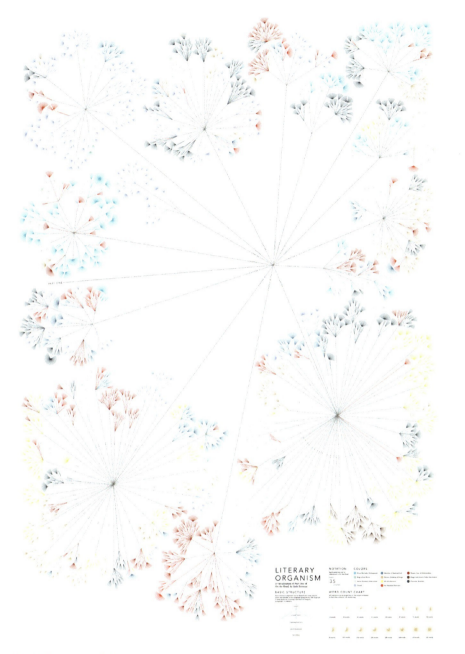

Fig. 22.4 *Writing Without Words*: *Literacy Organism*, Stefanie Posavec, 2009

SP: The reason I often translate text into form is that I am interested in the idea of compressing an entire work of literature into its essence or symbol. Normally reading text is a linear journey: the reader starts at the beginning of the book and follows the story until they reach the end. I enjoy creating project where I try to 'see' literature from an angle where we normally are unable to see while we are reading it.

DB: The close reading of your graphic creations opens up to a very broad interpretation of forms, in their strength and unity (e.g. *Murakami's 1Q84*, *Rythm Textures*, and also *Measuring Kraftwerk* projects)…
  Do you pay particular attention to the Gestalt laws/principles?

SP: I think you could consider me self-trained: I had a relatively brief design education on my BFA (due to the American approach to university education of taking a wide range of subjects) and during my MA, all of my development was self-initiated, so I only learned about the Gestalt laws/principles were this year! So, yes, I have been exploring these laws/principles in an intuitive, undefined sense, but haven't really explored these formally. I've mainly been creating my data projects based on my own ideas of which visual forms are readable and which aren't, and adjusting the design accordingly.

DB: Do you think that 'reading an image' requires higher cognitive effort, among others, than 'reading a text'?

SP: It would depend on the subject matter: some subjects are best communicated by text, and others are better communicated through image. One benefit of 'reading an image' is that you can very quickly get a view of the 'bigger picture', whereas with text the length of time that it takes for you to get the bigger picture is really dependent on how long it takes you to read to the end of the text:-)

DB: Do you consider that a picture says more than a thousand words (or 1,001 words are worth more than a picture, as the Chinese say:-)?

SP: I think a picture is worth a thousand words, but that might be because I find it easier to communicate using images than text!
Though really, I think my opinion on this might be the similar to the previous question: I think it's easier to quickly gain an overview of information in image form as opposed to text form.

DB: In the *Literary Organism* project, you put into images (using a hierarchical visualization model) and represent the formal and narrative structure of the work of literature entitled *On the Road*[2] written by Jack Kerouac (…)
  Could please precise what we can learn—or know more—about the story in this way?

---

[2] See Kerouac J (1957) On the Road. Viking Press, New York.

SP: The original intention of the *Writing Without Words* project was to explore systems of visualizing text that could be applied to any novel and used to aid comparison between different writers. So I would say that this method of visualizing a novel would be most functional and useful when one work was compared to other literary works, and a viewer could compare and contrast the different forms that different literary works take.

However, in the context of the On the Road, since I'm not really offering the viewer the opportunity in *Literary Organism* to compare Kerouac's writing to anything else, the goal of this project was mainly to showcase the hidden complexity, structure, and order that is found within a novel, and explore the idea of how something that communicates such a subjective, emotional experience has this beautiful, underlying orderly structure that holds it all together.

In hindsight, I would say that my interest in choosing to present the data in this way makes the *Literary Organism* piece less of a traditional data visualization and more of a data illustration, where I am visualizing the work in order to communicate my feelings about literature.

DB: For the On the Road project (and also others), your creative process start, as all data designers, with the step of gathering data. To do so, you use highlighters and pens to indicate graphically each part of the text (…)

Behold a simple and certainly very effective method that already gives a preview of the book structure (…)

Could you please explain in more details how you proceed to collect data?

SP: When I first work with a dataset, I like to 'get my hands dirty' work with the data by hand, using highlighters and pens to start categorizing the information and to begin gathering numerical data.

After I've begun to discover some interesting patterns that indicate what 'shape' the data will take in its final form, I'll either stop gathering data and have someone else automate the rest, or I will gather the rest of the data myself.

I often gather data by hand when I am looking to use data that a computer can't normally gather, or when I am working with data that has never been captured before.

I know I could use basic visualization tools to create rough visualizations for early analysis, for me the act of writing down content means it is more likely to stick in my mind… also, by going about the mundane task of gathering data, it opens up the creative side of my mind to think further about the project, in a similar way to when people say they come up with their best ideas when washing the dishes!

DB: Are there any concepts/ideas/things that can only be represented graphically by the means of images?

SP: 1. A person's face—The public is always shown photofits of criminals or missing people, as just writing a description of what someone looked like isn't really enough to capture the essence of a person.

2. The past—By this, I mean the reconstruction of extinct animals and eras looked like. I don't think that the research of the many people looking into various aspects of our past would be as impactful if it weren't made visual.

DB: It seems to me that your work is not unrelated to photography (metaphorically speaking). As Petteri Sulonen wrote: "the power of photography is to show us what is underneath the skin of the world: to discover that which is hidden in plain sight".[3] Is this not also the ultimate goal you want to achieve?

What is your relationship to photography and the image as a contemporary artist?

SP: Yes, I agree that my work is very similar to photography: a friend of mine recently introduced me as "someone who makes boring things interesting," and while it made me laugh, I agree with her! I like searching for patterns of data that I can use to highlight the idea that there are beautiful, interesting angles upon everything, even that which is boring or mundane.

DB: You create very beautiful data illustrations, or data graphics (e.g. 11-x-series and Sentence Length projects) that are distinguished by an extremely pure style focusing on essentials (…)

Do you think that the quality (or performance) of a data graphic depends on its expressiveness (simplicity, clarity, simplicity, accuracy and so on)? (Figs. 22.5, 22.6, 22.7 and 22.8).

SP: I agree that clarity, simplicity, and accuracy are keys to creating a quality data graphic. While many of my projects fall more into the category of 'illustration' instead of pure visualization, I still try to ensure that my projects are simple and clear, in order to communicate the data faithfully. Since I am pushing much of my data work into the subjective realm, it's important that I follow a few 'rules' to prevent the data from becoming unreadable.

DB: What are both rhetorical and aesthetic values of data graphics? Is there a relationship between the aesthetic qualities and the aesthetic value of data graphics?

SP: Yes, I think there is a relationship, as I think that people are more compelled to look at a data graphic if it is visually appealing, and then more people will receive the data graphic's message. But I agree that many data projects that are aesthetically-appealing might not have a strong message, or strong content, but I suppose this type of work falls in a similar category of beautifully-designed books, posters, advertisements, and all other design: all style, no substance.

DB: What are your creative tools (software tools and other antique tools like pens, paper, and highlighters)?

---

[3] See *Why Most Landscapes Suck*, URL, January 16, 2014: http://primejunta.blogspot.fr.

22 The Art & Craft of Portraying Data

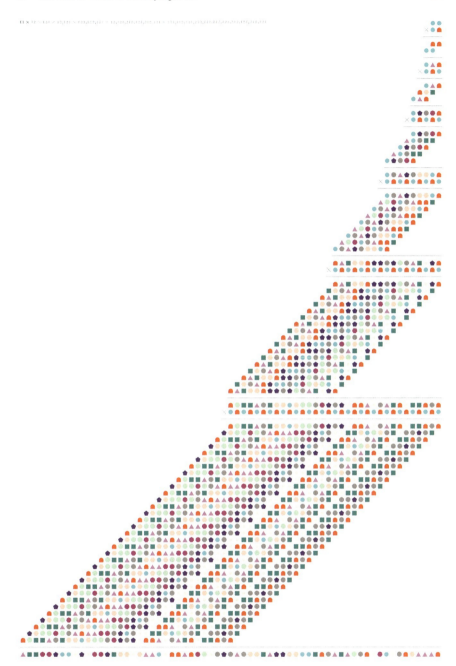

Fig. 22.5 *11-x-series*, Stefanie Posavec, 2012

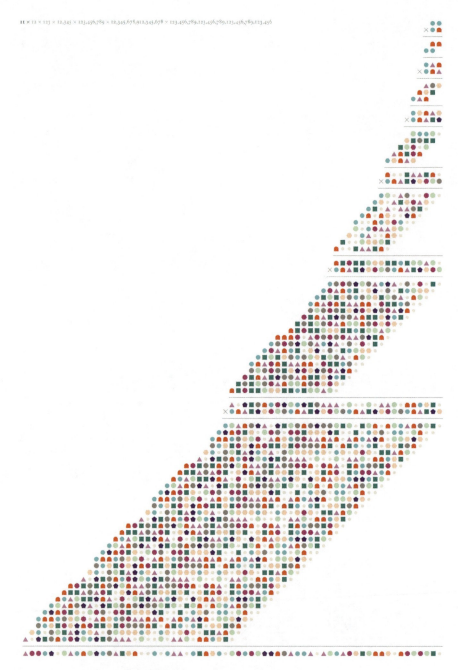

Fig. 22.6 *11-x-series*, Stefanie Posavec, 2012

22 The Art & Craft of Portraying Data 417

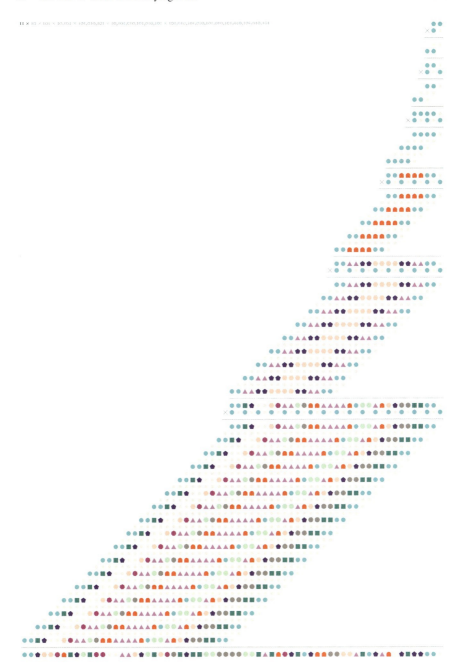

Fig. 22.7 *11-x-series*, Stefanie Posavec, 2012

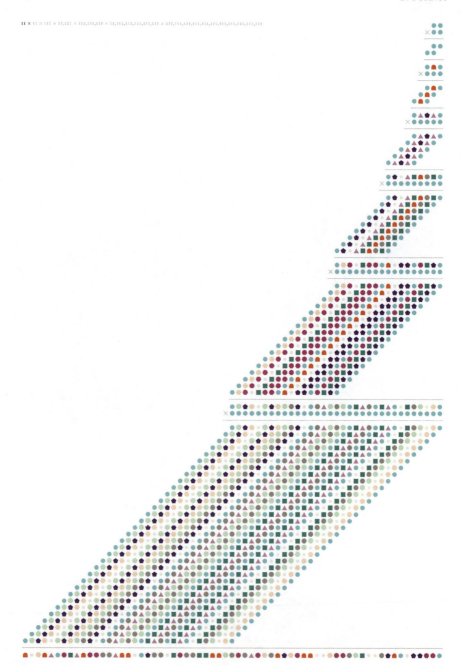

Fig. 22.8 *11-x-series*, Stefanie Posavec, 2012

SP: My creative tools include Adobe software (Illustrator, Photoshop, and InDesign), pens, pencils, paper, highlighters, sketchbooks, and the occasional use of a developer to write programs.

My main creative tool is my sketchbook: I take copious notes at the beginning of every data project, and it's only after sitting in a cafe with a sketchbook and a pencil that I begin a project properly.

DB: One of the remarkable singularities of your personal creative process is to do data design by hand. Are you attempting to develop (in a sense) a new form of craftsmanship: the Art & Craft of Portraying Data? (Fig. 22.9).

SP: I find the idea of the Art & Craft of Portraying Data appealing, and I have often described the difference between data visualizations/graphics created by hand and those which are generated using a program as having a similar relationship as a hand-knitted and a machine-knitted sweater have with each other: neither sweater is *better*, but each has different qualities.

While I accept that for many projects working with data by hand isn't suitable, I think there is a different aesthetic that comes from planning a data project by hand in that one isn't working with a computer's computing power, but with a human's computing power, and I think this affects the visual decisions that are made. Also, I like the fact that there is often space for "mess-ups" when you create work by hand, and this adds to the charm of final visual.

DB: What does it mean for you to design data graphics by hand? Is it a way to escape from predefined design models/patterns that stifle creativity?

SP: It actually started as more of a workaround for the fact that I don't code. I do work with developers now, but often for personal projects I am interested in the imperfections and distinctly different aesthetic quality that arises from creating something by hand.

But much of this comes from the fact that when I first started my career in design, I tried not to look at anyone else's work for fear that I would be influenced and I would begin "copying" someone else's approach. So in some cases, using the approaches that are already out there feels like cheating (even though it really isn't)!

DB: In another interview, you said that you only work with data that you are genuinely interested in, and also on topics that have caught your eye (…) You engage in with great interest and passion in all the topics you treat. When you start a new project, you usually read different books, articles and other specific documents in order to familiarize yourself with your subject and terminology. In addition, you often end up spending a lot of time exploring data in order to find new creative ways to depict them (…)

Why is it important for you to seize all topics/issues with such intensity (time, energy, courage)?

It seems also extremely important for you to be deeply concerned by the data you handle? If so, why?

Fig. 22.9 Highlighted book/poster, Stefanie Posavec, 2004

SP: I think much of it is part of my internal design strategy: if I can "fall in love" with a subject, then it's easier for me to come up with a design solution that has the emotional tone and concept that I am interested in looking for in my creative work. I am drawn to the idea of using data to communicate something poetic about a subject, and it's easier for me to add poetry to a graphic if I have an emotional connection.

Also, I tend to research and obsessively analyze a topic because I am quite cautious and indecisive when it comes to design decisions. I like to make sure that my design choices and conceptual choices are underpinned by research, logical thinking, and a watertight argument. It's for this reason that I find it quite difficult to create very spontaneous, loose design work: it's such a different type of design thinking than what I use when I work.

DB: Is this the price to pay to be able to unveil/reveal the hidden meaning of the data—thanks to sustained efforts and a considerable personal investment?

SP: I like challenges, and I like projects that have an element of human "toil" to them, they feel more "real" to me. I also like the idea that the amount of human toil needed to visualize a specific subject by hand is another way of communicating the scale and complexity of a dataset, and when this is combined with a visualization give the visualization a power that it might not have if it were automated.

DB: Would it be fair to say that you are trying to turn big amounts of raw data into rich insights that inspire people—to turn all this data into something insightful and valuable for us all?

SP: Yes, I think one of the main things I am trying to do is to look for the most inspiring, compelling patterns of data found in the world and to showcase them to a wider audience, hopefully using a visual approach that is appealing and accessible.

DB: How do you find original ways to present data graphically (to show data from new angles, which provide new ideas and richer understanding)?

SP: I'm not sure, but I think it's getting harder, mainly because there are certain shapes that specific datasets just won't be able to take (…) so there are naturally limitations on what you can do with form in the realm of data visualization.

I think much of the originality comes from the subject matter and the dataset that I've chosen to visualize: without compelling content, the final outcome will likely not be very successful.

DB: You create very high quality works that demand the utmost precision, and a very great skill (…)

Thus, you seem to continuously work with greatest attention to detail, the details that lead to perfection, always keeping in mind that perfection is no mere trifle—of course, all great designers know that design is in the details (…)

SP: Yes, I try to make sure that all the data that I use is accurate, and is represented accurately, even if that data is a secondary aspect of the piece, or that people will likely not be scrutinizing the finished product in such a fine-grained way. Even if no one will be using my visual as an academic data visualization, I need to make sure that the visual still truthfully portrays the data. It's that element of truthfulness that makes me try to make everything as perfect as possible.

However, I have noticed recently when speaking to skillful data visualizers that they aren't as worried when I am when a very small detail in my design isn't perfect: perhaps I should take my lead from them and relax slightly!

DB: What are for you the big challenges for data design?

SP: A challenge is to keep the work innovative and new, and explore new methods of visualization.

DB: What is your vision for the future of data design?

SP: I hope that data will be used more often as a raw design material by designers when creating all visual aspects of our world, from shoes to posters to cars (…) I hope that working with data isn't something that is on the periphery of design (as it feels it is today) but is fully integrated.

DB: How do you intend to tackle these challenges?

SP: In the world of data it's very easy for people to discuss the 'right' and 'wrong' ways of visualizing data (…) instead of being so hardline about these rules, I think the best way to innovate is to explore what happens when we break some of the 'rules' that other people champion, this is what will revitalize and invigorate the community.

Author Biography

Posavec Stefanie (Independent Designer & Data Illustrator) was raised in Denver, Colorado, and moved to London permanently in 2004 when completing her MA in Communication Design from Central Saint Martins College of Art & Design (2006). Her studies focused mainly on the visualization of literature, which led to time as a book cover designer for Penguin Press and the visualization of Stephen Fry's autobiography for his iPhone app My Fry in 2010.

After going freelance in late 2010, she focuses on projects ranging from data visualization and information design to designing book covers (or anything in between) for publishers and creative agencies. Her personal projects have been exhibited internationally, and this work tends to focus mostly on language and literature with an interest in hand-analyzed or handmade data visualizations.

For more information and contact:

http://www.stefanieposavec.co.uk
http://uk.linkedin.com/pub/stefanie-posavec/9/98a/44a
https://twitter.com/stefpos

Chapter 23
From Experience to Understanding

Benjamin Wiederkehr

David Bihanic's Interview with Benjamin Wiederkehr

David Bihanic (DB): As a way of introduction, could you please introduce yourself and describe your background (e.g. training, work experience, career path, etc.)?

How did the Interactive Things company come into being?

Benjamin Wiederkehr (BW): My name is Benjamin and the visual form has always my preferred way of thinking and talking about things. Based on this deeply rooted interest in the Arts and Visual Communication, I got started in graphic design and illustration. The start of my formal education was the preliminary course for creative expression at the School for Visual Arts Bern. While I enjoyed working with the static medium, I remember vividly the empowering feeling I got once I started writing code to create generative and interactive visuals. To advance my skills in this exciting area, I applied and got accepted to the Interaction Design program at the Zürich University of the Arts. My curriculum included a very broad variety of applications of human-computer interaction. Theoretical courses like technology philosophy were complemented by applied courses ranging from interface design to physical computing. During my studies I developed a very strong interest for projects in the fields of generative design and information visualization.

Apart from my passion for data-driven projects I furthermore discovered two principles that have since informed a lot of my decisions: First, the balance between user-centered design and technology-driven experimentation should never be neglected. Great pieces are based on the right technology to fulfill the user's expectations. Second, collaboration is essential everything I do. With the support

---

B. Wiederkehr (✉)
Interactive Things, Seefeldstrasse 307, 8008 Zürich, Swiss
e-mail: benjamin@interactivethings.com

and feedback from fellow practitioners, you can achieve results that soar high above what you can do on your own.

I learned a lot about the power of collaboration in joint projects with my classmates Christian Siegrist, Christoph Schmid, Jeremy Stucki, and Peter Gassner. They are the people with whom I run our company these days. The five of us all studied together and graduated from the ZHdK in 2008. Our ways parted for one year when everyone followed different career paths in big corporations, small design agencies, or academic institutes. One year later, in 2009, Christian, Jeremy any myself sat down together and formed the vision for our very own design studio. A place where real problems get solved. A space where people are free and able to explore, create and learn. A platform where designer and developer interlace and collaborate. This was the birth of Interactive Things, an interaction and information design studio.

DB: Your design projects are very wide-ranging: data design (data visualizations, infographics) but also user interface design and interaction design, mobile service design, webUX design, etc.

It seems highly important to you to ceaselessly strive to diversify your activity, to constantly expand your areas of work according to your wide range of skills, and also to renew your interest in your own work through connections to new collaborators (project partners, clients, etc.).

Is that right? What is your motivation behind this investment? Is it a way to make the most of your creativity?

BW: The broad variety of design projects that we are working on is rooted in our personal diversity of origin, expertise and experience. In that sense, diversity is less an activity, but rather a condition. On one side, our team is composed of people with backgrounds in graphic design, interaction design, motion design, information visualization, and computer science. On the other side, Interactive Things is set up to be a studio for creative research, production and advice where people actively collaborate. It's our believe that a person's creative potential is best unleashed when given the right amount of time, an inspiring and comfortable space, and a solid structure to reduce friction and distraction. The wide-ranging list of projects is simply the result of allowing our diverse group of practitioners to pursue their interests and to apply their skills to problems that are meaningful and impactful.

DB: Thanks to the crossdisciplinary skills nurtured by this investment, you are ideally positioned to blaze new creative paths—especially in the field of data design, which implies a transversal approach to design practice (crossover design) an approach that can best be described as cross fertilization (…)

How do you view this? Is your holistic approach to design (global design) to be regarded as a source of creative development and of enriching cross-fertilization?

BW: Looking at the work we are doing, you can easily distinguish two reoccurring themes: User-Centered Design for interfaces between humans and machines and Data-Driven Design for interfaces between humans and information. If you would analyze our work in more detail, you would be able to place them on a range between these two themes with the majority of projects landing somewhere in the middle where we have a strong overlap between user-centered and data-driven. For us, this is the space where our capabilities can have the biggest impact. This is where we can contribute not only on a per project-basis but hopefully to the design discipline as a whole. The intersection of User Experience Design and Data Visualization is a space of increasing importance considering the increasing amount of information available and the increasing ubiquity of computing capabilities. When we can create solutions that follow the best practices of these two fields, we are able to optimize interfaces for human capabilities while maintaining data integrity.

The competences from these two fields can even be applied to projects that do not include both set of requirements. When designing for interaction, keeping the best practices from information visualization in mind, helps us to create more readable and easier to understand interfaces. When developing for visualization, keeping the best practices from user experience design in mind, helps us to create more usable and more immersive systems. It's important to see this as a mutual enrichment that goes both ways. Not only will novel ways of interacting with visualizations emerge from this practice, but also novel interface types that following visualization design principles.

DB: Could you please tell me what your global design approach/method/process is? (You may refer here to your design practice process divided into four main steps: explore, define, design and implement).

BW: The practice in our studio can be described as research-led design. That said, we do not stick to any dogma on how to work on our projects, nor do we force ourselves to replicate the same process for every project. Different organizations, domains, goals, and circumstances demand for different means. On a big scale our design process can be described as a set of four phases: Discover, Define, Design, and Develop. This description is by far not our own invention. It has been established in design theory by much smarter people that we are and gets applied in design studios all around the words. Divergent thinking and convergent thinking is applied in alternation. Each of the phases can be broken down into activities focusing on individual key objectives of the phase. For each activity we have a set of methods that we can apply to conduct the activity. These can be well-established methods from the fields of ethnography, design research, cognitive science, information visualization, or computer science to more experimental or personal methods for acquiring knowledge, generating ideas, and developing concepts.

In the first phase we try to learn as much as possible about the project and discover its restrictions and opportunities. We explore the data and content, research the audience, analyze benchmarks, and gather requirements and stakeholder expectations. Our goal here is to build a shared understanding of the project and its context to inform our future decisions. Combining the insights from our exploration, we work closely with our client to establish a strong product strategy. This includes the basic concepts for technology, architecture, and user experience.

Collectively we specify what the product is going to be—and what it won't. We define a project plan and specify the product scope. We begin the design process by creating different approaches to each problem. Using prototypes and evaluation, we find the most effective way to present information and guide the user. Every design decision must prove its purpose and is executed with the highest attention to detail. In the last phase we implement the final application. This usually builds upon a series of prototypes that have previously been built as part of the design phase. Now is the time for the final rounds of evaluation with users, testing thoroughly in terms of usability, stability and security, and optimization for deployment. It's important to note that the phases can't be seen as closed steps. They much rather represent a scaffolding for conducting a series of related and dependent activities. Therefore, the phases are interconnected and get repeated in iterations where necessary and appropriate (Fig. 23.1).

DB: (Like the Studio NAND for instance) Does your creative process imply a global view and a transversal approach to the technical tools, methods, uses and their interrelations?

BW: Our process is less a strict recipe for how we do our daily work. It's much more a useful framework that we use to plan our projects, allocate the resources and manage the communication with our clients. We are open to change it as necessary, and refine it whenever possible. Inside the process framework we place the technical activities to advance our solutions. These activities are the things that inform our thinking, render tangible results, and eventually solve the problems at hand. Clearly described tools and methods are the means by which we conduct each activity. Each activity has a set of methods that we can use and it's a conscious decision which method we select to do the job. Examples for such methods are Contextual Inquiry for conducting user research or Rapid Prototyping for evaluating visualization methods. And as with all tools humans have ever used, they influence the characteristics of the end result. That's why selecting the right tools for the task at hand is crucial. The tool should be optimized for both the problem domain as well as the problem solver. In that sense, we need to consider the balance between the designer's expertise and experience as well as the content, context, and user of the solution.

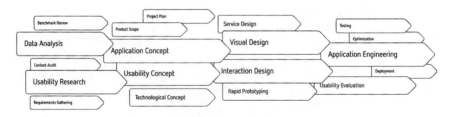

Fig. 23.1 Global design process

DB: How do you address specific data design issues? What creative process (method and tools) do you use to turn complex data into clear graphics (and/or interactive graphics)?

What is your own design-oriented data visualization approach (principles, method, and process)?

BW: As a design studio with the focus not only on Data Visualization, but also on User Experience, it's evident that we follow fundamentally a user-centered design process. When I look closer at the individual phases that we typically go through when designing visualizations, I see ten steps that are connected and get repeated as necessary:

1. *Inform*: We inform everyone on the team about the project in its entirety.
2. *Prepare*: We research, acquire, clean, and format the data.
3. *Explore*: We learn about the structure and texture of the data and what answers we will be able to draw from it.
4. *Discover*: We connect with the data and find stories, patterns, correlations, and challenges.
5. *Sketch*: We test out ideas visually based on the data to see if our ideas for the visual form are applicable and comprehensive.
6. *Question*: We ask questions about our sketches to verify if the selected visualization methods truly provide valuable insights.
7. *Design*: We refine and finalize the visual and functional design of the visualizations and interfaces.
8. *Develop*: We implement a flawlessly working application.
9. *Evaluate*: We ensure that the solution is understandable, readable, usable and useful and make corrections and optimizations where necessary.
10. *Deliver*: We conclude the project by publishing and documenting it appropriately.

DB: Do you consider yourself a sort of present-day adventurer who sets out to discover the insights hiding in data (or as Moritz Stefaner said: "(…) to find the truth and beauty in data")?

BW: I personally enjoy the mental picture of a present-day adventurer sailing out to explore new frontiers and bringing back precious and rare discoveries. But in reality, we understand our work more as craftsmanship. In most projects that we are involved in, we help the end user to understand information more intuitively, to analyze patterns more efficiently, or to conduct tasks more easily. We see ourselves less in the role of the adventurer himself, but rather as the drawers of maps, builders or tools and architects of platforms that makes such adventures possible.

DB: What makes a good data visualization (as refers to aesthetic and functional qualities)?

BW: First, it's important to note that "good" can mean anything under different circumstances. The quality of a data visualization should only be evaluated by

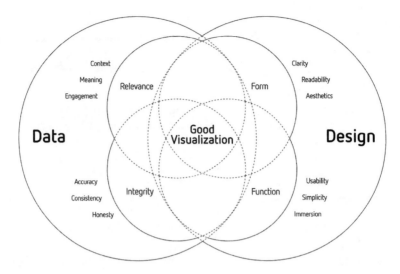

Fig. 23.2 Qualities of successful visualization

considering the target audience for it, the context where it's deployed, and the goals the author hopes to achieve with it. When we evaluate visualizations we typically do this with four dimensions grouped into Data and Design.

Data Relevance examines the importance of the data for the specific context of the user. The data must hold substantial meaning for the user and must motivate her to actively engage with the visualization. Data Integrity evaluates the accuracy, consistency, and honesty of the data source and its provider. Design Form analyzes if the data is presented appropriately. The presentation needs to be clear, readable and holding up to high aesthetic standard. Data Function assesses the overall usefulness as a combination of usability, simplicity and immersion (Fig. 23.2).

DB: In your various projects, you proved that there is no competition between aesthetics and function (…) You showed instead that the combination of aesthetic and functional qualities confers several benefits (…)
Is it still one of your main targets: design with aesthetics and function?

BW: At the studio, we refer to the design of an artifact as the elegant combination of form and function. That's why we never see aesthetic requirements in competition with functional requirements. They need to be neatly integrated to render a good design solution. Design can be seen as the combination of engineering, which traditionally is focused on functionality, and art, which traditionally is focused on content, context and aesthetics. Becoming overly attached to one or the other can hurt the overall elegance of a design solution. It might leave the user with something that works, but is undesirable or something that's beautiful but not useful. The goal should be to combine these different requirements and create a solution that's both useable and desirable.

Fig. 23.3 *Social Progress Index*, Interactive Things (Client: Social Progress Imperative), 2013

DB: In your statistical data-visualization applications, you design graphical visualization models that especially increase perception of relationships between data. To do so, you don't usually starting from scratch but often from existing models (like bar charts, pie charts, scatter plots and so on) that you strive to adapt/transform in accordance with the specificities of each data set (cf. *Social progress index, Human development index 2.0, Wide* projects, etc.)

Could you please tell me exactly how you do that? (Fig. 23.3).

BW: The visualization methods that have proven to be successful in making data understandable are plentiful. Our goal with our work is not necessarily to invent new ones, although we certainly are not shy with experimentation. We put our efforts into finding the right, not the most innovative, solution to a problem. "Right" can only be evaluated by considering closely the characteristics of the user and the data as well as the context in which the system is being used. Then again, we do not consciously look for predefined and ready-made visualization methods that can be applied to any problem. We much rather analyze the data that we want to visualize and then encode the different dimensions using visual and behavioral attributes that can be processed pre-attentively by the human brain. This way, we make best use of the user's capabilities as a human and simplify and deepen comprehension of the information that we represent.

DB: Do you consider that old/classical graphical models for data representation (drawn by Florence Nightingale, William Playfair and many others) are no longer efficient to make huge datasets easy to visualize and explore?

Consequently, is it necessary to invent new ways of representing new data sets (for each dataset, a specific presentation mode)?

~ For this, many predefined visualization models can be used (trees, networks, and diagrams) but they will not be as effective as the dedicated modes ~

BW: I agree that for vast, dynamically changing or even user generated data sets, existing visualization methods do not suffice. The sheer size of a data set doesn't match with methods that originate from hand drawings. That said, a lot of the most innovative looking visualization build on top of very traditional representation models. But with the introduction of algorithms we are now able to advance them into much more sophisticated visualization types that are well-suited to satisfy today's visualization needs. Force-directed networks, large-scale triangulation, or edge bundling of circular network edges are a few examples among many.

To answer the question about the necessity of inventing specific presentation models for each new data set, we must first agree on what specific means in this regard. I think that the specificity of a visualization method doesn't necessary mean that we have to invent a whole new visual vocabulary. Instead, every single decision we make along the process of encoding the information visually, audibly or behaviorally, is an increase of specificity. Such modifications always need to support the data set at hand and they cannot be transferred between different visualizations and maintain their effectiveness. I see this circumstance being mainly rooted in the nature of the data sets. They vary significantly in quality, quantity, structure and texture of its content. Consequentially, a visualization that has been refined to bring out the most from a specific data set, will always be more intuitive to read than one that simply has been applied without considering the characteristics of the underlying data (Figs. 23.4 and 23.5).

DB: You opened an online laboratory area that allows you to share insights, ideas, inspirations and knowledge, also to dialogue with people around your research topics, etc.

More broadly, do you intend to develop a sort of open design lab for crowd-sourced brainstorming and discovery?

BW: The Lab at Interactive Things provides us with a platform to do different things that wouldn't fit otherwise into our daily routine. First, it's the home of our academic research projects from our own studies as well as collaboration with research institutions. Second, it's also where we exhibit our pro-bono work done as support for the Open Government Data task force in Switzerland. Third, we use the lab to document our speaking engagements and publications. This mostly includes links to audio and video footage, notes and assets, or complete documentation website.

It's important to us to share our ideas freely as we strongly believe in the benefits of that can be gained from community feedback and support. Open Sourcing our insights, ideas, experiments, trials and errors, so to speak. A lot of the projects we have conducted inside the Lab were set up as collaboration with other practitioners.

23 From Experience to Understanding

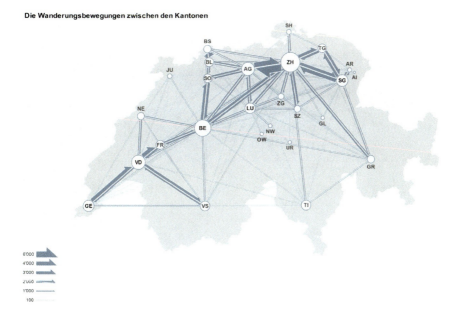

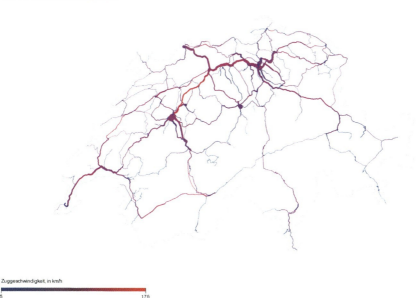

Fig. 23.4 *Swiss Maps*, Interactive Things (Client: Neue Zürcher Zeitung), 2013. Die Wanderungsbewegungen zwischen den Kantonen (*top*). Wie der Schienenverkehr pulsiert (*bottom*)

Therefore it's only right to share them with the broader community instead of keeping it behind closed doors.

DB: Would this be a new way to combine practice-based research with experimental design (especially in data design)? How do you view this?

BW: In all honesty, if so then it's less intentional that it may seem. The lab is where we are able to try out things that we can't during client projects. This includes novel approaches to how we structure a project, innovative visualization methods, tackling data that we don't get our hands on otherwise, experimenting with emerging technologies, and so on. It's important to all of us to stay curious, attentive and proactive. Experimentation is the best way for us to keep up the excitement for our field and not to get worn down due to constraints that we may encounter in client projects. But true experimentation is only possible if people are willing and free to fail. If the pressure of producing something functional, learning something useful or researching something worthwhile is too high, people are more likely to not even start experimenting. To prevent this, barriers to get started should be low and consequences sandboxed in this playground. We do our best effort to keep lab projects lightweight and provided with the right environment and timeframe to grow and flourish.

DB: Do you work with academic/scientific laboratories/institutes? If so, what collaborative relationships have you established with them? What is your overall work methodology? How do you proceed in setting common or shared objectives?

BW: We maintain a close relationship with the Zürich University of the Arts and collaborate with them on many different strings. For one, we are involved in the curriculum of the Interaction Design program and teach technical courses where we share our expertise in designing and implementing interface and visualization systems. We are also involved in the newly setup design incubator of the ZHdK that helps young designers to establish themselves and their practice in the industry outside of the academic environment. These are young professionals that strive to become design-driven social entrepreneurs providing services or building products that might have meaningful impact on the local community or society at large. Our role here is to mentor some of the incubates with our experience from starting out as an independent design studio.

Beside our collaboration with ZHdK, which is a very natural connection due to our academic upbringing, we also work with a range of other educational institutes like Zürich University of Applied Science (ZHAW), the University of St. Gallen (HSG), or the Swiss Federal Institute of Technology in Zurich (ETHZ). One important collaboration that I wouldn't like to miss out, is the work we do together with the Swiss National Science Fund. They are the national institution for funding scientific research and closely observe the research activity in Switzerland. In this collaboration our involvement comes full-circle as we use visualization methods to explore, evaluate and communicate data about the state of research fields, funding, and activities.

Fig. 23.5 *Ecoplace Dashboard*, Interactive Things (Client: AXA), 2012

DB: You made a website dedicated to a series of inspirational interviews with famous data designers. If you don't mind, I would like to ask you a question you ask them: "What role does collaboration with others play for your projects?" (Fig. 23.6).

BW: Thanks for mentioning Substratum—it's a project that I am fascinated about and it has been a big honor to work on it with so many people I admire. I'm more than happy to answer this specific question and I see collaboration taking place in three main forms in our studio:

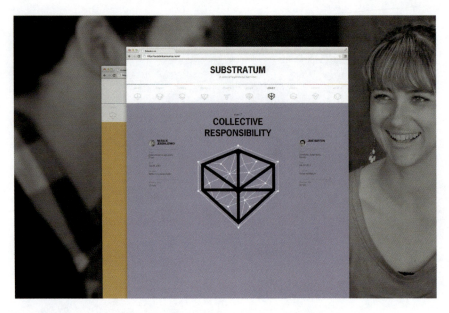

Fig. 23.6 *Substratum*, Interactive Things, 2011

Most importantly, the collaboration with professionals from deep inside of an expert domain is crucial to the success of our work. Being able to help them answer their questions quickly and precisely requires us to learn as much as possible about what they're doing and how they're doing it. We need to understand what's meaningful to them and why. The more domain-specific the information or interactions, that we need to design for are, the more closely do we need to observe and learn from the experts. On the other hand, it's also fascinating to collaborate with people who are not interested in and who do not understand what we're doing. It helps you remember what it was like to be more naïve and unknowing of what's possible. It helps you to keep a healthy, user-centered view towards technology. This way, we remind ourselves that the user isn't interested in the technology, but in the experience it delivers. Lastly, our process highly benefits from the collaboration between professionals from the same field. If you are working with people who are really good in the same domain as you, your ability to communicate is greatly enhanced. References to existing methods, technology, works or people are understood and act representative for concepts that would require much longer descriptions for laymen. It's a shared vocabulary, which means we can communicate in a really high bandwidth way.

DB: You recently led a workshop in collaboration with "Neue Zürcher Zeitung" focused on data-driven storytelling.

What creative methods (and tools) do you use in those workshops to raise the awareness of a maximum number of people?

BW: These workshops have a slightly different focus than typical design workshops and is highly inspired by the work of journalists. The main objective of the Data-Driven Storytelling workshop is to learn how a data-driven journalism team, consisting of a Journalist, an Interactive News Developer and an Interactive News Designer, works together in a situation of limited time and information available. The general structure of the workshop is to identify one meaningful message out of a generic agency news report. Then, connecting this message with relevant data collected from credible and verifiable sources. Finally, developing the visual and functional concept to tell a captivating story with supporting interactive visualizations and multi-media components. The results are generally sketches and descriptions that are presented at the end of these workshops that can between take 2-6 h of time. It's a fast-paced program and requires attention and courage from the participants to make a successful presentation. At the same time, it's a realistic situation most journalists find themselves in on a regular basis. After extraordinary events take place, the time to reflect, research, ideate, conceptualize, produce and deploy is very short and yet, the pressure for a fast publication is very high.

The program of these workshops have been developed in collaboration with the Neue Zürcher Zeitung with whom we have a continuous collaboration developing data-driven journalistic pieces. The insights that we share with our participants are mostly lessons that we have learned during this challenging but highly rewarding collaboration.

DB: In your statistical data visualization projects, you offer users to view, slice and interact with data sources, and thus to make helpful observations about it (with the objective of updating and developing their own knowledge).

In this way, do you want to help people discover their own insight? (Figs. 23.7 23.8 and 23.9)

BW: Visualization can be applied with different goals in mind and increasing knowledge about a data set certainly is one of the most important and one of the most wide spread use cases. In contrast to applications with the goal of communicating certain insights about a data set or with the goal of telling a compelling and statistically supported story, applications that strive to advance the users knowledge always benefit from free exploration and evaluation. If the presented facts become verifiable through interactivity and self-experimentation, they are perceived as being trustworthy. Discoveries become repeatable and subject to personal interpretation. This way, the insights are no longer only the insights of the creator of the application. Instead they become part of the personal experience of the user. We as humans are much more favorable of integrating facts into our knowledge when we understand from where they originate.

DB: Interactive data visualizations are a great way to engage and inform large audiences. They offer users an overall and comprehensive view about different data sets and also allow them to focus on interesting parts and details (in reference to the Visual Information Seeking Mantra: Overview first, zoom and filter, then details-on-demand).

What other major benefits do they offer users?

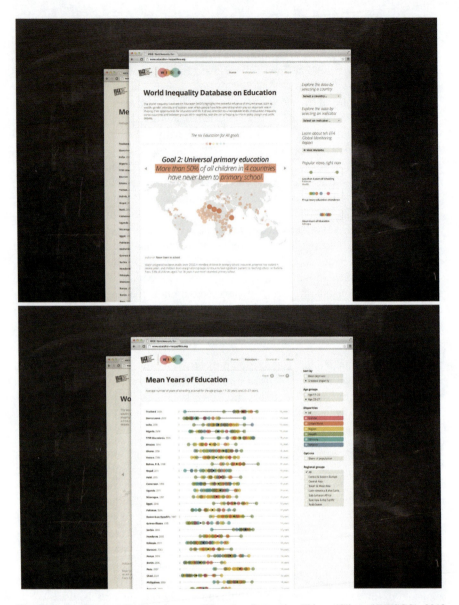

Fig. 23.7 *World Inequality Database on Education*, Interactive Things (Client: UNESCO), 2012

BW: Wow, this question has been answered before by much smarter people than me but let me describe a concept for interactive visualizations that influences a lot of decision I make in my work. In his paper *The Eyes Have It: A Task by Data Type Taxonomy for Information Visualizations* Ben Shneiderman introduced the visual

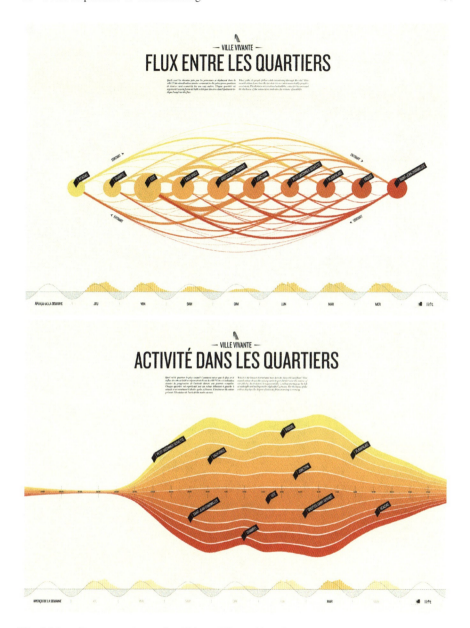

Fig. 23.8 *Ville Vivante*, Interactive Things (Client: City of Geneva), 2012

information seeking mantra that goes like this: Overview first, zoom and filter, then details-on-demand. An interactive visualization should provide the user with an overview of the entire collection first. Then, the user starts focusing on the relevant items and dismisses the irrelevant ones. After finding the items of interest, the user

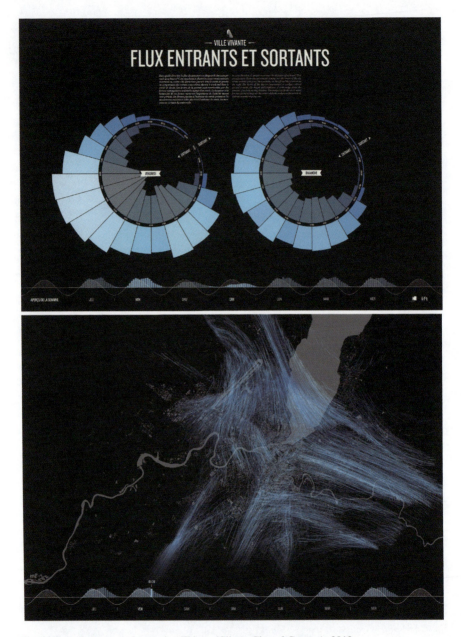

Fig. 23.9 *Ville Vivante*, Interactive Things (Client: City of Geneva), 2012

can get more detailed information on demand. In the transition from a static to an interactive visualization, also the manner of consumption shifts from author-driven to reader-driven without a predescribed ordering. That doesn't mean that interactive visualizations can't follow a narrative, though. Instead, it introduces new types of visual narratives with both, a clear storyline and the possibility for exploration. Edward Segel and Jeffrey Heer wrote an excellent paper about such approaches called "Narrative Visualization: Telling Stories with Data."

DB: When designing data visualizations, you also renewed the way we can interact with data... In fact, your work has significantly improved our understanding of the crucial role of interaction in data visualization (...)

Do you think interaction can lead to a better understanding of relationships between data (and also to a better perception of data structures)?

BW: Technology supported data collection, processing and representation already empowers user to scrutinize data and gain insights from it. Using information visualization that leverages the principles of visual perception as well as user interaction has proven to be indispensable in understanding large and complex data sets in many areas. Because the user's context plays a vital role in how information is being perceived, finding the appropriate configuration of content, presentation and interaction is key for the application to be usable and useful. The direct and directed interaction between the human and the visualization can aid understanding by connecting the users' actions to the visualizations' reactions. Such interaction models become increasingly impactful when applied to facetted views that provide multi-dimensional views on the same data set. This way, the user is able to focus his manipulations on one dimension yet is provided with a multitude of different views.

DB: Nate Agrin and Nick Rabinowitz said in a recent article that most of the time real data is ugly: "(...) In practice, when dealing with most real-world data sets, expect to spend up to 80 per cent of your time finding, acquiring, loading, cleaning and transforming your data."

Is it crucial to have good and rich data sets to develop successful visualizations?

BW: What a great point! Consistent and coherent data is indeed crucial to tell meaningful stories or to build comprehensive exploratory instruments. The foundation needs to be reliable to have meaningful impact. And while I fully agree to the importance of finding, acquiring, cleaning and formatting data, I must say that in our daily work, we do not have such a heavy emphasis on these tasks. Luckily, we mostly have our client delivering the data in a well-prepared format for us. We often collaborate with public research institutions or private companies that are responsible for collecting and storing their own data. Therefore, finding the right data source and then working towards a useful data set is a matter of hours, not weeks.

DB: A data designer reveals relationships in data sets that are not evident from crude data (...)

Do these discoveries depend heavily on the designer's intuition?

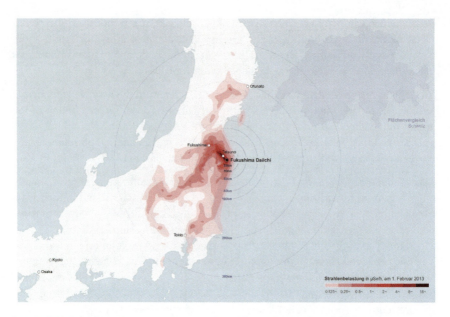

Fig. 23.10 Fukushima, Interactive Things (Client: Neue Zürcher Zeitung), 2013

BW: I do not see intuition as being the best driving-factor to make big discoveries. Considering the amount of data that we are handling in our projects, relying on our intuition would be too big of a risk to find groundbreaking insights. Instead, we rely on our curiosity and our critical thinking. When we are curious, we keep searching for interesting, relevant, and exciting relevant information. When we are critical, we keep questioning our findings, assumptions, and interpretations. The collaborative process that we use to keep looking and keep questioning is what helps us make meaningful discoveries (Fig. 23.10).

DB: What major challenges do data designers face today? In your opinion, what are the big challenges for data design?

BW: The challenges we face today when designing data visualizations are plentiful and multi-facetted. A few of them are very evident and recognizable on first sight: How do we handle big, dynamically changing data? How do we secure that our underlying data is respecting the privacy of people and organizations while still providing an honest look on structures, patterns, trends, and outliers? How do we work towards universally accessible systems? I would like to mention a topic that has become more relevant in recent years and where advancements could allow us to reach more people with our visualizations than ever before.

In the past years we have observed a fast and far spread adoption of computing devices that no longer stand on our desks, but instead lie in the palms of our hands. Our smartphones have become incredibly powerful and have taken over the role of the computer for a broad variety of tasks. The exploration, evaluation, and communication of quantitative information is no exception. We have seen mobile-compatible applications that provide complex data analysis systems. We have built such tools ourselves and by doing so, learned that mobile visualizations need to look and function significantly different to be useable and useful. It's important to note that simply adapting the visualization design to a smaller screen size barely scratches the surface of the challenges that we are confronted with or of the possibilities we are provided with. It's true that we have to consider the technological boundaries of mobile devices of which screen size is one. Limited computing power, data storage capacity, or bandwidth all play roles when implementing mobile visualization systems. But there's one major challenge that technological advancements will not solve for us and that's the altered user context. An application that is being used on the go has a completely different set of requirements than a desktop application. Such applications are used in short intervals making it necessary to provide insights quickly and intuitively. They are mostly used single handed and interaction metaphors have changed from point-and-click to gestural interactions. But not only the form and function of a visualization needs to be adapted to the mobile context. With shifted requirements the content that is presented to the user needs to be adapted as well. It needs to be relevant to her location, adequate to her attention span and actionable for her behavior.

DB: How do you intend to tackle these challenges?

BW: In our work on truly mobile-optimized visualization designs, we strive to optimize along the three axes of content, form and function. Each decision for how to handle these dimensions should be informed by observing why, how, when, and where mobile applications are being used by the target audience. By conducting user research upfront and user evaluation throughout the design and prototyping process, we can do our best to optimize for this increasingly important use case. Only by verifying our assumptions with real users are we able to leave the path of pre-existing patterns to discover novel ways of displaying and interacting with information. I strongly believe that the mobile user context demands for such novel ways. The demand is what pushes our work towards overcoming the above mentioned challenges. Furthermore, there's also a force that pulls us towards this vision and that's the available technology. The capabilities of mobile devices might be still limited in some areas, but they are liberating in others. Considering the advanced sensor technology that these devices host, it's easy to imagine a completely new set of interaction methods. Location-aware filtering, full-body navigation, multi-user collaboration through interconnected devices, just to name a few of the things that are being evaluated.

I come back to our previously mentioned role as drawers of maps, builders of tools and architects of platforms that make adventures in data exploration possible. In this regard, we have to invest in advancing our toolset if we want to catalyze discoveries beyond the frontiers.

Author Biography

Benjamin Wiederkehr (Interaction-UX Designer, Co-founder and Managing Director of Interactive Things) is an interaction designer with a focus on information visualization and interface design. He is founding partner and managing director of Interactive Things, a User Experience and Data Visualization design studio he established in 2010 together with Christian SiegristSIEGRIST Christian and Jeremy StuckiSTUCKI Jeremy. The Zurich based team designs and implements interactive applications and visualizations for private and public sector organizations and companies including the UNDP, UNESCO, WEF, UEFA and National Geographic.

Benjamin is also part of the Open Government Data task force in Switzerland and helps to facilitate open access to government data for everyone. On Datavisualization.ch, Benjamin and his co-authors provide insights into their research and working process as well as they document topical use cases in the field of data visualization. His work and research interests center around persuasive technology, information and knowledge visualization and emerging interaction principles.

For more information and contact:

http://www.interactivethings.com
http://benjaminwiederkehr.com
http://ch.linkedin.com/in/benjaminwiederkehr
http://datavisualization.ch
https://twitter.com/wiederkehr

Author Index

A

Abrioux, Yves, 26
Adorno, Theodor, W., 39
Agrin, Nate, 441
Akten, Memo, 297
Albers, Joseph, 42, 97
Alexander, Ruth, 239
Almossawi, Ali, 153
Anderson, Chris, 112
Antico, Susanna, 125
Antonelli, Paola, 300
Aristotele, 75
Armengaud, Marc, 123
Armengaud, Matthias, 123
Ash, Michael, 238
Asmussen, Jörg, 239, 247
Auge, Marc, 121
Auster, Paul, 78
de Azua, Felix, 120

B

Babwahsingh, Michael, 113
Bach Jean, Sebastian, 393
Bad, Charlotte, 284
de Balzac, Honoré, 121
Balsamo, Anne, 209
Barabasi, Lazlo, 161
Bari Corbero, Jordi, 124
Basili, Victor R., 142
Battiston, Stefano, 243
Bederson, Benjamin B., 24
Benjamin, Walter, 121
Bergson, Henri, 28
Bernardi, Marco, 86
Bernhard Beus, Hans, 247
Berque, Augustin, 123
Berry David, M, 207
Berthoz, Alain, 40
Bertin, Jacques, 29, 32, 295

Beschizza, Rob, 112
Biddulph, Matt, 194
Bihanic, David, 27, 47, 48, 343, 355, 379, 393, 407, 425
Binx, Rachel, 283, 284
Birt, Arlene, 253, 254
Bohnett, David, 224
Borgman L. Christine, 210
Bostock, Mike, 302
Bourriaud, Nicholas, 332
Boyd, Danah, 366
Boyer, Bryan, 288
Brand, Christoph, 247
Bret, Victor, 159
Brouillet, Denis, 26
Browning, Tyson R., 146
Bruegel, Pieter, 239
Brunet, Roger, 32
Buchanan, Richard, 210
Buckminster Fuller, Richard, 394
Burdick, Anne, 209

C

Cage, John, 81, 82, 83
Cairo, Alberto, 103, 106
Calle, Sophie, 332
Cameron, Andy, 317
Capriles Radonski, Henrique, 107
Caroll Douglas J., 24
Carreras, Genis, 43
Carter, Shan, 62, 302
Case, Ryan, 343, 353
Cassidy C. David, 176
Castells, Manuel, 122
Caviglia, Giorgio, 207, 317
Cawthon, Nick, 386
Cerda, Ildefons, 121
Cezanne, Paul, 70
Chabris, Christopher, 114

Chang, Jih-Jie, 24
Chang, Kai, 309
Chavez, Hugo, 107
Chekhov, Anton, 319
Chen, Alexander, 42
Chianchetta, Alessandra, 123
Cho Young, Sang, 43, 46
Chomsky, Noam, 26
Chopra, Deepak, 259
Chouvel, Jean-Marc, 29
Cialdini, Robert, 275
Ciampaglia, Giovanni Luca, 402, 403
Cicourel, Aaron, 27
Citraro, Dino, 175
Ciuffi, Davide, 86
Clark, Jeff, 309
Cleveland S. William, 148
Confino, Jo, 260
Cope Brinton, Willard, 8
Corner, James, 121
Covey, Stephen, 270
Crawford, Kate, 366
Csikszentmihalyi, Mihály, 361

D
Danna, Serena, 86
Defleur Margaret H., 111
Delaunay, Boris, 287
Deleuze, Gilles, 238, 247
Denis, Michel, 24
Depraz, Nathalie, 30
Dewar, Mike, 86
Diakopoulos, Nick, 178
Dickens, Charles, 121
Dickerman, Leah, 79
Dokic, Jérôme, 26
Draghi, Mario, 239, 241, 243
Drucker, Johanna, 207, 208, 209, 211, 215
Dufrenne, Mikel, 39
Dunbar Kevin N., 176

E
Eames, Charles, 62, 64
Eco, Umberto, 40
Ehmann, Sven, 265
Epler, Matthew, 42
Eppinger, Steven D., 146
Escher, Maurits Cornelis, 393

F
Fellow, Tow, 178
Felton, Nicholas, 287, 343, 345, 346, 347, 348, 349, 350, 353
Few, Stephen, 113, 114, 260
Fiedler, Steffen, 355
Fisher, Eric, 122
Fisher, Max, 369
Flores, Fernando, 27
Forester de Rothschild Lynn, 239
Foucault, Michel, 235, 237
Fragapane, Federica, 86
Freire, Juan, 134
Friedman, Ken, 210
Friedman, Milton, 247
Froese, Tom, 31
Fry, Benjamin, 27, 385
Fuller, Jack, 114
Funk Kirkegaard Jacon, 239

G
Galbraith, Kate, 305
Gansterer, Nikolaus, 78
Gassner, Peter, 426
Gatzer, Werner, 247
Gauchat, Gordon, 177
Girardin, Fabien, 189, 190
Gladwell, Malcolm, 70
Glattfelder James B., 243
Godel, Kurt, 393
Gold Matthew K., 207
Goldacre, Ben, 111
Gottschall, Jonathan, 262
Griesemer James R., 194
Groch-Begley, Hannah, 103
Guerra, Stefania, 86
Guinea Montalvo, Pietro, 86

H
Haigh, Alison, 43
Hansen, Holger, 239
Harasser, Karim, 79
Harre, Niki, 273
Harris, Jonathan, 300, 306, 320, 321, 322, 323, 324, 325, 326, 327, 329, 330, 331, 333, 334, 335, 336, 337
Hascher, Xavier, 29
Hashemi, Mahmoud, 49

Hatton, Les, 142
Hayles Katherine N., 208
Healey Christopher, 26
Heer, Jeffrey, 295, 441
Heisenberg, Werner, 176
Hemment, Drew, 356, 357, 358, 360
Heraclitus, 162
Herndon, Thomas, 238
Hidalgo Cesar A., 152, 153
Hidalgo Cesar A, 151
Hofstadter Douglas R., 393
Hosken, Ben, 380
Huff, Darrel, 104, 105
Hwang, Sha, 283, 284, 286

I

Ikeda, Ryoji, 36
Ilyas, Shahee, 42

J

Jaeger Richard M., 114
Jenkins, Nicholas, 214
Johnson, Mark, 26
Jonathan, Harris, 317

K

Kahneman, Daniel, 114
Kamvar, Sepandar, 300, 306, 320, 321, 322, 323, 324
Kan Stephen H., 142
Kandinsky, Vassily, 80
Kapitza, Nicole, 41
Kapitza, Petra, 41
Kaufman, Leslie, 271
Kepes, Gyorgy, 123
Kerouac, Jack, 414
Klein, Julian, 364
Klein, Naomi, 247
Koch, Vaughn, 286
Koch-Weser, Caio, 239
Koestler, Arthur, 40
Kosslyn, Stephen M., 113, 114
Kotler, Arnold, 260
Krueger, Myron, 317
Krum, Randy, 106
Kruskal B. Joseph, 24
Kuang, Cliff, 369
Kuniavsky, Mike, 196
Kurzban, Robert, 114

L

Lakoff, George, 26
Lamping, John, 27
Lang, Andrew, 92
Lang, François, 26
Langacke,r Ronald, 26
Laporte, Stephen, 49
Latour, Bruno, 26
Lautrey, Jacques, 28
Lave, Jean, 27
Laville, Frédéric, 27
Leibniz Gottfried W., 28
Lenay, Charles, 30
Lev, Nadya, 284
Levin, Golan, 318, 319, 336
Ljungberg, Jan, 372
Loh, Jonas, 355
Lopez Pena, Javier, 199
Lozano-Hemmer, Rafael, 306
Lumley, Thomas, 111
Lunenfeld, Peter, 209
Lupi, Giorgia, 62, 79, 86

M

Maduro Moros, Nicolás, 107, 108
Maeda, John, 40, 410
Majno, Francesco, 86
Malamed, Connie, 261
Malevich, Kasimir, 80
Manovich, Lev, 213, 403
Mantegari, Glauco, 86
Marsh, Bill, 8
Marsh, Sarah, 239
Martinez Diez, Pablo, 119, 124
Maslow, Abraham, 268
Masud, Luca, 215
Mccabe, Thomas J., 143
Mccandless, David, 5
Mccandless, David, 43
Mccarty, Willard, 209
Mcchesney, Robert W., 111
Mcclelland, James, 26
Mccormack, Alan D., 143
Mceachren, Alan M., 114
Mcgann, Jerome, 212, 215
Mcghee, Geoff, 305
Mcgonigal, Jane, 272
Mckenzie-Mohr, Doug, 259, 267, 270
Meadows, Donella H., 261
Medcraft, James, 298
Meeks, Elijah, 214

Merkel, Angela, 243
Meyer, Cheryl L., 178
Mondrian, Piet, 80
Monmonier, Mark S., 105
Moretti, Franco, 212
Morin, Edgar, 40
Muir, John, 260
Murray, Scott, 37, 38, 214, 305
Myin, Erik, 27

N

Nakamura, Yugo, 36, 318, 319
Narboni, Roger, 125
Nhat Hanh, Thich, 260
Nichols, John, 111
Nightingale, Florence, 431
Nin, Anais, 176
Nix, Markus, 122
Norman, Donald A., 26
Norton, Quinn, 285
Nowviskie, Bethany, 215
Nussbaum, Bruce, 361
Nye, Joseph S., 247

O

O'regan, J. Kevin, 27
Oberman, Michelle, 178
Opalka, Roman, 346
Ortiz, Santiago, 27, 161, 165, 168, 169, 170, 300, 301
Owens, Trevor, 211

P

Parakh, Vasundhara, 43
Parakh, Vasundhara, 45
Park, Haeyoun, 303
Patterson, E. Thomas, 111, 114
Paulos, John Allen, 111
Perec, George, 77
Peres Galdos, Benito, 121
Perricone, Barry T., 142
Petitot, Jean, 29
Piaget, Jean, 26
Picasso, Pablo, 70, 332
Plato, 236, 267
Playfair, William, 431
Pope, Rob, 212
Popova, Maria, 79
Posavec, Stefanie, 407, 408, 409, 412, 413, 417, 418, 419, 420, 422

Poupyrev, Ivan, 367, 368, 374
Presner, Todd, 209

Q

Quadri, Simone, 86
Quayola, Davide, 297
Quealy, Kevin, 303

R

Rabinowitz, Nick, 441
Raciti, Elisa, 86
Rajoy, Mariano, 110
Ramsay, Stephen, 209
Ratto, Pierenrico, 86
Reas, Casey, 318
Rees, Kim, 175
Rehn, Olli, 238, 239, 240, 242
Reinhart, Carmen, 238, 239
Reith, Lord, 139
Rezner, John, 224
Riekoff, Christian, 367, 368
Ritte,l Horst W. J., 236
Riva, Matteo, 86
Robinson, Andrew, 262
Rockefeller, David, 239, 240
Rockwell, Geoffrey, 213
Rogers, Richards, 210
Rogers, Simon, 390
Rogers, Timothy T., 26
Rogoff, Kenneth, 238, 239, 240, 242
Rosenblueth, Arturo, 237
Rosenkrantz, Jessica, 289
Rosling, Hans, 399
de Rosnay, Joël, 403
Rossi, Gabriele, 86
Rumelhart, David E., 26

S

Sachs, Goldman, 240, 247
Samuels, Lisa, 212, 215
Santamaria-Varas, Mar, 119, 124
Schank-Smith, Kendra, 193
Schauble, Wolfgang, 238, 239, 242
Schmid, Christoph, 426
Schnapp, Jeffrey, 209
Schoch, Christof, 211
Schon, Donald Alan, 210, 216
Schosslera, Philipp, 368
Schreibman, Susan, 207, 208
Segel, Edward, 295

Sennett, Richard, 120
Sferlazza, Roberta, 86
Shen, Vincent Y., 142
Shere, David, 103
Shermer, Michael, 111
Shneiderman, Ben, 305, 385, 438
Shore, James, 142
Shumaker, John Abraham, 114
Siegrist, Christian, 425, 426
Simons, Daniel, 114
Simpson, Bart, 62
Simpson, Lisa, 62
Sinclair, Stéfan, 213
Slavin, Konstantin, 365
Smith, William, 259
de Sola-Morales, Manuel, 120
Spence, Robert, 27
Spiers, Adam, 31
Srnicek, Nick, 247
Star, Susan Leigh, 194
Stefaner, Moritz, 27, 34, 35, 83, 356, 357, 358, 359, 360, 380, 393, 396, 397, 398, 399, 401, 402, 403, 404, 405, 429
Stein, Ben, 248
Stein, Courtney, 176
Straw, Will, 123
Stucki, Jeremy, 425, 426
Sturtevant, Daniel Joseph, 141, 144
Sulonen, Petteri, 416

T

Taraborelli, Dario, 402
Teather, David, 243
Teisl, Mario F., 272
Tharp, Twyla, 60, 77
Thiel, Stephan, 355
Thompson, D'Arcy Wentworth, 160
Thoreau, David Henri, 77
Thorp, Jer, 27
Tiravanija, Rikrit, 332
Touchette, Hugo, 199
Tufte, Edward, 5, 29, 104, 126, 148, 275
Tulp, Jan Willem, 379, 383, 384, 388, 390

Twyman, Luke, 43
Twyman, Luke, 44

U

Uboldi, Giorgio, 207
Usherwood, Zoe, 239

V

Valery, Paul, 33
Van der rohe mies, Ludwig, 33
Van doesburg, Theo, 39
Vande moere, Andrew, 386
Verdi, Giuseppe, 61, 69
Viegas, Fernanda, 27
Vijgen, Richard, 221, 229
Vitali, Stefania, 243
Von klaeden, Eckard, 247
Vygotski, Lev, 26

W

Wagner, Richard, 61
Wainer, Howard, 105
Warnow, Christopher, 235, 240, 242, 244, 245, 246
Wattenberg, Martin, 27
Webber, Melvin M., 236
Wiederkehr, Benjamin, 425
Wiener, Norbert, 237
Williamson, Christopher, 305
Windolf, Paul, 243
Winograd, Terry, 27
Wolff, Michael, 112
Wopperer, Wolfgang, 372
Worthington, Philip, 302, 304
Wu, Eric, 286

Z

Zapponi, Carlo, 43
Zapponi, Carlo, 45
Zetsche, Dieter, 247
Zook, Matthew, 190, 191